integrated
urban
models
2

integrated urban models 2

SH Putman

new research and applications of optimization and dynamics

Pion Limited, 207 Brondesbury Park, London NW2 5JN

© 1991 Pion Limited

All rights reserved. No part of this book may be reproduced in any form by photostat microfilm or any other means without written permission from the publishers.

British Library Cataloguing in Publication Data
A CIP catalogue record for this book is available from the British Library.

ISBN 0 85086 126 8

Printed in Great Britain by Page Bros (Norwich) Limited

Preface

Some years ago I read a biographical account of Louis Agassiz, the great Swiss naturalist who travelled to the United States of America in 1846 and accepted a post at Harvard in 1847 (Mc Cullough, 1977). He remained there, a major figure in the history of natural science, until the end of his life in 1873.

In my opinion there are many parallels between the study of natural science as it was in the second half of the nineteenth century, and the study of urban systems in the second half of the twentieth century. More and more data are accumulating; considerable effort is being spent on the description of observed phenomena; various theories are proposed, and often dismissed, with inadequate examination; theories are ranged against each other with little or no empirical support; there is only limited consensus amongst researchers in the field; and a great deal of additional work still lies ahead.

McCullough (page 9) describes the way in which Agassiz would start prospective students on their course of work with him.

> "His initial interview at an end, Agassiz would ask the student when he would like to begin. If the answer was now, the student was immediately presented with a dead fish—usually a very long-dead, pickled, evil-smelling specimen—personally selected by 'the master' from one of the wide-mouthed jars that lined his shelves. The fish was placed before the student in a tin pan. He was to look at the fish, the student was told, whereupon Agassiz would leave, not to return until later in the day, if at all.
>
> "Samuel Scudder, one of the many from the school who would go on to do important work of their own (his in entomology), described the experience as one of life's memorable turning points.
>
>> 'In ten minutes I had seen all that could be seen in that fish ... Half an hour passed—an hour—another hour; the fish began to look loathsome. I turned it over and around; looked it in the face—ghastly; from behind, beneath, above, sideways, at three-quarters view—just as ghastly. I was in despair.
>>
>> 'I might not use a magnifying glass; instruments of all kinds were interdicted. My two hands, my two eyes, and the fish: it seemed a most limited field. I pushed my finger down its throat to feel how sharp the teeth were. I began to count the scales in the different rows, until I was convinced that that was nonsense. At last a happy thought struck me—I would draw the fish, and now with surprise I began to discover new features in the creature.'
>
> "When Agassiz returned later and listened to Scudder recount what he had observed, his only comment was that the young man must look again.
>
>> 'I was piqued; I was mortified. Still more of that wretched fish! But now I set myself to my task with a will, and discovered one new thing after another ... The afternoon passed quickly; and when, toward its close, the professor inquired:
>>
>> "Do you see it yet?"
>>
>> "No," I replied, "I am certain I do not, but I see how little I saw before."'
>
> "The day following, having thought of the fish through most of the night, Scudder had a brainstorm. The fish, he announced to Agassiz, had symmetrical sides with paired organs.
>
>> '"Of course, of course!" Agassiz said, obviously pleased. Scudder asked what he might do next, and Agassiz replied, "Oh, look at your fish!"'

"In Scudder's case the lesson lasted a full three days. 'Look, look, look,' was the repeated injunction and the best lesson he ever had, Scudder recalled, 'a legacy the professor has left to me, as he has left it to many others, of inestimable value, which we could not buy, with which we cannot part.'"

Agassiz's approach embodies a philosophic point of view which I believe is most relevant to the study of urban systems. We must endeavour to balance theory development with extensive empirical testing and analysis. This is the context in which the work presented here should be taken.

Stephen H Putman
Philadelphia, Pennsylvania
Autumn 1990

Acknowledgements

There was a major difference in the process used in creating this book, as compared with that used for its predecessor. In the last book I reported on the results from a long series of Federally supported research projects which, in total, came to nearly $1 000 000. Direct Federal funding of the research described in this book came to less than 10% of what was spent on the work reported in my previous book. In an era of little or no Federal government support for research on urban transportation and location interactions, progress in the field becomes much more dependent upon coordination of smaller projects and the work of graduate students. As such, acknowledgement is due to many individuals.

Federal support was used, indirectly, and amongst other tasks, to cover the development of the Washington, DC, data and many of the computer experiments which were done with it. Fred Ducca and Ed Weiner at the US Department of Transport were very supportive of that project. Rod Green at Howard University was helpful as the director of the project for which some of the work in this book was a subtask.

Alan Clark and Jerry Bobo of the Houston–Galveston Area Council permitted the use of the Houston data, developed as part of work done for them in my role as their consultant, in the experiments described here.

Quite a number of projects were done by students in the Department of City and Regional Planning at the University of Pennsylvania which related to the work presented in this book. It is difficult to say which of their efforts was most important, and so the students are mentioned here in alphabetical order.

Mohammed Almaani did a PhD dissertation in which, in part, he examined the performance of assignment algorithms.

Chin-Hsiang Chiu quite apart from doing a PhD dissertation, the results of which are not described herein, did the rather nasty job of digitizing and checking the Washington area highway network.

Hsun-Jung Cho, who's PhD dissertation is not discussed here either, digitized both the Washington and the Houston maps used in many of the figures of the book, and ran the Florian problem discussed in chapter 8.

Majid Enani worked with Ziad el Mously in doing some of the intramodel equilibrium runs necessary to produce the chaotic behavior graphs shown in chapter 12.

Robert Keith wrote, as part of his MCP (Master of City Planning) Professional Project, the early revisions of the mode-split program discussed in conjunction with the computer experiments described in chapter 9.

Manuel Marin, as part of his MCP Professional Project, did some of the first computer experiments to investigate intramodel equilibrium.

Hugh Miller did a PhD dissertation which was one of the first theoretical examinations of the intramodel equilibrium question.

Kazem Oryani wrote a PhD dissertation, examining the behavior of the TOPAZ model and comparing its outputs to a spatial interaction model (DRAM).

Acknowledgements

Vincent Patterson did an MCP Professional Project which examined the performance of assignment algorithms on the Archerville data, and which resulted in early versions of some of the figures in chapter 6.

David Stiff and Osman Shahenshah collaborated on a project which further examined some of the equilibrium experiments described in chapter 11.

It is perfectly obvious that computers played a central role in the work discussed in this book. The IBM Corporation, as a part of its Threshold Grant to the University of Pennsylvania, donated the PC and the plotter which I used as a remote terminal, word processor, and graphics generating device throughout the writing of this work.

Most of the computer experiments described in this book were done on an IBM 3090-200 E at the David Rittenhouse Computing Facility (DRCF) of the University of Pennsylvania. Special thanks are due to Roy Marshall the Director of the Center, and to Mike Kearney, Tom Denier, and Mike Guilfoyle who helped out when FORTRAN, graphics, or SAS problems emerged that were resistant to my efforts to solve them.

Kathleen Crossin of the Department of City and Regional Planning did some of her first word processing on portions of the manuscript for this book. From there she ploughed right through the worst of it with a great generosity of spirit.

I have undoubtedly forgotten someone. To them, as well as to the many people who sometimes found me inattentive or preoccupied as I worked on the material described here, my thanks for your forbearance.

Stephen H Putman
Philadelphia, Pennsylvania
Autumn 1990

To Mary

Her strength, wisdom, and love
have helped make everything possible.

Contents

1	**Introduction and overview**	
1.1	Introduction	1
1.2	Purpose	2
1.3	Setting this book in context	4
1.4	The plan of this book	6
2	**Location and transportation model examples**	
2.1	Introduction	8
2.2	The data	8
	2.2.1 The zones and land use	9
	2.2.2 Location of socioeconomic activity	10
	2.2.3 Transportation facilities	10
2.3	Employment location	13
2.4	Residence location	17
2.5	Trip distribution and mode split	19
2.6	Trip assignment	24
2.7	A note on simple numerical examples	29
2.8	Conclusion	30
3	**Optimization and optimizing models**	
3.1	Introduction	31
3.2	Optimality	31
	3.2.1 The objective function	31
	3.2.2 The constraints	32
3.3	Linear programming	32
	3.3.1 Graphical solution method	32
	3.3.2 The simplex method	36
	3.3.3 The shortest-path problem	41
	3.3.4 The minimum-cost flow problem	44
3.4	Nonlinear programming: unconstrained	48
	3.4.1 Algebraic methods	48
	3.4.2 Search methods: estimation of parameters of location models	51
3.5	Nonlinear programming: constrained	55
	3.5.1 The nonlinear minimum-cost flow problem	55
	3.5.2 The Frank–Wolfe algorithm	58
4	**Location models in optimizing frameworks: 1**	
4.1	Introduction	63
4.2	A land-use plan design model	63
4.3	The Herbert–Stevens model: original forms	65
4.4	The Herbert–Stevens model: application issues	70
4.5	The Herbert–Stevens model: numerical examples	73
4.6	The Herbert–Stevens model: further developments	81
4.7	Numerical examples of the modified forms of the Herbert–Stevens model	87

5	**Location models in optimizing frameworks: 2**	
5.1	Introduction	95
5.2	Technique for optimal placement of activity in zones: TOPAZ	95
5.3	TOPAZ: a numerical example	100
5.4	TOPAZ: a numerical example incorporating dispersion	106
5.5	A comment on sensitivity analysis	111
5.6	Some conclusions	113
6	**Transportation models in optimizing frameworks: 1**	
6.1	Introduction	116
6.2	Unconstrained assignment	116
6.3	Capacity-constrained assignment	117
6.4	Equilibrium assignment	119
6.5	Discussion of solution procedures for the UE problem	123
6.6	Numerical experiments with assignment algorithms	124
6.7	Numerical experiments with assignment algorithms: larger networks	132
7	**Transportation models in optimizing frameworks: 2**	
7.1	Introduction	139
7.2	Properties of the Frank–Wolfe solution to the UE problem	139
7.3	The stochastic user equilibrium (SUE) problem	140
7.4	Solution procedures for the SUE problem	146
7.5	Congestion functions	151
7.6	Levels of network congestion	153
7.7	The effects of levels of detail	155
7.8	System calibration and empirical verification	157
7.9	A note on the effects of starting points	158
7.10	Final tests of large-scale networks	159
7.11	Conclusion	161
8	**Simultaneous location–transportation models: 1**	
8.1	Introduction	162
8.2	A simple combined location and trip-assignment model	162
8.3	A simple combined model: numerical examples of intractable solutions	166
8.4	Reformulating and solving the combined model: the Florian approach	171
8.5	Reformulating and solving the combined model: the Evans approach	177
8.6	The Evans approach: a numerical example	179
8.7	The MSA approach: a numerical example	184

8.8	The MSA approach: a numerical example using a spatial interaction model	186
8.9	Conclusions	188
9	**Simultaneous location–transportation models: 2**	
9.1	Introduction	190
9.2	Alternate model system configurations	190
	9.2.1 Variable activity location with fixed transport cost	191
	9.2.2 Fixed activity location with variable transport cost	192
	9.2.3 Variable activity location with variable transport cost: one mode	192
	9.2.4 Variable activity location with variable transport cost: two modes	194
	9.2.5 Variable activity location with variable transport cost: equilibrium	194
9.3	The Washington and Houston data sets	195
9.4	Test 1. Variable activity location with fixed transport cost	202
	9.4.1 Houston	203
	9.4.2 Washington	208
	9.4.3 Summary	209
9.5	Test 2. Variable activity location with variable transport cost: one mode	214
	9.5.1 Houston	215
	9.5.2 Washington	221
	9.5.3 Summary	226
9.6	A comment on sensitivity to data and assumptions	226
9.7	Test 3. Variable activity location with variable transport cost: two modes	233
9.8	The equilibrium solution of the problem of variable activity location with variable transport cost	242
9.9	Summary	247
10	**Equilibrium solutions to location models**	
10.1	Introduction	253
10.2	Solution methods for individual models	253
	10.2.1 EMPAL	253
	10.2.2 DRAM	254
	10.2.3 LANCON	256
	10.2.4 NETWRK	258
10.3	Solving the linked set of models	258
10.4	Experiments with model solution procedures: DRAM	259
10.5	Solution methods for linked models: DRAM and NETWRK	265
10.6	Chaotic behavior of a deterministic system	273
	10.6.1 DRAM and NETWRK: 1	273
	10.6.2 DRAM and NETWRK: 2	278
10.7	Conclusion	281

11	**Preliminary development of dynamic spatial models**	
11.1	Introduction	284
11.2	Adjustment to equilibrium as a modelling framework	285
11.3	Some preliminary experiments: Archerville with fixed travel costs	291
11.4	More preliminary experiments: Archerville with variable travel costs	298
11.5	Intramodel equilibrium experiments: Houston and Washington	300
11.6	Very preliminary tests of an adjustment to equilibrium structure	304
11.7	Conclusion	312
12	**Conclusion: next research steps for dynamic spatial models**	
12.1	Introduction	314
12.2	Supply–demand interaction in spatial models	314
12.3	Supply–demand interaction in programming models	316
12.4	Location–transportation equilibrium: structures and solutions	318
12.5	Dynamics of comprehensive integrated model structures	321
12.6	Calibration of systems of models	322
12.7	Additional calibration problems for PAMEQ structures	324
12.8	Theory development and data analysis	325
12.9	Conclusion	326

Appendix

A.1	Introduction	328
A.2	Calibration results	329
	A.2.1 DRAM, by household type	329
	A.2.2 EMPAL, by employment type	332
	A.2.3 LANCON	334

References 337

Author Index 347

Subject Index 349

1

Introduction and overview

1.1 Introduction

This book is as much a complement as it is a successor to *Integrated Urban Models* (Putman, 1983a). That book was an account of the development and testing of a set of mathematical models of the interrelationships between transportation, location, and land use. The research described there began in 1971. Prior to that time there had been virtually no work done towards quantifying the forwards *and* backwards linkages between the metropolitan patterns of employment and population location and the networks of transportation facilities which connected them.

There had, of course, been considerable development of mathematical models of location (land use) and of mathematical models of transportation. Even so, the models were often improperly specified and their workings only poorly understood. The interactions between such models were largely unexplored. In the twelve years' research described in *Integrated Urban Models* considerable work was done to develop and test improved, *practical*, versions of the models and to examine and understand their interactions.

One of the results of that research was the development of a set of computer versions of the mathematical models. Taken together these are known as the Integrated Transportation and Land Use Package (ITLUP). The employment location model (EMPAL) and the residential location model (DRAM) are two of the major components of ITLUP. The overall model system was the first operational integrated transportation and land-use model package. The ITLUP package and its component models have been tested with data for nearly four dozen different urban/metropolitan regions worldwide. EMPAL and DRAM, in particular, have seen more actual agency policy forecasting applications than any other spatial models. They have been used in regions as different as Seattle, Houston, Washington (DC), Los Angeles, Taipei, and Sarajevo.

Another of the results of that same research was the evolution of the various components of ITLUP into a computerized 'test bed', as well as the development of a substantial collection of metropolitan area data sets. It was thus possible to engage in a continuing series of experiments. New theoretical questions could, *relatively* quickly, be tested against actual data sets. Results, successful or unsuccessful, could be confirmed by repeating tests on numerous data sets.

Not long after the ITLUP research began there started an extensive and impressive development of new theory regarding individual model structures as well as linked model structures. One major theme of these developments was that of discrete choice models, often with logit formulations—though these models were first used with reference to mode choice, they were later extended to location choice as well. Another theme has been that of equilibrium analysis, and along with it, some considerable speculation on dynamics. A third theme has been the notion of recasting spatial interaction models, derived from entropy maximization approaches, as mathematical programming models of, say, consumer surplus maximization. Simultaneously, and in a sense cutting across the other lines of endeavor, there has also been some work in other disciplines on model systems and on more general issues of system complexity.

All this, the development of the modelling 'test bed', the acquisition of data archives, the lessons learned and questions raised by repeated model applications, and the new progress in theory development, set the stage for the work described here.

1.2 Purpose

The purpose of the work presented in this book is to describe some of the more important theoretical developments, as well as the numerical experiments and empirical comparisons which were made to evaluate them. This has led to the resolution of some of the outstanding issues from previous integrated model work, as well as pointing the way for future research.

The problem of assigning trips to networks has received considerable research attention in the past decade. In particular, the theoretical development of the user-equilibrium assignment model along with algorithms for its solution, has attracted a good deal of attention. The stochastic multipath assignment model, developed a few years earlier was still considered by some to be a valid approach as well. There was work towards integrating the two concepts into the stochastic user-equilibrium method. An important goal in this book is to compare these methods, as well as the more traditional all-or-nothing, incremental assignment, and quantal assignment methods. As well as dealing with new developments in the theory, these tests help to resolve some questions remaining from the *Integrated Urban Models* work, where only very limited tests were possible, and none of those were made with the newer model constructs.

The problem of 'simultaneous' trip distribution and trip assignment has also received considerable attention in the past decade. The problem of achieving an equilibrium solution to ITLUP was left unresolved in *Integrated Urban Models*. These are two closely related problems. A major goal in this book was to resolve the ITLUP equilibrium problem. A theoretically sound and eminently practical solution to the ITLUP equilibrium problem is given in chapter 8. What is more, its development illuminates several important model system issues as well as setting the stage for additional work. In particular, many questions regarding the meanings of, and differences between, intermodel and intramodel equilibrium solutions become more clearly defined.

The initial development of the solution to the ITLUP equilibrium problem, as well as the solutions to the user-equilibrium and stochastic user-equilibrium assignment problems, was through a mathematical programming approach.

Another goal in this book was to discuss and explore aspects of the mathematical programming approach for both activity location and trip assignment. The trip assignment work has already been mentioned. In addition, many numerical experiments were done with mathematical programming models of location. Although some points stand out rather clearly, for example, for 'realistic' results to be obtained a dispersion term must be included in mathematical programming models of activity location, more questions were raised than were resolved. Overall, however, mathematical programming, or an optimizing framework, was shown to be a potentially useful approach to model development, notwithstanding a number of difficulties which are discussed in the text.

Time after time during the experimental work and the writing of this book the issue of complexity arose. In some of the more recent theoretical works in this field the reader is confronted with dizzying cascades of ever more complex equations. Even in this book where the emphasis is deliberately placed on the practical consequences of the developments in the theory some fairly complex model structures are presented. There are two worrisome points in this regard. First, there is the issue of obtaining the data necessary to test and evaluate such complex structures. The field must beware of becoming too well described by a remark attributed to Mark Twain: "There is something fascinating about science—one gets such wholesome returns of conjecture out of such a trifling investment of fact." Second, there is the possibility that even as data become available, the systems and their requisite solution procedures may become so complex as to make it quite difficult to understand what is happening in any case. These issues will have to be faced time and time again as work in this field continues.

Much of the theory development in location and transportation modelling is done for greatly simplified examples and test cases. Frequently one theory is proposed and is subsequently deposed by another, with neither of them ever having been subjected to any empirical testing. It has been said that, "in the social sciences there may not be any simple problems" (Broad, 1983, page C1). This assertion is certainly true for the types of problem addressed by use of the models discussed here. It is *not*, however, a justification for ignoring the empirical testing of theory. Theory development in the complete absence of empirical testing is speculation, not science. The purpose and goals in this book have been stated. The method used here to accomplish them is a constant interplay of theory development and description, with numerical experiments and empirical testing. It seems likely that future progress in this field will become increasingly dependent upon this approach. The systems which we wish to understand are too complex for simple approaches. Even if the 'laws' which govern our systems are found to be relatively simple (which most researchers doubt will be the case), the hundreds, or thousands, of variables necessary to describe the systems cannot be managed without the help of some form of computer model. It may even be the case that after learning to master the complexity of these systems, simpler underlying laws will be discovered. For the present, the combination of empirical analysis with theory development which is made possible by computer modelling seems to be the most productive approach.

1.3 Setting this book in context
Throughout the 1970s when this author was doing the research described in *Integrated Urban Models*, other researchers were also working on the general problem of integrating transportation and location or land-use models. It is unfortunate that rather little of that work is available in book form, though the reader who cares to take the effort can find some of it in journal articles and various published working papers. The purpose in the next few paragraphs is to give some guidance as to where this other work may be found, while at the same time giving a very general notion of how it relates to the material in this book.

Several researchers have worked on integrated transportation and land-use models which are, in very general terms, rather like the author's ITLUP package. Echenique and his associates have developed a model package called MEPLAN which has seen several actual agency applications (Echenique, 1985). Mackett spent quite some time on the development of the LILT package, which has been extensively tested on data for the city of Leeds in England (Mackett, 1980). A third major effort of this sort was conducted by Wegener, who used data for the city of Dortmund, West Germany (Wegener, 1986).

The recently completed report on the work of the International Study Group on Land-Use/Transport Interaction (ISGLUTI) describes these efforts as well as several others (Webster et al, 1988). Additional efforts described there include those by Floor and de Jong (1981), Lundquist (1973), Nakamura et al (1983), and Sharpe and Karlquist (1980). In addition, there is a useful recent review by Berechman and Small (1988).

Two books have recently been published (after most of this one had been written), both of which are about integrated transportation and land-use modelling. One, by de la Barra (1989), unfortunately gives very little information on the author's actual models, with the bulk of the book being a review of theories which form the basis for integrated modelling. The other book, by Kim (1989), is focused solely on nonlinear mathematical programming formulations of the combined location-distribution-assignment problem as discussed here in chapter 8. Kim also presents considerable discussion of the mathematical programming approach and descriptions of several solution procedures for these models. These two books have in common a total lack of reference to each other and a virtually complete lack of reference to any of the published work on integrated models such as, for example, that reviewed in the ISGLUTI report. This is regrettable, as it is by now quite clear that no individual researcher's efforts are going to solve the integrated transportation and location modelling problem.

The ISGLUTI report makes an excellent starting point for a review of a number of the integrated model systems which have seen substantial empirical testing, and the list of references in the report is invaluable. Each of these modelling efforts has its own differences from and similarities to the others. All attempt, generally, to link models of activity location and models of the assignment of trips to transportation networks. In some cases an attempt is made to represent explicitly both demand and supply in activity location, whereas in other cases these phenomena are subsumed into a single model-process. To some extent, these model system configurations evolved partly in response to the theory–data interaction. The availability of a particular data source for the development of a model inevitably helps shape the model. This serves to explain some of the differences between the models. Other differences result from the perspectives of the individual developers of the models and simply underscore the fact that there remain many unsolved problems in this general area and many possible approaches to solving them. Common to all of the modelling efforts discussed in the ISGLUTI report, ITLUP included, is a constant testing of hypotheses against real data. This results in a rather persistent and productive cycling between model development and analysis of data. In general, the output of this process is an increased understanding of the phenomena being studied, and, one hopes, improved models.

The particular virtue of integrated transportation and location models is their ability to represent a much broader range of phenomena than can be done by either type of model in isolation. Most of these model systems are run in a recursive mode, with the output of one run of the model being the input to the next, and with the output of one time period being input to the next. This approach, along with the location–transportation interaction, results in integrated models being uniquely able to address indirect and induced response effects. Thus, for example, a policy which might be shown by a transportation model to reduce travel cost, could be shown by an integrated model system to have the longer-term effect of increasing travel cost. This contradictory result would be a consequence of the spatial relocations produced by the initial travel-cost reduction, and the possible tripmaking increase attendent upon those spatial relocations.

The work of Bertuglia et al (1987) comes from a somewhat different perspective. In chapter 7 of that work Wilson articulates a very sophisticated framework "which contains the main submodels of an urban and regional model system". Although there is no attempt at empirical analysis in that discussion, the consideration of the elements of the framework is a useful goad towards further model development and empirical testing. Wilson discusses this framework again in a somewhat later paper (1989). These discussions, along with those, say, of Bertuglia et al (1987) in the last chapter of their book, hold out possibilities, or possible paths, for future work.

The work described in *Integrated Urban Models* is in the same class of work as that discussed in the ISGLUTI report. That class of integrated models are the only reasonable option for agency application at present. The work described in this book takes that of its predecessor as a starting point. In order to explore questions of equilibrium, research is described here which addresses the virtues and the faults of mathematical programming formulations of location models, transportation models, and integrated models. This is the point of connection to work such as that of Boyce et al (1983), Kim (1989), Prastacos (1983), Sharpe et al (1984), and Wilson et al (1981). At the same time, the research described here is firmly grounded in empirical analysis and verification, and thus retains its intimate connection to the previous development of ITLUP. A third facet of the material given here is the beginning of what will undoubtedly be a lengthy effort to incorporate the work done in economics on disequilibrium and partial adjustment models (Bowden, 1978a; 1978b; Fisher, 1983; Maddala, 1983). Many of the comments on various methods discussed and tested here, some given *en passant*, will be of use to agency or consultancy practitioners. The other purpose is to try to define some major paths of endeavor for the research that will take us into the twenty-first century.

1.4 The plan of this book

After this introductory chapter, there are two chapters to set the context. In chapter 2 simple numerical examples of location, distribution, mode split, and assignment models are given. These are the types of models which are discussed and extended throughout the remainder of the book. In chapter 3 a brief introduction to optimization is given and, again through the use of simple numerical examples, some of the concepts and algorithms which will be used later in the book are illustrated.

Chapters 4 and 5, and 6 and 7, are parallel pairs. In the first two, chapters 4 and 5, models of the location of activities (residents, in these discussions) cast in terms of optimizing frameworks are presented. Chapter 4 begins with an examination of the Herbert–Stevens model, a linear programming formulation. Then in both discussion as well as by use of simple numerical examples the formulation is converted to a nonlinear programming model incorporating dispersion. In chapter 5 the process is repeated with the TOPAZ model, including dispersion, showing its relationship both to the Herbert–Stevens model on one hand and to entropy-derived spatial interaction models on the other.

Chapters 6 and 7 are about the assignment of trips to networks. In chapter 6 the traditional all-or-nothing assignment procedure is discussed, followed by multipath assignment and, finally, user-equilibrium assignment. The methods are described, discussed, and compared in theory as well as in simple numerical examples and larger-scale problems. In chapter 7 the discussion of assignment algorithms with stochastic user-equilibrium formulations is continued. A number of other issues are examined, including volume/delay functions and levels of congestion. Finally, two larger networks are used to compare all the assignment algorithms.

Chapters 8 and 9 turn to the combined (simultaneous) activity location and trip assignment problem. Description of the problem begins in chapter 8 in which the mathematical programming formulation of the location problem given in chapter 5 and that of the user-equilibrium assignment problem in chapter 6 are drawn together. Both in equation form and by use of simple numerical examples the difficulties of solving this problem are explored. Then two approaches to the solution of the combined problem are described and, again, illustrated by use of simple numerical examples. Finally, the algorithm is extended, with use of a procedure used to solve the stochastic user-equilibrium assignment problem, to solve a combination of a spatial interaction model with user-equilibrium trip assignment. In chapter 9 a series of extensive numerical experiments is described. These begin with simple linked models of location and with fixed transportation costs. Next, variable transportation costs are introduced, and linked location and trip assignment models are solved. Finally, with use of the algorithm developed in chapter 8, these models are solved for an equilibrium solution to the combined location, land use, and trip assignment problem.

In chapter 10 the issue of equilibrium solutions both to individual models and to linked sets of models is taken up. First the question of what these solutions mean in the context of the model structures is dealt with. Numerical experiments are done to show the effects of alternative structural configurations. The concept of equilibrium solutions as components of a partial adjustment model of location and transportation is introduced, and numerical examples are used as illustrations. In chapter 11 these notions are extended to the full-scale Houston data set, and a further set of tests and discussions is presented. The problem of estimating the supply-side and demand-side aspects of these models is introduced, tying the formulations of chapter 4 back into the overall system.

In chapter 12 a set of four general areas for further research is discussed. The first of these is the topic of supply-side representation in location models, along with reconciliation of supply and demand. The second topic is that of equilibrium structures, meaning, and solutions. Third, the set of issues surrounding the problem of calibrating complex systems of models is looked at. The last topic is that of the interaction between data analysis and theory development.

Location and transportation model examples

2.1 Introduction
The purpose in this chapter is to provide a set of simple numerical examples of the major types of model which will be dealt with in this book. Examples are given of employment and residence location models, a mode split model, and procedures for assigning trips to networks. Some readers will be familiar with this material and may wish simply to check the notational conventions established here and move on. Other readers will find this material a complete, albeit brief, introduction to the types of models that will be elaborated in later chapters.

2.2 The data
For the numerical examples presented in this chapter there will be a common data set. To the extent that it is possible, this same data set will be used for all the numerical examples presented throughout the book. This will make possible the comparison of results from many different models and procedures over a common base. The example used will be a five zone region. Before proceeding it should be noted that this will give occasional problems of discontinuity due to the 'lumpiness' of a data set with so few zones. Yet to increase the number of zones to the thirty or so which are

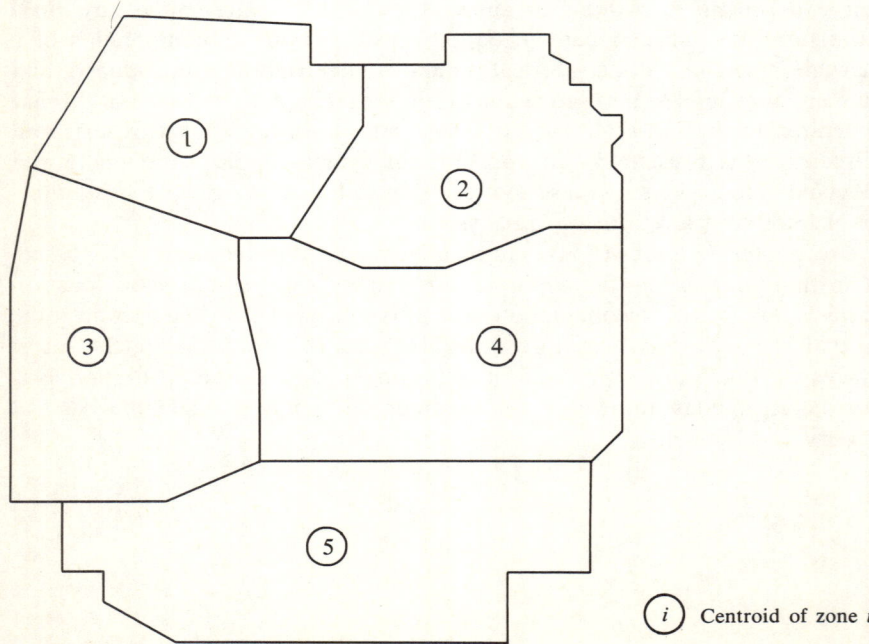

Figure 2.1. Archerville: general configuration.

required to ensure against the occurrence of such discontinuities would defeat the idea of a simple numerical example. Further, note that in subsequent uses of the example data it will occasionally be necessary to add additional variables. This will be done in such a way as to remain consistent with the original data as presented here.

2.2.1 The zones and land use

A map of the example region which, for convenience, can be referred to as Archerville, is shown in figure 2.1. Each of the zones has a centroid, located as shown in figure 2.1. Land use in the region is divided into four classes: (1) residential, (2) commercial, (3) industrial, and (4) vacant. The land use, by type, in each zone, along with the total area of each zone, is given in table 2.1.

Table 2.1. Archerville: land-use and socioeconomic data.

	Zone					Total
	1	2	3	4	5	
Land-use (area)						
Residential	2.5	3.0	1.0	2.5	1.5	10.5
Commercial	1.0	1.0	2.0	1.0	0.5	5.5
Industrial	1.0	1.5	3.0	0.0	0.0	5.5
Vacant	0.5	0.7	0.8	4.5	6.6	13.1
Total	5.0	6.2	6.8	8.0	8.6	34.6
Number of employees (by type of employment)						
Commercial	150	200	100	200	100	750
Industrial	150	150	400	0	0	700
Total	300	350	500	200	100	1450
Number of households (by socioeconomic status)						
Lower-income	200	300	150	100	50	800
Higher-income	100	50	50	300	150	650
Total	300	350	200	400	200	1450
Population	860	1050	585	1030	515	4040

Calculation of the employee-to-household conversion matrix

Cross-tabulation

	LI	HI	Total
Commercial	400	350	750
Industrial	400	300	700
Total	800	650	1450

Conversion matrix

	LI	HI	Total
Commercial	0.533	0.467	1.00
Industrial	0.571	0.429	1.00

Note: LI, low-income household; HI, high-income household.

2.2.2 Location of socioeconomic activity

Both employment and population will be subdivided into types for this example (though not all the illustrative calculations will use all the information about activity type). The employment types will be: (1) commercial, consisting of both retail and service types of activity, and (2) industrial, consisting of durable and nondurable manufacturing, as well as wholesaling, transportation, and public utilities. The population will be classified according to household income, into low-income households and high-income households. It will be assumed that for this example there is no unemployment in the region, and that there is one worker (employee) per household. Further, it is assumed that there are 3.1 and 2.4 persons per household for the low-income (LI) households and high-income (HI) households, respectively. The region's socioeconomic data are given in table 2.1 along with the land-use data.

It is also necessary to calculate the employee-to-household conversion matrix. The use of this matrix will become clear later in this chapter. The calculation of the conversion matrix is shown at the bottom of table 2.1.

2.2.3 Transportation facilities

Archerville has two transportation modes: the private automobile on streets and highways, and a public transit system. The highway system has five two-way load links to connect each zone centroid to the network. It has an additional sixteen two-way network links. These twenty-one two-way links are equivalent to forty-two one-way links. The links intersect at five load nodes and eleven network connection nodes, for a total of sixteen nodes. The network is located on an arbitrary (X, Y) coordinate system, and the (X, Y) coordinates of all the nodes are given in table 2.2. The twenty-one two-way links are given in table 2.3 along with the nodes they connect, their uncongested travel 'costs', and their design capacities in terms of numbers of trips. The links are symmetric with respect to costs and design capacities. The highway system is shown in figure 2.2.

Table 2.2. Archerville: coordinates of the highway system nodes.

Node	Coordinate		Node	Coordinate	
	X	Y		X	Y
1	24	141	9	78	129
2	81	136	10	56	79
3	17	78	11	78	81
4	88	81	12	24	78
5	48	23	13	12	60
6	28	132	14	72	42
7	20	117	15	48	36
8	20	100	16	27	27

Table 2.3. Archerville: characteristics of the highway system links.

Link	'From' node	'To' node	Cost	Capacity
1	1	6	1.0	200
2	2	9	1.0	200
3	3	12	1.0	200
4	4	11	1.0	200
5	5	15	1.0	200
6	6	7	2.0	50
7	6	9	4.0	50
8	6	12	5.0	50
9	7	8	2.0	50
10	8	12	2.0	50
11	9	10	5.0	50
12	9	11	5.0	50
13	9	12	7.0	50
14	10	11	1.0	50
15	10	12	2.0	50
16	11	14	4.0	50
17	12	13	2.0	50
18	12	14	6.0	50
19	13	16	4.0	50
20	14	15	2.0	50
21	15	16	2.0	50

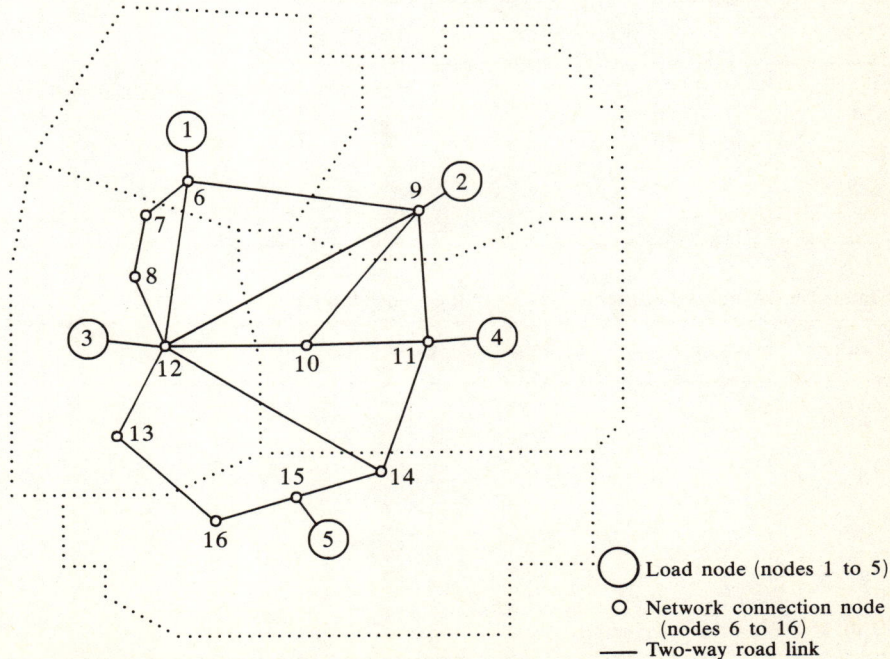

Figure 2.2. The highway system in Archerville.

The transit system in Archerville, formerly the Archerville–Betseyburg Transit Company, is now publicly owned and operated, and consists solely of trolley cars operating along their own right-of-way. As was the case for the highway network, there are five two-way load links to connect the zone centroids to the network. There are an additional five two-way network links. These ten two-way links in the transit network are equivalent to twenty one-way links. The links intersect at five load nodes and five network connection nodes, for a total of ten nodes in the transit network. This transit network is located on the same (X, Y) coordinate system as the highway network. The (X, Y) coordinates of the transit system nodes are given in table 2.4. The ten two-way links are given in table 2.5 along with their travel 'costs'. The transit network is assumed to be capable of dealing with any trip load without an increase in link cost. Thus it is not necessary to specify the capacities of the transit links. The Archerville transit system is shown in figure 2.3.

This completes the definition of the basic items of data needed for the simple numerical examples which follow. It will be necessary, within the context of each of the examples, to define further data items. To avoid confusion these are left to be defined where needed.

Table 2.4. Archerville: coordinates of the transit system nodes.

Node	Coordinate		Node	Coordinate	
	X	Y		X	Y
1	24	141	6	28	132
2	81	136	7	78	129
3	17	78	8	24	78
4	88	81	9	78	81
5	48	23	10	48	36

Table 2.5. Archerville: characteristics of the transit system links.

Link	'From' node	'To' node	Cost
1	1	6	1.5
2	2	7	1.5
3	3	8	1.5
4	4	9	1.5
5	5	10	1.5
6	6	7	5.5
7	6	8	6.0
8	7	8	8.5
9	8	9	4.5
10	9	10	8.0

Location and transportation model examples

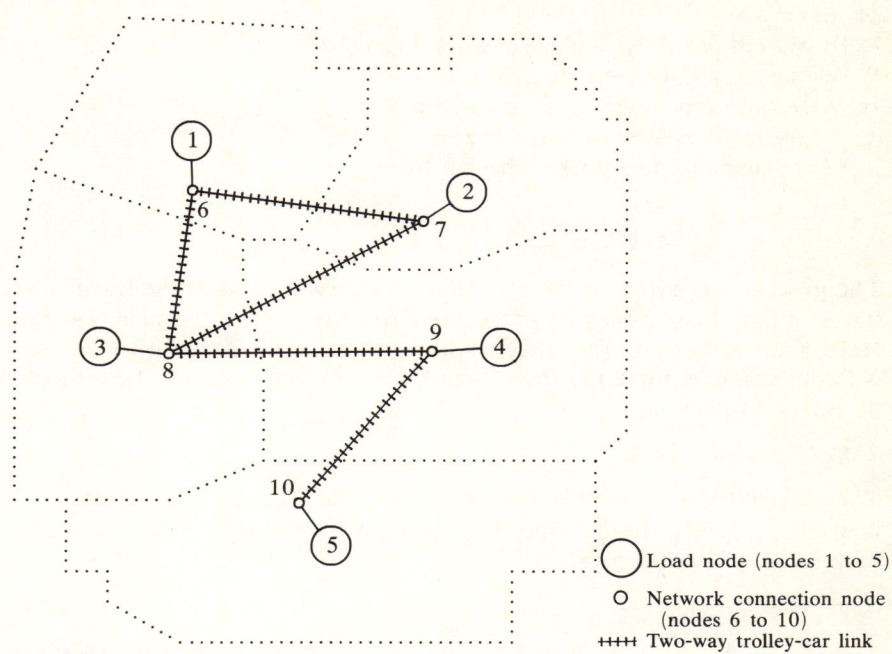

Figure 2.3. The transit system in Archerville.

2.3 Employment location

Both of the location models to be discussed in this chapter, one for employment and the other for residence, are variations on spatial interaction models as derived according to entropy maximizing principles. A great deal has been written about the derivation of these models, with perhaps the two major works being the books by Wilson (1974) and Batty (1976). For the most part there has been rather little published material on models of industrial employment location, with the bulk of the attention being given to spatial interaction models of retail and/or service employment location. A model designed for the purpose of describing intraurban location of all employment types was proposed in a previous book by this author (Putman, 1983a, chapter 6). Subsequent applications and modifications have shown that this model, EMPAL, does a good job of forecasting employment location (PSCOG, 1986; Putman, 1981). EMPAL is a modified form of spatial interaction model and, in simplified form, will be used for the first of the numerical examples to be presented here.

The general form of the model is what is usually called a singly constrained spatial interaction model. In equation form this is,

$$E_j = \sum_i P_i A_i W_j \mathrm{f}(c_{ij}) , \qquad (2.1)$$

where
E_j is the employment, at place of work, in zone j;
P_i is the population (residents) in zone i;
W_j is an 'attractiveness' measure for zone j;
c_{ij} is the travel cost from zone i to zone j;
A_i is a 'balancing' factor which has the form

$$A_i = \left[\sum_k W_k \, \mathrm{f}(c_{ik}) \right]^{-1} . \tag{2.2}$$

The precise definitions of the attractiveness measure and of the travel-cost function have been discussed in the book previous to this (Putman, 1983a). Here, some rather arbitrary specifications will be made in order to proceed with the example. First, the travel-cost function will be taken to be a simple declining exponential:

$$\mathrm{f}(c_{ij}) = \exp(-\beta c_{ij}) . \tag{2.3}$$

The attractiveness measure will be taken to be the product of the appropriate land use (of type n) in the zone, $L_{j,n}$, and the zone's total employment, E_{jT}, each raised to a parameter:

$$W_j = L_{j,n}^{\alpha} E_{jT}^{\delta} . \tag{2.4}$$

Now, as the Archerville data set contains two types of employment, the final equation form, for each employment type, will be

$$E_{j,n} = \sum_i P_i \left[L_{j,n}^{\alpha_n} E_{jT}^{\delta_n} \exp(-\beta_n c_{ij}) \right] \left[\sum_k L_{k,n}^{\alpha_n} E_{kT}^{\delta_n} \exp(-\beta_n c_{ik}) \right]^{-1} . \tag{2.5}$$

The parameters α, β, and δ are assumed to be empirically determined by a procedure known as model calibration (Putman, 1983a, chapter 7). This procedure is, in principle, similar to the fitting of models to data by multiple regression methods, with the difference that here the equations are nonlinear.

There is yet another set of questions regarding the time dimension, or sequencing, of the variables in this model. In the complete formulation of EMPAL there is a second term on the right-hand side of equation (2.5) which is the lagged (time $t-1$) amount of type n employment in zone j. The left-hand side of equation (2.5) was taken to be employment of type n in zone j at time t. Other variables on the right-hand side were all taken to be at time $t-1$, under the assumption that, with time periods on the order of five years each, the use of lagged variables provided a reasonable lag in the response of employment location to changes in the other variables.

It should be noted that with the simplified formulation of the model given here the computed value of $E_{j,n}$ on the left-hand side of equation (2.5) will not be properly scaled to a set of exogenously provided employment totals for the region for time t. In a more complete formulation the necessary scaling would be incorporated into the structure of the equation. In this example it must be done after the fact, by simple prorating of the results calculated in equation (2.5) to whatever exogenous control totals are provided.

The last question to be answered before doing some calculation is that of the zone-to-zone costs. The highway and transit networks for Archerville are given above, but not the zone-to-zone costs. Later in this chapter the questions of mode split, trip assignment, and composite cost calculation will be discussed. At this point there is something of a problem of how to start the process. Eventually it will be shown how the location of activities and trip assignment are interrelated, and the problem of finding suitable starting points for the algorithms will be discussed. For the moment the simplest answer is to use the costs for the shortest paths through the Archerville highway network. Referring to figure 2.2 and table 2.3, it can be seen, for example, that the shortest path from zone 1 to zone 5 is from node 1 via nodes 6, 12, 13, 16, and 15, to node 5. The length, in cost, of this path is 15 units. The shortest path from zone 2 to zone 5 is by nodes 2, 9, 11, 14, 15, and 5, with a total cost of 13 units. The matrix of zone-to-zone costs calculated along the shortest paths of the Archerville highway network is given in table 2.6.

For these calculations the following parameter values are assumed, where type 1 employment is 'commercial', and type 2 employment is 'industrial', and the parameters are defined in equations (2.3) and (2.4).

$\alpha_1 = 0.90, \quad \alpha_2 = 1.50,$
$\delta_1 = 0.50, \quad \delta_2 = 0.70,$
$\beta_1 = 0.10, \quad \beta_2 = 0.05.$

Now, substituting these values into the equations, we find the attractiveness of zone j for the location of type 1 employment is given by

$$W_{j,1} = L_{j,1}^{0.9} E_{jT}^{0.50},$$

and for type 2 employment by

$$W_{j,2} = L_{j,2}^{1.50} E_{jT}^{0.70}.$$

Using the Archerville data substituted into the employment model equations, we obtain the employment forecasts, by type and by zone as shown for 'run 1' in table 2.7. Note that the regional totals were left

Table 2.6. Archerville: the shortest zone-to-zone costs on the uncongested highway network.

Zone	Zone				
	1	2	3	4	5
1	1	6	7	10	15
2	6	1	9	7	13
3	7	9	1	5	10
4	10	7	5	1	8
5	15	13	10	8	1

unchanged from the base year. Had the regional totals been, say, 10% greater than the base year then the forecasts for each zone would also have been 10% greater. If the location of both employment types were more sensitive to transportation cost so that $\beta_1 = 0.20$ and $\beta_2 = 0.10$, then the employment forecasts would be those for 'run 2' in table 2.7. If, alternatively, the sensitivity to transportation cost had been unchanged but the industrial and commercial employment were less sensitive to the agglomeration economies represented by the E_{jT} term, with $\delta_1 = 0.30$ and $\delta_2 = 0.50$, then the results for 'run 3' in table 2.7 would be obtained.

The results of run 2 show a slight decrease of commercial employment in zone 3, the 'core' zone, with corresponding small increases in commercial employment in the surrounding zones. Industrial employment shows a similar but even smaller shift from zone 3, the 'core', to zone 2.

For run 3 there was slightly greater dispersion of both employment types, away from the 'core' zone to the surrounding zones. Both sets of results are quite reasonable, especially so, given the greatly simplified equation structure of the model, but they only begin to suggest the possibilities for experiment even with simple numerical examples such as these. Note also that the runs point up an error introduced in the simplified form of the model that does not exist in the full form. With this model form, equation (2.5), when a zone has for either or both employment types zero employment to begin with, and thus no land use of that type either, there will never be any employment of that type allocated to the zone. This is not a serious failing with respect to these simple examples, but it does show the kinds of difficulties which can arise even when one is only trying to do simple examples such as these.

Overall this simple model behaves reasonably, with perhaps a tendency to disperse employment to all zones more than would be realistic. This could be remedied by the inclusion of a constraint variable in the attractiveness measures. Simple policy experiments could be done with this model alone, or with connections to the additional models which are described later in this chapter.

Table 2.7. Archerville: employment location forecasts.

Zone	Run 1		Run 2		Run 3	
	comm.	ind.	comm.	ind.	comm.	ind.
1	129	65	131	65	133	70
2	153	140	164	147	152	147
3	317	495	292	488	293	483
4	117	0	124	0	130	0
5	34	0	39	0	43	0
Total	750	700	750	700	750	700

Note: comm. commercial, ind. industrial.

2.4 Residence location

A simple residence location model can be developed along the same lines followed for the employment location model. The residence location model, too, is a modified form of spatial interaction model and has also seen considerable application. The full model, DRAM, is described in detail in *Integrated Urban Models* (Putman, 1983a, chapters 7 and 8). The general form of the equation is

$$N_i = \sum_j E_j B_j W_i \mathrm{f}(c_{ij}) , \qquad (2.6)$$

where, in addition to the notation already defined,
N_i is the number of households, at place of residence, in zone i; and
B_j is a 'balancing' factor of the form

$$B_j = \left[\sum_k W_k \mathrm{f}(c_{kj}) \right]^{-1} . \qquad (2.7)$$

Again, the precise definitions of the attractiveness measure and of the travel cost function have been discussed in *Integrated Urban Models*. Here too, for the numerical example, arbitrary definitions will be made. The same declining exponential travel cost function of equation (2.3) will be used, with different values of β. The attractiveness measure will be taken to be the product of residential land in the zone and one plus the percentage of each of the household types in the zone. All three of these variables are raised to a parameter. Thus

$$W_i = (L_{i\mathrm{R}})^\gamma (1 + n_{i1})^\sigma (1 + n_{i2})^\rho , \qquad (2.8)$$

where

$$n_{i1} = \frac{N_{i1}}{\sum_h N_{ih}} = \frac{N_{i1}}{N_{i1} + N_{i2}} . \qquad (2.9)$$

With all the terms gathered, the equation form for each household type is

$$N_{ih} = \sum_j E_{j\mathrm{T}} [L_{i\mathrm{R}}^{\gamma_h} (1 + n_{i1})^{\sigma_h} (1 + n_{i2})^{\rho_h} \exp(-\beta_h c_{ij})]$$

$$\times \left[\sum_k L_{k\mathrm{R}}^{\gamma_h} (1 + n_{k1})^{\sigma_h} (1 + n_{k2})^{\rho_h} \exp(-\beta_h c_{kj}) \right]^{-1} . \qquad (2.10)$$

The parameters, γ, σ, ρ, and β are also assumed to have been empirically estimated by model calibration. The question of temporal definition appears here too, and again the simplest assumption is made. The left-hand side of equation (2.10) is taken to represent households of type h in zone i at time t. All variables on the right-hand side of equation (2.10) are taken to be at time $t-1$. It is again worth noting that the question of these time subscripts is a difficult one. Considerable discussion will be devoted to this in later chapters of the book.

It is also worth commenting on the use of the variable $E_{j\mathrm{T}}$ on the right-hand side of equation (2.10). In a more complete version of the model better results could be gained by using the employment types described here, multiplied by the conversion matrix given in table 2.1. Thus a substitute for $E_{j\mathrm{T}}$ would be made, of the form

$$E_j = \sum_l \phi_{lh} E_{j,l} \,, \tag{2.11}$$

where

ϕ_{lh} is the conversion rate from employees of type l to heads-of-households of type h.

This conversion procedure will not be used in the example given here, but will be used in examples later in the book.

For the numerical example the highway costs along the shortest path given in table 2.6 will again be used. For employment by zones, the run 1 forecasts from the employment model example as given in table 2.7 will be used. The assumed values of parameters will be

$\gamma_1 = 0.90,\quad \gamma_2 = 1.50,$
$\sigma_1 = 2.00,\quad \sigma_2 = 0.20,$
$\rho_1 = 0.90,\quad \rho_2 = 3.00,$
$\beta_1 = 0.20,\quad \beta_2 = 0.10.$

Now, as before, the attractiveness of zone i for the location of type 1 (LI) households may be written, and substituting the numerical values of the parameters gives us

$$W_{i1} = L_{i\mathrm{R}}^{0.90}(1 + n_{i1})^{2.00}(1 + n_{i2})^{0.90} \,.$$

For type 2, it is

$$W_{i2} = L_{i\mathrm{R}}^{1.50}(1 + n_{i1})^{0.20}(1 + n_{i2})^{3.00} \,.$$

Assuming regional forecasts of 850 LI households and 725 HI households, and taking the run 1 employment location forecasts (from the employment model numerical example), we may substitute the Archerville data into the residence model equations, yielding the household forecasts, by type and by zone, as shown in table 2.8.

The results from run 1 were produced with the use of the data and the parameter values given above. As for the employment model examples, further calculations were done for different parameter values. For run 2 the β values were reduced for both LI and HI households—to 0.10 and 0.05, respectively—to reflect a lessening of sensitivity to transport cost by both household types. Both household types showed a more dispersed pattern of location, with the largest changes being the increased numbers of households locating in zone 5, Archerville's most 'suburban' zone. In run 3, the β values were restored to their original values, but the σ and ρ values were changed to represent reductions in the apparent separation of the household types; thus σ was 1.20 (HI) and 0.90 (LI), and ρ was 1.00 (HI) and 2.10 (LI).

Use of these values results in a decrease of segregation of household types, with the concomitant substantial increase of dispersion of both household types amongst all zones. As was the case with the employment model example, these few sets of results only begin to suggest the possibilities for experimentation which exist, even with these simplified model forms. More such experiments will be described in later chapters of this book. In the next section of this chapter, the examination of the linkage from the location models to the transport models is begun.

Table 2.8. Archerville: household location forecasts.

Zone	Run 1		Run 2		Run 3	
	LI	HI	LI	HI	LI	HI
1	207	148	219	144	202	175
2	246	120	266	119	219	175
3	181	40	133	34	165	53
4	170	324	167	314	209	250
5	46	93	65	114	55	72
Total	850	725	850	725	850	725

Note: LI low income, HI high income.

2.5 Trip distribution and mode split

Having shown simple models for employment location and household (residence) location, the next step is to deal with the transport-related impacts of these locations. In particular, the concern here is with the trips made by the employees and residents located by the two location models. To simplify matters for these examples, only the home-to-work trips will be considered, and it will be assumed that there is one home-to-work trip per employee. With these assumptions it is an easy matter to extract the home-to-work trip matrix from the household location calculations.

First, note that the household location model shown in equation (2.6) is actually a trip-end summation of a zone-to-zone trip estimation procedure. Thus,

$$T_{ij} = B_j W_i f(c_{ij}) , \qquad (2.12)$$

where

T_{ij} is the number of trips by residents from place of residence in zone i to place of work in zone j.

Further, to get the total number of households, assuming one worker per household, we evaluate

$$N_i = \sum_j T_{ij} . \qquad (2.13)$$

Finally, it should be clear that, in the numerical example above, two such matrices are generated, one for the LI tripmakers and one for the HI tripmakers. These can be added together to get a matrix of work trips. These trip matrices, for run 1, are shown in table 2.9.

The next problem is that some of these trips are on the highway network and some are by transit. Thus a method to split the trips by mode will be necessary. The most frequently used model for mode split is the logit model. Many different formulations exist, and these will be discussed in greater detail in a later chapter. Here, a simple model of the disaggregate type will be used, and will be of the following general form, without zone subscripts,

$$p_{m,h} = \exp U_{m,h} \left(\sum_m \exp U_{m,h} \right)^{-1}, \tag{2.14}$$

where

$p_{m,h}$ is the probability that a traveller type h (for example, head of type h household) will use mode m;

$U_{m,h}$ is the 'utility' of mode m to traveller type h.

Table 2.9. Archerville: home-to-work trip matrices.

Zone of residence	Zone of work					Total
	1	2	3	4	5	
Low-income households						
1	65	32	102	7	1	207
2	30	110	87	17	3	247
3	9	8	154	9	2	182
4	8	19	109	31	4	171
5	2	4	25	5	10	46
Total	114	173	477	69	20	853
High-income households						
1	35	29	75	7	2	148
2	18	41	52	8	2	121
3	4	5	29	2	1	41
4	31	57	195	34	7	324
5	9	15	55	8	6	93
Total	97	147	406	59	18	727
All households						
1	100	61	177	14	3	355
2	48	151	139	25	5	368
3	13	13	183	11	3	223
4	39	76	304	65	11	495
5	11	19	80	13	16	139
Total	211	320	883	128	38	1580

Location and transportation model examples

The 'utility' measure is usually some linear multivariate expression incorporating variables describing the tripmakers and the costs and service level of the transport modes. For these examples a somewhat simplified utility measure will be used, of the form

$$U_{m,h} = a + b_1 I_h + b_2 \Delta c ,\qquad(2.15)$$

where
I_h is the household income of traveller type h,
Δc is the difference between travel costs on the different modes,
a, b_1, b_2 are empirically estimated parameters.

In cases where exactly two modes are being considered it is necessary that the probabilities for the two modes, for each ij zone pair, sum to 1.00. This being true, the algebra allows for some simplification of the model and its estimation. If mode 1 is taken to be automobile (highway) and mode 2 to be transit, then $U_{1,h}$ is arbitrarily set to zero and $U_{2,h}$ is estimated by equation (2.15). This produces the following equations. First, for just two modes, equation (2.14) expands to

$$p_{1,h} = \frac{\exp U_{1,h}}{\exp U_{1,h} + \exp U_{2,h}},\qquad(2.16.1)$$

and

$$p_{2,h} = \frac{\exp U_{2,h}}{\exp U_{1,h} + \exp U_{2,h}}.\qquad(2.16.2)$$

Now, if $U_{1,h} = 0$, then

$$p_{1,h} = \frac{1.0}{1.0 + \exp U_{2,h}},\qquad(2.17.1)$$

and

$$p_{2,h} = \frac{\exp U_{2,h}}{1.0 + \exp U_{2,h}},\qquad(2.17.2)$$

which is precisely what would be obtained if it were required that

$$p_{1,h} + p_{2,h} = 1.0 ;\qquad(2.18.1)$$

and thus

$$p_{1,h} = 1.0 - p_{2,h} .\qquad(2.18.2)$$

Substituting equations (2.16) into equations (2.18) will yield equations (2.17). This simplification allows such a model to be fitted to data for only one mode, thus reducing its application cost. For these numerical examples the following specific form will be used, where $U_{m,h,ij}$ is the utility of mode m to traveller of type h travelling between zones i and j.

$$U_{1,h,ij} = 0.0 ,\qquad(2.19.1)$$

and

$$U_{2,h,ij} = 1.3 - 0.01 I_{i,h} + 0.40 \Delta c_{ij} , \qquad (2.19.2)$$

where
- $I_{i,h}$ is the average income of type-h residents in zone i (in hundreds of dollars),
- Δc_{ij} is equal to $c_{1,ij} - c_{2,ij}$,

The dependent variables will be $p_{1,h,ij}$ and $p_{2,h,ij}$, the probabilities that a type-h resident in zone i will choose mode 1 (or mode 2) for the journey to a place of work in zone j.

For example, from table 2.6 the highway cost from zone 2 to zone 3 is 9 units. From the transit network in figure 2.3 and table 2.5, the minimum cost of travel by transit from zone 2 to zone 3 is 11.5 units. The entire matrix of zone-to-zone transit costs is given in table 2.10. If it is assumed that the average income of LI households is $15 000 and that of HI households is $25 000 it is possible to calculate the probability that an individual, of either type, will use either mode. For example, from zone 2 to zone 3 $\Delta c_{ij} = 9 - 11.5 = -2.5$, so for employed heads of LI households

$$\begin{aligned} p_{1,1,23} &= \{1.0 + \exp[1.3 - 0.01\,(150) + 0.40\,(-2.5)]\}^{-1} , \\ &= [1.0 + \exp(-1.20)]^{-1} , \\ &= (1.301)^{-1} , \\ &= 0.769 . \end{aligned}$$

Thus between zone 2 and zone 3, 77% of employed heads of LI households will use automobiles and 23% will use transit. For employed heads of HI households the figures are 90% and 10% respectively.

Table 2.10. Archerville: shortest zone-to-zone path costs on transit network.

Zone	Zone				
	1	2	3	4	5
1	1.5	8.5	9.0	13.4	21.5
2	8.5	1.5	11.5	16.0	24.0
3	9.0	11.5	1.5	7.5	15.5
4	13.5	16.0	7.5	1.5	11.0
5	21.5	24.0	15.5	11.0	1.5

To continue the example, the work trips of each of the two income groups may now be split into transit trips and highway trips by use of equations (2.17.1), (2.17.2), (2.19.1), and (2.19.2). The matrices of highway trip probabilities are given in table 2.11. Applying these trip probabilities to the trip matrices of table 2.9 and then summing the trips by mode gives us the trips by mode shown in table 2.12. Just as with the two location models, it is possible to do simple sensitivity tests of this mode split model by varying one or more parameters and observing the consequences. If, for example, sensitivity to the difference between transit cost and highway cost were to be greater, as expressed by the parameter b_2 increasing from 0.40 to 0.60, then there would be modest increases in highway trips and modest decreases in transit trips for most of the origin–destination pairs in Archerville. The first row of the highway trips matrix of table 2.12 would become 69, 54, 150, 13, and 3. The first row of the transit matrix would become 31, 7, 26, 1, and 0. A decrease in b_2 to a value of 0.2, marking a slight reduction in sensitivity to the modal cost differences, would result in a modest shift to increased

Table 2.11. Archerville: highway trip probabilities.

Zone	Zone				
	1	2	3	4	5
Low-income travellers					
1	0.60	0.77	0.73	0.83	0.94
2	0.77	0.60	0.77	0.98	0.99
3	0.73	0.77	0.60	0.77	0.92
4	0.83	0.98	0.77	0.60	0.80
5	0.94	0.99	0.92	0.80	0.60
High-income travellers					
1	0.80	0.90	0.88	0.93	0.98
2	0.90	0.80	0.90	0.99	1.00
3	0.88	0.90	0.80	0.90	0.97
4	0.93	0.99	0.90	0.80	0.92
5	0.98	1.00	0.97	0.92	0.80

Table 2.12. Archerville: work trips by mode.

Zone	Zone				
	1	2	3	4	5
Highway (mode 1)					
1	67	51	140	12	3
2	40	99	113	24	4
3	10	10	115	9	2
4	35	75	260	46	9
5	10	18	77	11	11
Transit (mode 2)					
1	33	10	36	2	0
2	9	52	25	0	0
3	3	2	67	2	0
4	3	1	45	19	1
5	0	0	4	2	5

transit trips and decreased highway trips. Clearly, more such tests could be done in order to more fully understand the behavior of this type of model. For the purposes of this chapter, however, the next item to be discussed is the fourth model procedure in the set: trip assignment.

2.6 Trip assignment
The last step of the procedures being described in this chapter is to assign the trips to the networks. This procedure takes the appropriate matrix of trips and assigns (routes) them to paths on the links of the network. If there were no maximum capacities associated with the links then the procedure would be relatively uninteresting, being simply a matter of finding the shortest path through the network for each zone-to-zone pair and then merely assigning the trips to the shortest paths. What makes the matter interesting, and of considerable importance to the concept of integrated location and transportation analysis, is the fact that the presence of these trips on the network results in link congestion. This congestion, in turn, results in changes in the link travel times and costs. These altered network characteristics, themselves, imply a potential rearrangement of locating activities, new trip matrices, and so on.

In this numerical example it has been assumed that the transit system has no capacity limits and therefore does not show changes in transport times or costs as a function of numbers of transit trips. The highway network, however, does have capacity specifications for each link, as given in table 2.3. The following functional form will be taken to describe the consequences on travel cost of congestion on a specific highway network link.

$$t_\ell = t_\ell^{(0)} \left[1 + \alpha \left(\frac{x_\ell}{k_\ell'} \right)^\beta \right], \tag{2.20}$$

where

t_ℓ is the 'congested' travel time or cost over link ℓ,
$t_\ell^{(0)}$ is the travel time or cost for free flow, that is, the travel time or cost on the uncongested link,
x_ℓ is the flow (volume) over link ℓ,
k_ℓ' is the 'design' or 'practical' capacity of link ℓ,
α and β are parameters whose usual values are 0.15 and 4, respectively.

Thus in any case where the link flow exceeds the 'design' capacity of that link, the link travel cost will increase as per equation (2.20). There are several algorithms which have been used to deal with this problem, all of which are discussed in considerable detail in chapter 6 and chapter 7 of this book. In this chapter only a few of these will be tested, for the purpose of showing the consequences of their use.

The first algorithm is called the 'all-or-nothing' (AON) algorithm. With this, the shortest path from each zone to each other zone is found, and all the trips going between those zones are assigned to the links on that path.

The presence of these trips on the 'shortest' path may congest the path's links to such a degree that it is no longer shortest, but in the AON algorithm no notice is taken of this fact. The travel costs over the congested network may be appreciably different from the original, unloaded network, travel costs.

Consider, for example, the minimum path from zone 1 to zone 5, which was found earlier in this chapter to be along links connecting nodes 1, 6, 12, 13, 16, 15, and 5, respectively. (Note: this path will now be expressed as 1-6-12-13-16-15-5; similarly for other paths.) The total cost of travelling this path, when there is no congestion, is 15 units. A glance at the matrix of highway trips in table 2.12 will show only 3 trips from place of residence in zone 1 to place of work in zone 5. As no link in the highway network has a capacity less than 50, one might be tempted to think that there would be no change in travel cost from zone 1 to zone 5 due to congestion. Yet there may well be trips between other origin-destination pairs that use some of the same links as are used by the trips from zone 1 to zone 5. First, all trips leaving zone 1 must use the 'connector' link from zone 1 to node 6. There are 206 trips from zone 1 to other zones, thus producing a slight degree of congestion on the connector link (note that intrazonal trips, that is, zone 1 to zone 1, are dealt with apart from the interzonal trips, that is, those from zone 1 to all other zones). The link from node 6 to node 12 carries 155 trips, 3 going from zone 1 to zone 5, 12 going from zone 1 to zone 4, and 140 going from zone 1 to zone 3. Thus the high congestion on the link from node 6 to node 12 is mainly due to trips from zone 1 to zone 3, but must be endured, under this assignment procedure, by those going from zone 1 to zones 4 and 5 as well. Tracing the flows on all the links enables us to calculate the consequences of congestion throughout. After these trips are assigned and the new congested costs calculated, the originally shortest path from zone 1 to zone 5 has a length of 84 units (owing primarily to the high level of congestion on the link from node 6 to node 12). This is no longer the shortest path from zone 1 to zone 5. The new shortest path is 1-6-7-8-12-13-16-15-5, and has a length of 16 units. Tracing these new shortest paths through the entire congested network gives the revised matrix shown in table 2.13. In addition to noting that many of the

Table 2.13. Archerville: shortest zone-to-zone path costs on an AON congested highway network.

Zone	Zone				
	1	2	3	4	5
1	1	7	19	11	16
2	6	1	23	7	13
3	8	9	1	5	10
4	22	18	29	3	10
5	16	17	21	13	1

zone-to-zone path costs are greater than those through the uncongested network as given in table 2.6, we must also note that the cost matrix is no longer symmetric. This can have a substantial impact on the estimates of employment and residence location.

The fact that the AON assignment procedure produces such uneven assignments of trips to the network links is clearly a problem. An alternative procedure called stochastic multipath (SM) assignment is an attempt to remedy this problem. In SM assignment the algorithm is used to investigate all reasonable paths between a given origin–destination pair and then to assign trips to these paths, pro rata, according to their costs. The exact scheme for allocating the trips to the paths is a matter of discussion in the literature and will be further examined in chapter 6. It suffices here to say that one such scheme uses a logit formulation similar to that of equation (2.14), where in this case the utilities are simply the path costs, and all travelers are assumed to have the same utility function. Trips from zone 1 to zone 5 for example, could use any of a number of paths such as
path 1: 1-6-7-8-12-14-15-5 (16 units);
path 2: 1-6-7-8-12-13-16-15-5 (16 units);
path 3: 1-6-12-14-15-5 (15 units);
path 4: 1-6-12-13-16-15-5 (15 units);
path 5: 1-6-9-11-14-15-5 (17 units).
Other paths are also possible. Note the number in parentheses at the end of each path description is the path cost on the uncongested network. The logit path split model would be of the form

$$p_{ij}^{(p)} = \exp(-\theta c_{ij}^{(p)}) \bigg/ \sum_k \exp(-\theta c_{ij}^k) , \qquad (2.21)$$

where
$p_{ij}^{(p)}$ is the probability of taking path p from zone i to zone j,
$c_{ij}^{(p)}$ is the cost of taking that path,
The parameter θ has the role of determining the degree of dispersion of trips over the reasonable paths [k in equation (2.21) denotes summation over the set of reasonable paths]. When θ equals 0, the trips are distributed evenly over all reasonable paths. As θ becomes greater than 1.00 the trips tend to be distributed primarily to the minimum path. If the five paths enumerated above are taken to be all the reasonable paths from zone 1 to zone 5, then the probabilities of taking each path (assuming for illustration's sake that $\theta = 0.5$) will be:

$$p_{1\,5}^{(1)} = \exp(-0.5 \times 16)[\exp(-0.5 \times 16) + \exp(-0.5 \times 16)$$
$$+ \exp(-0.5 \times 15) + ...]^{-1} ,$$

and so on, to give $p_{1\,5}^{(1)} = 0.169$, $p_{1\,5}^{(2)} = 0.169$, $p_{1\,5}^{(3)} = 0.279$, $p_{1\,5}^{(4)} = 0.279$, $p_{1\,5}^{(5)} = 0.104$.

The trips from zone 1 to zone 5 would be allocated to these paths accordingly. After the table 2.12 highway trip matrix is assigned to the Archerville highway network by using the SM procedure, the shortest path from zone 1 to zone 5 is the same as that for AON assignment, and so is its total cost. This is owing to the small numbers of trips involved. The overall zone-to-zone costs for the SM congested network are somewhat different from those of the AON congested network, and are given in table 2.14. Note that two sets of results are given, for $\theta = 0.5$ in table 2.14(a), and for $\theta = 0.1$ in table 2.14(b). Individual network links also show different flows and congestion levels. Again note that the congested highway cost matrix is not symmetric.

In both cases, the fact that once trips are assigned to a highway network link they cannot later be reassigned to a less congested link, is both unrealistic, and may tend to distort the final zone-to-zone cost estimates. A third type of trip assignment procedure, called user-equilibrium (UE) assignment, deals with this problem by doing what is essentially a reassignment of trips to less congested links. As will be discussed in chapter 6, the purpose of these reassignments is to achieve an equilibrium of trips on the network links. The overall zone-to-zone costs for the congested network resulting from UE assignment is given in table 2.15. These costs are noticeably greater (for most zone-to-zone pairs) than those from either AON or SM assignment results. At first glance this might appear to be an unwanted result. In fact, what has happened is that the trips were more evenly spread over the network links as the UE algorithm reassigned trips to less congested paths. There are 42 one-way links (twice the 21 two-way links) in the network. The AON algorithm leaves 12 links unused. The SM algorithm

Table 2.14. Archerville: shortest zone-to-zone path costs under stochastic multipath assignment for the congested highway network.

Zone	Zone				
	1	2	3	4	5
(a) $\theta = 0.5$					
1	1	7	19	11	16
2	7	1	22	7	13
3	7	9	1	5	10
4	17	12	26	3	10
5	16	17	22	9	1
(b) $\theta = 0.1$					
1	1	13	19	11	16
2	12	1	26	7	13
3	7	9	1	5	10
4	19	9	26	3	10
5	16	14	22	9	1

uses all links, but assigns rather low volumes to links on paths longer than the minimum path (alternate runs with different values of θ change this somewhat). The UE algorithm uses all but 5 of the links.

Again considering the path from zone 1 to zone 5, the minimum path through the UE congested network uses the same links as were used for both the AON and the SM congested networks. The path length resulting from both AON and SM assignments was 16 units, whereas the path length for the UE solution is 25 units. The path which had been minimum before trip assignment had lengths (cost) of 84, 23, and 27 units following trip assignment by the AON, SM, and UE algorithms respectively. For paths with more highly congested links the AON and SM results tend to be closer together in magnitude (SM larger than AON in this case) and the UE result notably smaller.

As mentioned above, there is much more to be said about trip assignment algorithms in a later chapter of this book. What is important here is: (1) regardless of which trip assignment algorithm is used, the congested network costs are greater than the 'empty' network costs, and (2) the congested network cost matrix is never symmetric. Both these results have important implications for the location model forecasts. Clearly if any of the

Table 2.15. Archerville: shortest zone-to-zone path costs under user-equilibrium assignment for congested highway network.

Zone	Zone				
	1	2	3	4	5
1	1	7	28	12	25
2	14	1	34	7	30
3	7	9	1	5	10
4	44	32	57	3	27
5	31	33	37	9	1

Table 2.16. Archerville: forecasts of employment and household location with use of travel costs on the AON congested network.

Zone	Employment		Households	
	comm.	ind.	LI	HI
1	163	92	134	185
2	183	197	200	146
3	158	411	456	131
4	189	0	35	151
5	57	0	25	112
Total	750	700	850	725

Note: comm. commerical; ind. industrial; LI low-income; HI high-income.

matrices of congested network costs were substituted into the employment or residence location models, there would be a rearrangement of spatial patterns. Table 2.16 shows the results of recalculating the employment and residence location using the AON congested network costs. A comparison with the original results obtained using the travel costs on the uncongested network, given in tables 2.7 and 2.8, shows significant differences. If these forecasts were used, along with the travel costs on the congested network, to calculate work trips and the modal splits, the results would differ from the originals too. If these new highway-using work trips were then assigned to the Archerville highway network, new levels of congestion would result. There are several important issues related to both the theory and the mechanics of such linked models, and these will be discussed throughout the remainder of the text. For the purposes of this chapter the point is that the characteristics of the transportation facilities affect the region's spatial patterns of activities, and these spatial patterns, in turn, affect the characteristics of the region's transportation facilities.

2.7 A note on simple numerical examples

Before moving on to the concluding section of this chapter a small precautionary note is in order. Small numerical examples of models such as those presented here suffer from a property perhaps best described as 'discreteness' of observations. This property results in 'lumpy' forecasts because there are too few zones and/or too few locators to allocate, and because there are too few network links and/or too few trips to assign to them. As an example, assume there are two equally attractive zones and only three residents to be allocated between them, or assume two network links of equal length and five trips to be allocated. The difference between a 2 : 1 allocation of residents to zones and a 1 : 2 allocation is, in percentage terms, substantial. So too, for the difference between a 3 : 2 or a 2 : 3 split of trips to paths.

This is not to say that small numerical examples are not useful—indeed, for pedagogical purposes they are unequalled. Yet in their design, that is, in the selecting of variables, abbreviated functional forms, parameters, and so on, one must carefully consider the purposes for which they are being developed. Everything must work out as intended, the 'discreteness' problem must not be allowed to interfere with the exposition. Setting up a small numerical example to run through a sequence of several models is not a trivial task. For research purposes such examples must be approached with great caution. Care must be taken to ensure that conclusions ascribed to model structure are not actually artifacts of the specific example. This point will be mentioned again; but especially in a text where there will be regular shifting from small examples to more realistic problems, the point is sufficiently important to be made specifically.

2.8 Conclusion

The four types of model presented in this chapter are the building blocks for the work which follows. Although there will be many modifications and rearrangements there will be similar elements throughout. A familiarity with the material presented here will allow a ready progression into the developments which follow.

On a substantive note, the matter of integrated transportation and location model systems should be raised here. In later chapters many different conceptualizations of such integrated systems will be presented and discussed. The four models presented here (with the addition of a procedure for producing a composite zone-to-zone transportation cost which takes both highway and transit costs into account) form the principal ingredients of an integrated system. The connection of the congested network costs back to the location models is the primary feedback loop. It is clear, however, that having made this connection a whole new series of questions is raised regarding the starting points, equilibrium conditions, algorithms, and so on of the solution. These questions are the driving force behind much of what will be presented in subsequent chapters. In addition to analyses of the performance of such systems of linked models, new formulations of simultaneous models, some presented within mathematical programming frameworks, will be described and evaluated. Before all that though, some further groundwork must be laid. In the next chapter, again by use of small numerical examples, the concepts of optimality and techniques of mathematical programming necessary for further model developments to follow will be introduced.

Optimization and optimizing models

3.1 Introduction

The purpose in this chapter is to introduce some of the principal concepts of optimization methods as they apply to land-use and transportation analysis. It is not intended as a replacement for a thorough understanding of the area, but as a refresher or an introduction. The hope is that by emphasizing those concepts and techniques of greatest importance at the expense of others, equally interesting, but less directly relevant, enough can be conveyed to give the reader the necessary background from which to continue in this text as well as others.

The next section will contain a brief review of the concepts of optimality in general terms. This will be followed by a discussion of linear programming, including some numerical examples in which the Archerville data are used. In the fourth section, other nonlinear optimization techniques in general will be discussed, and use will be made of the Archerville data for a small numerical example. In the fifth section, nonlinear programming will be discussed, and again the Archerville data will be drawn upon.

3.2 Optimality

Virtually everyone has an intuitive concept of the meaning of optimality. Here a brief review of its more formal mathematical definition will be given. Before starting, however, it should be noted that there are several excellent treatments of mathematical programming (Bradley et al, 1977; Killen, 1983; Lapin, 1981). This chapter should by no stretch of the imagination be considered an equal substitute for one of those books. The interested reader is strongly urged to go to such texts for additional material on these topics. This chapter will serve to provide the background for the material which follows in the rest of the book.

In discussing optima and optimality it is useful to do an immediate classification of types of problem. Perhaps to begin with, the most useful will be a two-by-two classification. First the functional forms being dealt with can be either linear or nonlinear. Second, the problem can be either constrained or unconstrained. A third matter of great importance is that the functional forms involved may be continuous or discontinuous. This latter classification will be dispensed with forthwith by requiring that all problems considered here have functional forms which are continuous in the ranges being considered.

3.2.1 The objective function

An optimum can be either a maximum or a minimum. For the most part it is irrelevant to the discussion as to whether a function is to be maximized or minimized as most problems can be converted from one type to the other by some simple algebraic changes of sign. In some cases, there will be questions about shapes of function which may affect such issues. These will be mentioned as they arise. With a strictly linear objective function the problem is quite straightforward, being simply a matter of getting as far from the origin as possible (in the case of maximization). In the multidimensional case the situation is the same. In general the minimizing or maximizing of a linear objective function is rather uninteresting unless there are constraints on one or more of the variables. With nonlinear objective functions, even without constraints, the problems can be considerably more difficult. This will depend upon the mathematical properties of the function itself, as will be seen in the discussion of nonlinear optimization later in this chapter. Thus although it is true that one can postulate any sort of algebraic function as the objective function in an optimizing problem, one really should be quite careful in their formulation, as there can be vast differences in the ease of solution of such problems as a result of what might seem to be rather modest differences in the formulation of the objective functions.

3.2.2. The constraints

Optimization problems can be either constrained or unconstrained. Unconstrained linear optimization is uninteresting. Unconstrained nonlinear optimization can be quite tricky. Constraints can themselves be either linear or nonlinear. A linear objective function with linear constraints gives a linear programming problem. If either or both the objective function and the constraints are nonlinear the problem may or may not be solvable by mathematical programming techniques. In many cases problems with nonlinearities are solved by developing linear approximations. Thus as a general introduction to mathematical programming, the next section of this chapter begins with a discussion of linear programming.

3.3 Linear programming

It would be presumptuous to even suggest that just a portion of a chapter could provide a complete introduction to this topic. There are many excellent texts which may be consulted. The intent here is to give a sense of what is being done by such methods and how they are relevant to this book's concerns with transportation and activity location. In linear programming both the objective function and all the constraints must be linear. The sense of a linear programming approach may best be conveyed with a simple example and a graphical solution.

3.3.1 Graphical solution method

Consider the following information on a health-care clinic for low-income families. Two strategies were to be investigated. The first involved sending people employed by the clinic on a house-to-house tour of the area being served. These employees would discuss health problems and when necessary they would urge people to come to the clinic, would offer transportation, and so on. This program could result in a reduction of deaths in the population of 3 persons per 5 000.

Optimization and optimizing models

The second strategy involved a smaller number of clinic employees who would make telephone calls to families in the clinic service area and then attempt similar conversations, though not face-to-face, to those held by the house-to-house interviewers. This program could result in a reduction of 1 death per 5000.

Obviously the costs for the two programs were rather different. The clinic estimates that the costs for the house-to-house program were $1000 per 5000 population reached, whereas the costs for the telephone program were $200 per 5000 population reached.

Suppose that the clinic has a $100000 budget for this program. It is now possible to write the problem in equation form. First, the goal of the clinic is to maximize the number of deaths prevented. There are 3 prevented per 5000 population for the house-to-house program (H) and 1 per 5000 population prevented with the telephone program (T). Thus in equation form the goal is expressed by

maximize $e = 3X_H + X_T$,

where X_H and X_T are the number of house-to-house and telephone programs, respectively, and e is the effectiveness. The costs are $1000 for one unit (reaching 5000 people) of the H program and $200 for one unit (reaching 5000 people) of the T program. Thus

total cost = $1000X_H + 200X_T$.

There is a $100000 budget maximum, so

$1000X_H + 200X_T \leq 100000$.

Last, for mathematical consistency we require that both X_H and X_T are nonnegative. Thus we have

$X_H \geq 0$ and $X_T \geq 0$.

This problem may be depicted graphically as in figure 3.1. First calculate the maximum number of telephone programs, X_T, which can be done within the budget. That is

$$\frac{\$100000}{\$200} = 500.$$

The maximum number of house-to-house programs, X_H, that can be done is

$$\frac{\$100000}{\$1000} = 100.$$

The budget constraint is shown in figure 3.1 as a straight line connecting these two extremes. The region of feasible values of X_H and X_T is the finite area bounded by the budget-constraint line and the two axes and is shown shaded in figure 3.1. Next the program effectiveness may be put on the graph as follows. Start at the point $X_H = 100$ on the X_H axis. At this point

$X_T = 0$, so by substituting into the objective function, we obtain

$$e = 3X_H + X_T = 3 \times 100 + 0 = 300 .$$

Move next to the X_T axis. On the X_T axis the value of X_H is 0 so, substituting into the above equation, we have

$$e = 300 = 3 \times 0 + X_T ,$$

and thus $X_T = 300$. This point is then connected to the $X_H = 100$ point on the X_H axis (the line '$e = 300$' in the figure 3.1). Another line can be drawn for $e = 400$. This line intersects the X_T axis at the point $X_T = 400$. Now, to calculate the X_H axis intercept, we use

$$e = 400 = 3X_H + 0 ,$$

and thus $X_H = 133$. Last, the line for $e = 500$ can be drawn in the same fashion:

$$e = 500 = 3X_H + 0 ,$$

and so $X_H = 167$.

Now, looking at the graph, we see that the line for $e = 300$ falls entirely within the 'feasible region', whereas the line for $e = 500$ touches the feasible region only at the point when $X_T = 500$ and $X_H = 0$. This point is the optimal (or best) solution to the problem of how to allocate the health clinic's funds in such a way as to maximize the effectiveness of the program in reducing deaths in the low-income population.

Suppose, now, that there is an additional constraint: owing to a shortage of trained interviewing personnel the clinic can only do 75 of the house-to-house programs. The problem is now written as

$$\text{maximize } e = 3X_H + X_T ,$$

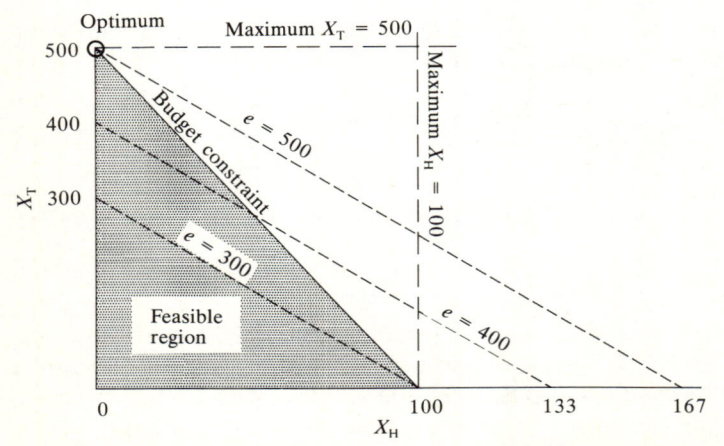

Figure 3.1. Solution to health-clinic problem with budget constraint.

subject to

$$1000X_H + 200X_T \leq 100\,000,$$

and to

$$X_H \leq 75,$$

and

$$X_H \geq 0 \text{ and } X_T \geq 0.$$

The result of adding this new constraint is to reduce the size of the feasible region, as shown in figure 3.2. However, as can be seen in this figure, as the reduction in the feasible region does not contact the line for $e = 500$, the constraint has no effect on the optimal solution.

Next, suppose the office-personnel situation restricts the health clinic to no more than 250 telephone programs. Now the problem is written

maximize $e = 3X_H + X_T$.

Subject to

$$1000X_H + 200X_T \leq 100\,000,$$

and to

$$X_H \leq 75,$$
$$X_T \leq 250,$$

and

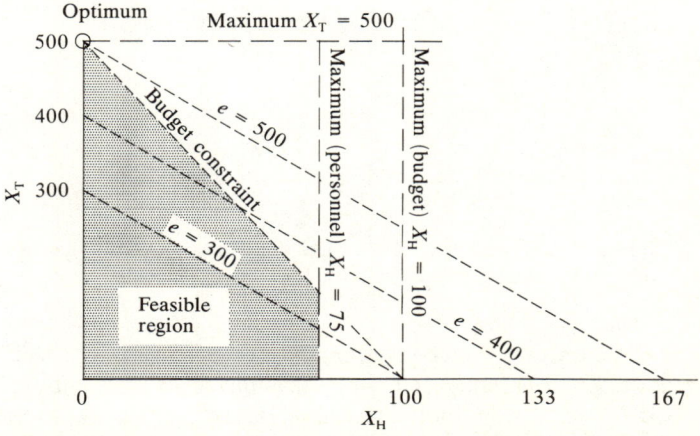

$$X_H \geq 0 \text{ and } X_T \geq 0.$$

Figure 3.2. Solution to health-clinic problem with budget constraint plus interviewing-personnel constraint.

The feasible region is thus reduced as shown in figure 3.3, and the line for $e = 500$ no longer touches the feasible region. As the optimal solution must fall within, or just touch the edge of, the feasible region, it is clear that this new constraint will change the problem solution. With $X_H = 75$, the clinic will use $75 000 of the budget for house-to-house programs. This leaves $25 000 for telephone programs. The effectiveness of this plan is found by first calculating that they can do $25 000/$200 (125) telephone programs. So, we may now calculate the effectiveness:

$$e = 3X_H + X_T = 3 \times 75 + 125 = 350 \ .$$

Suppose that, next, X_T is set to 250, thus requiring $50 000 of the budget for telephone programs. This leaves $50 000 for house-to-house programs. Now, this means there are a possible $50 000/$1 000 (50) house-to-house programs, so the effectiveness is given by

$$e = 3X_H + X_T = 3 \times 50 + 250 = 400 \ .$$

Thus this is a more effective use of the total budget and is indeed the optimal solution subject to the constraints. This graphical solution method, although pedagogically useful, cannot be used when there are more variables than can conveniently be expressed graphically (for example, n-dimensional graph paper is still a rather expensive commodity). To solve more realistic problems a more comprehensive approach is necessary.

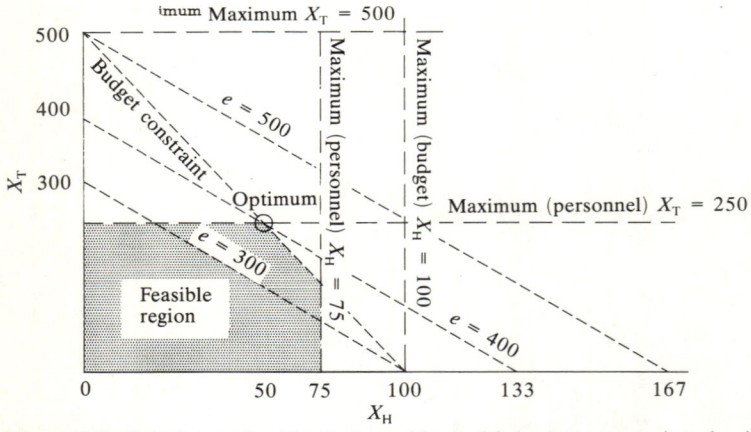

Figure 3.3. Solution to health-clinic problem with budget constraint plus interviewing-personnel and telephoning-personnel constraints.

3.3.2 The simplex method
The simplex method for solving linear-programming problems is, in effect, an algebraic procedure for efficiently examining feasible problem solutions until the optimum is found (if possible). Its success depends upon, in addition to a proper formulation of the problem, the convenient fact that the optimal solution will be at a vertex of the feasible region. A vertex is one of

the points where two constraints intersect or where a constraint intersects one of the axes. The simplex method is an efficient procedure for examining in turn each of the vertices in the problem to see if it is the optimal solution.

To apply the simplex method to a linear programming problem, the problem must be formulated in a particular way, called a canonical form. Any linear programming problem can, by a series of transforming steps, be put in canonical form.

The first step is to ensure that all the original variables in the problem (also known as the 'decision' variables) have a sign constraint. This is normally done for most variables in the nonnegativity constraints, such as the requirements that both X_H and X_T be greater than or equal to zero in the sample problem discussed above. If any decision variable in the original problem is not so constrained then it must be replaced by a pair of new variables with unit coefficients and opposite signs. This must be done both in the objective function and in the constraint equations. To illustrate, consider the following small problem:

maximize $Z = 2X_1 + 3X_2 + X_3 - 5X_4$,

subject to

$X_1 + X_4 \leq 5$,

$3X_2 - X_3 \leq 4$,

and

$X_1, X_2, X_3 \geq 0$.

Note that there is no constraint on the sign of X_4. Thus X_4 must be replaced by two new variables, X_5 and X_6:

$X_4 \rightarrow X_5 - X_6$.

The problem thus becomes

maximize $Z = 2X_1 + 3X_2 + X_3 - 5X_5 + 5X_6$,

subject to

$X_1 + X_5 - X_6 \leq 5$,

$3X_2 - X_3 \leq 4$,

and

$X_1, X_2, X_3, X_5, X_6 \geq 0$.

The second step is to change all inequality constraints to equalities by the addition of variables called 'slack' or 'surplus' variables. Consider the following constraint:

$3X_2 - X_3 \leq 4$;

with the addition of the slack variable X_7, this constraint becomes

$3X_2 - X_3 + X_7 = 4$.

Note that X_7 does not get included in the objective function; thus if it has some numerical value other than 0 it will not contribute to the value of the objective function. The slack variable(s) will be subject to nonnegativity constraints.

Now consider the constraint

$X_2 + 5X_3 \geqslant 6$;

here a surplus variable, X_8, must be added to yield

$X_2 + 5X_3 - X_8 = 6$.

Note that surplus variables have coefficients of -1, and that they too are not included in the objective function. Surplus variables will also be subject to the nonnegativity constraint.

The third step is to find any constraints with a negative value on the right-hand side of the equation and multiply the entire equation by -1. This will result in no equation having a negative value on the right-hand side. Consider the constraint

$X_1 + 3X_2 \geqslant -10$:

First convert it to an equality by adding a surplus variable:

$X_1 + 3X_2 - X_9 = -10$.

Next multiply the equation through by -1 to eliminate the negative right-hand side:

$-X_1 - 3X_2 + X_9 = 10$.

The simplex algorithm, as mentioned above, begins at one of the vertices of the solution space (the solution space forms what is known in linear algebra as a simplex, hence the name of the algorithm), and examines each vertex, in turn, for an optimum solution to the problem. Obviously, to begin the process there must be a starting point—the initial feasible solution. In many cases the initial feasible solution will simply be 'all decision variables are equal to zero'. In some cases, however, finding an initial feasible solution can be a bit of a trick. These problems and their solutions are discussed in all books on mathematical programming and need not be treated here. There can also be cases where there is no feasible solution or no unique optimum. These problems, too, are dealt with in books on the topic and need not be discussed here.

The health-clinic problem solved graphically above was

maximize $e = 3X_H + X_T$,

subject to

$$1000X_H + 200X_T \leq 100\,000,$$
$$X_H \leq 75,$$
$$X_T \leq 250,$$

and

$$X_H, X_T \geq 0.$$

This can be put in canonical form with no change in the objective function and with the constraints converted to

$$1000X_H + 200X_T + X_1 = 100\,000,$$
$$X_H + X_2 = 75,$$
$$X_T + X_3 = 250,$$

and

$$X_H, X_T, X_1, X_2, X_3 \geq 0.$$

The simplex solution yields the same result as the graphical solution:
$e = 400$, $X_1 = 0$,
$X_H = 50$, $X_2 = 25$,
$X_T = 250$, $X_3 = 0$.

Further, the values of the slack variables tell us about the optimal solution in relation to the constraints. X_1 is 0, showing that the budget is fully utilized (the first constraint is the budget constraint). X_3 is 0, showing that the maximum allowable number of telephone interviews is being done (the third constraint). X_2 is 25, indicating that twenty five more house-to-house programs could be done if there were budget available to do them.

It is possible to do a great deal more analysis of such simplex solutions. One of the principal methods involves the analysis of what is known as the dual program solution. The set of equations solved above may be referred to as the primal problem. Every primal problem has an associated dual problem. The dual problem can be created by, in effect, transposing the primal.

Consider the general primal problem:

maximize $Z = \sum_j C_j X_j$, $j = 1, n$,

subject to

$$\sum_j a_{ij} X_j \leq b_j, \quad j = 1, n, \quad i = 1, m,$$

and

$$X_j \geq 0.$$

Then the corresponding dual problem is

$$\text{minimize } V = \sum_i b_i Y_i, \quad i = 1, m,$$

subject to

$$\sum_i a_{ij} Y_i \geq c_j, \quad j = 1, n, \quad i = 1, m,$$

and

$$Y_i \geq 0.$$

Now consider the health-clinic problem written in the form shown above with all coefficients included even when they are 0 or 1:

$$\text{maximize } e = 3X_H - 1X_T,$$

subject to

$$1000X_H + 200X_T \leq 100\,000,$$
$$1X_H + 0X_T \leq 75,$$
$$0X_H + 1X_T \geq 250,$$

and

$$X_H, X_T \geq 0.$$

Then the dual will be

$$\text{minimize } V = 100\,00\,Y_1 + 75\,Y_2 + 250\,Y_3,$$

subject to

$$1000\,Y_1 + 1\,Y_2 + 0\,Y_3 \geq 3,$$
$$200\,Y_1 + 0\,Y_2 + 1\,Y_3 \geq 1,$$

and

$$Y_1, Y_2, Y_3 \geq 0.$$

To solve the dual problem by use of the simplex algorithm slack variables need to be added to the two constraints. These may be called Y_4 and Y_5. The solution produced by the simplex algorithm is

$V = 400,$ $Y_3 = 0.04,$
$Y_1 = 0.003,$ $Y_4 = 0.00,$
$Y_2 = 0.00,$ $Y_5 = 0.00.$

Note first that at their optimum values e and V are equal, both being 400. What is perhaps more interesting is the meaning of the variables Y_1, Y_2, and Y_3 in the dual. These are, in effect, opportunity costs. They give, at the optimal solution to the dual, the amount of increase in the health clinic's overall effectiveness, e, which would result from the availability of one more unit of

Y_1, Y_2, or Y_3. Thus Y_1, from the budget constraint of the primal problem, shows that each additional dollar of budget, if spent in the most efficient way, would yield an increase of 0.003 units of clinic effectiveness. Y_2, from the constraint on the number of house-to-house interviews, shows that no resources should be devoted to such interviews until all possible staff resources are used on telephone interviews. The value of Y_3 shows that additional staff resources, if they were available, should be devoted to telephone interviews, and would result in an increase in the clinic's effectiveness of 0.4 units per unit of telephone-interview program performed. Of course, in this simple problem these facts may have been evident by inspection, but in a larger sized problem it is much more difficult to learn these things in such a way. Thus the ability of the dual to produce information on these opportunity costs or, shadow prices, can be invaluable to the policy analyst.

Last, it should be noted that by simple algebraic manipulations the solution to the primal can give the solution to the dual, and vice versa. For computational purposes the results are equivalent. Yet, it is a fact that the simplex algorithm works more efficiently on problems with a greater number of objective function variables than constraints. Thus if one is faced with a problem having rather few objective function variables, but rather a lot of constraints, its solutions may be more efficiently produced by solving the dual. This fact will be quite useful in problems to be dealt with later in this book. Next, to illustrate the use of the simplex algorithm, some transport network problems will be solved for Archerville

3.3.3 The shortest-path problem

A frequently encountered problem in transportation and location analyses is that of finding the shortest path from one node to another over the links of a network. This is a problem which can be considered as a linear programming problem. The general form is

$$\text{minimize } Z = \sum_{i,j} c_{ij} x_{ij} , \tag{3.1}$$

subject to

$$\sum_j x_{ij} - \sum_k x_{ki} = \begin{cases} 1, & \text{if } i \text{ is the origin,} \\ -1, & \text{if } i \text{ is the destination,} \\ 0, & \text{otherwise,} \end{cases} \tag{3.2}$$

and

$$x_{ij} \geq 0, \quad \forall i, j , \tag{3.3}$$

where

c_{ij} is the cost of traversing link i, j,
x_{ij} is the flow (number of trips) on link i, j.

The objective function is straightforward, being simply to minimize the sum of the link cost multiplied by the trips incurring that cost. The principal constraint equations [expressions (3.2)] are a set of flow–balance relationships which ensure that the flows balance at each network node. Thus for each node the total flow of trips into the node minus the total flow of trips out of the node must equal the net trips supplied (or demanded) at the node. The last set of constraint equations [expression (3.3)] are simply the nonnegativity requirements which prohibit negative flows. It can readily be seen that writing the shortest-path problem in this form yields a good-sized problem. In the objective function the number of terms equals the number of links in the network. There must be a constraint equation for each network node [equation (3.2)], and a constraint equation for each network link [inequality (3.3)].

To give a sense of what is involved, the Archerville highway network can be used to illustrate the expression of a minimum-path problem in terms of a linear program. Note that here one can deal only with the shortest path for links with fixed travel costs. Thus the problem is formulated for a given set of link volumes and link costs. The costs may have been derived as a function of volumes, but all must remain constant in the linear programming example. Further, finding the shortest path in this way requires that the problem be separately formulated and solved for each origin–destination pair. For zone 1 as origin and zone 5 as destination the problem has the following form:

$$\begin{aligned}
\text{minimize } Z = &\; x_{1\,6} + x_{2\,9} + x_{3\,12} + x_{4\,11} + x_{5\,15} + x_{6\,1} + 2x_{6\,7} + 4x_{6\,9} + 5x_{6\,12} \\
&+ 2x_{7\,6} + 2x_{7\,8} + 2x_{8\,7} + 2x_{8\,12} + x_{9\,2} + 4x_{9\,6} + 5x_{9\,10} + 5x_{9\,11} \\
&+ 7x_{9\,12} + 5x_{10\,9} + x_{10\,11} + 2x_{10\,12} + x_{11\,4} + 5x_{11\,9} + x_{11\,10} \\
&+ 4x_{11\,14} + x_{12\,3} + 5x_{12\,6} + 2x_{12\,8} + 7x_{12\,9} + 2x_{12\,10} + 2x_{12\,13} \\
&+ 6x_{12\,14} + 2x_{13\,12} + 4x_{13\,16} + 4x_{14\,11} + 6x_{14\,12} + 2x_{14\,15} \\
&+ x_{15\,5} + 2x_{15\,14} + 2x_{15\,16} + 4x_{16\,13} + 2x_{16\,15}\,. \quad (3.4)
\end{aligned}$$

Note that this objective function has forty two terms, one for each one-way link in the network. The coefficient of each term is the free-flow link cost. Next the flow-balance constraints may be written, one for each network node. Recall that in this case, node 1 is to be the origin or source node and node 5 is the sink or destination node.

Minimize Z subject to

$$x_{1\,6} - x_{6\,1} = 1\,, \quad (3.5.1)$$

$$x_{2\,9} - x_{9\,2} = 0\,, \quad (3.5.2)$$

$$x_{3\,12} - x_{12\,3} = 0\,, \quad (3.5.3)$$

$$x_{4\,11} - x_{11\,4} = 0\,, \quad (3.5.4)$$

Optimization and optimizing models

and

$$x_{5\,15} - x_{15\,5} = -1 , \qquad (3.5.5)$$

$$x_{6\,1} - x_{1\,6} + x_{6\,7} - x_{7\,6} + x_{6\,9} - x_{9\,6} + x_{6\,12} - x_{12\,6} = 0 , \qquad (3.5.6)$$

$$x_{7\,6} - x_{6\,7} + x_{7\,8} - x_{8\,7} = 0 , \qquad (3.5.7)$$

$$x_{8\,7} - x_{7\,8} + x_{8\,12} - x_{12\,8} = 0 , \qquad (3.5.8)$$

$$x_{9\,2} - x_{2\,9} + x_{9\,6} - x_{6\,9} + x_{9\,10} - x_{10\,9} + x_{9\,11} - x_{11\,9} + x_{9\,12} - x_{12\,9} = 0 , \qquad (3.5.9)$$

$$x_{10\,11} - x_{11\,10} + x_{10\,12} - x_{12\,10} + x_{10\,9} - x_{9\,10} = 0 , \qquad (3.5.10)$$

$$x_{11\,4} - x_{4\,11} + x_{11\,9} - x_{9\,11} + x_{11\,10} - x_{10\,11} + x_{11\,14} - x_{14\,11} = 0 , \qquad (3.5.11)$$

$$x_{12\,3} - x_{3\,12} + x_{12\,6} - x_{6\,12} + x_{12\,8} - x_{8\,12} + x_{12\,9} - x_{9\,12}$$
$$+ x_{12\,10} - x_{10\,12} + x_{12\,14} - x_{14\,12} + x_{12\,13} - x_{13\,12} = 0 , \qquad (3.5.12)$$

$$x_{13\,12} - x_{12\,13} + x_{13\,16} - x_{16\,13} = 0 , \qquad (3.5.13)$$

$$x_{14\,11} - x_{11\,14} + x_{14\,12} - x_{12\,14} + x_{14\,15} - x_{15\,14} = 0 , \qquad (3.5.14)$$

$$x_{15\,5} - x_{5\,15} + x_{15\,14} - x_{14\,15} + x_{15\,16} - x_{16\,15} = 0 , \qquad (3.5.15)$$

$$x_{16\,13} - x_{13\,16} + x_{16\,15} - x_{15\,16} = 0 . \qquad (3.5.16)$$

Note that the first five of these constraint equations are for the load nodes. Equation (3.5.1) is for the origin node and equation (3.5.5) is for the destination node. It should be clear, too, that the equations for the other load nodes simply serve to keep the program solution from attempting to use them.

The last set of constraints is in fact forty two inequalities, one for each link, each of the form

$$x_{ij} \geq 0 . \qquad (3.6)$$

Referring to the map of the Archerville highway system (figure 2.2), it can be seen that there are two minimum paths from node 1 to node 5, with equal lengths of 15 units. In terms of the notation used here these paths would be:

path 1: $x_{1\,6} - x_{6\,12} - x_{12\,14} - x_{14\,15} - x_{15\,5}$;

path 2: $x_{1\,6} - x_{6\,12} - x_{12\,13} - x_{13\,16} - x_{16\,15} - x_{15\,5}$.

Thus the solution to the linear program of equations (3.4), (3.5), and (3.6) should show either of the above sets of x values equal to 1, and all others equal to 0. In point of fact, actual computer runs of the above problem give path 2 as the optimal solution, as the sequence of steps in the simplex algorithm results in the vertex which represents path 2 being reached first, and no other path offers any improvement (reduction) in path cost. It is worth noting that even though there are hundreds of thousands of vertices on the simplex defined by this linear program, the simplex algorithm, being a very efficient search procedure, investigated just a few before finding the optimum. Actually, only one other complete path was found before the optimal path. The previous, nonoptimal path was

path 0: $x_{1\,6} - x_{6\,9} - x_{9\,11} - x_{11\,14} - x_{15\,5}$.

The total cost of path 0 is 17 units. Several other vertices (seven to be exact) are then examined before the algorithm finds path 1 and terminates with an optimal solution.

If it were necessary to find the shortest path from node 1 to node 4, then it would be necessary to reformulate the problem and rerun the simplex algorithm. The objective function [equation (3.4)] would remain unchanged. Constraints incorporating the flow-balance relationships for the nonload nodes [equations (3.5.6) to (3.5.16)] would remain unchanged, as would the nonnegativity constraints of equations (3.6). The constraints defining the flow-balance relationships for the load nodes would change, and would become

$$x_{1\,6} - x_{6\,1} = 1 , \qquad (3.7.1)$$

$$x_{2\,9} - x_{9\,2} = 0 , \qquad (3.7.2)$$

$$x_{3\,12} - x_{12\,3} = 0 , \qquad (3.7.3)$$

$$x_{4\,11} - x_{11\,4} = -1 , \qquad (3.7.4)$$

$$x_{5\,15} - x_{15\,5} = 0 . \qquad (3.7.5)$$

Thus to trace all the minimum paths from each node to each other node would require that twenty different linear programs be formulated and solved. Although the formulations would only differ in the load-node constraint equations, this would still represent a substantial effort, not to mention what would be involved in a realistically sized problem. Other procedures have been developed for such calculations which are used in full-scale applications.

3.3.4 The minimum-cost flow problem

Another type of linear programming problem is that known as the minimum-cost flow problem. It is also known as the 'transportation problem' of linear programming. As an example, the Archerville data may again be used. There are a known number of households of each income group living in each zone and a known number of employees of each type working in each zone. The illustrative residence location model described in chapter 2 is a way of calculating where employees will live and work. Implicitly there is a zone-to-zone matrix of home-to-work trips as described in chapter 2 also. Suppose, now, that one takes as given the location of households shown in table 2.1 and the location of industrial employment given in the same table. By first assuming that there will be one employee per household and then applying the employee–household conversion ratios given in table 3.1, one may calculate the number of industrial employees residing in each zone. This number is given in table 3.1. The number of industrial employees working in zones 1 to 5 is 150, 150, 400, 0, and 0, respectively. With the models described in chapter 2 the distribution of employees at the place of residence to place of work was done according to equations (2.10) and (2.12). The assignment of these trips to the network links was done by one of the assignment algorithms discussed in chapter 2 as well.

Here it is possible to consider the proposal that each employee chooses a place of work such that the total travel cost for all employees is minimized. If network link capacities and the consequent congestion are ignored for this illustration, the problem may be stated as a linear program:

$$\text{minimize } Z = \sum_{i,j} c_{ij} x_{ij} , \qquad (3.8)$$

subject to

$$\sum_j x_{ij} - \sum_k x_{ki} = \begin{cases} O_i, & \text{if } i \text{ is the origin,} \\ -D_i, & \text{if } i \text{ is the destination,} \\ 0, & \text{otherwise,} \end{cases} \qquad (3.9)$$

and

$$x_{ij} \geqslant 0, \qquad \forall i,j , \qquad (3.10)$$

where
O_i is the net number of trips leaving node i,
D_i is the net number of trips arriving at node i.

Equations (3.8), (3.9), and (3.10) are the general minimum-cost flow problem. If there were specified link capacities k'_{ij} for each link, which could not be exceeded, then another set of constraints would be substituted for equations (3.10). This new set of constraints would be of the form

$$k'_{ij} \geqslant x_{ij} \geqslant 0, \qquad \forall i,j . \qquad (3.11)$$

Table 3.1. Archerville: the number of industrial employees residing in each zone in the base year.

Zone	Low-income household residents	High-income household residents	Industrial employees resident	Industrial employment	Additional workers required
1	200	100	146	150	4
2	300	50	173	150	−23
3	150	50	98	400	302
4	100	300	188	0	−188
5	50	150	95	0	−95

Note: 0.500 industrial employees per low-income household; 0.462 industrial employees per high-income household for this example.

Note that the objective function here is the same as that for the shortest-path problem given in equation (3.1). The constraints [equations (3.9)] are a set of flow-balance relationships similar to those of equations (3.2).

As the intrazonal travel costs are 1.0 for each zone, clearly the first consideration is that all employees residing in any zone first be assigned to jobs in that zone. Referring to table 3.1, we can see that zone 1 requires 4 additional workers, and zone 3 requires 302. Zones 2, 4, and 5 have surplus workers to the amount of 23 188, and 95, respectively. Thus equations (3.9) become, for this case:

$$x_{1\,6} - x_{6\,1} = -4 \, , \tag{3.9.1}$$

$$x_{2\,9} - x_{9\,2} = 23 \, , \tag{3.9.2}$$

$$x_{3\,12} - x_{12\,3} = -302 \, , \tag{3.9.3}$$

$$x_{4\,11} - x_{11\,4} = 188 \, , \tag{3.9.4}$$

$$x_{5\,15} - x_{15\,5} = 95 \, . \tag{3.9.5}$$

The remainder of equations (3.9) are the flow-balance equations and because the network is unchanged they are identical to equations (3.5.6) to (3.5.16). The links were left unconstrained and thus only the nonnegativity constraints [equations (3.10)], are added to the problem. The link flows produced by the simplex algorithm are shown in figure 3.4.

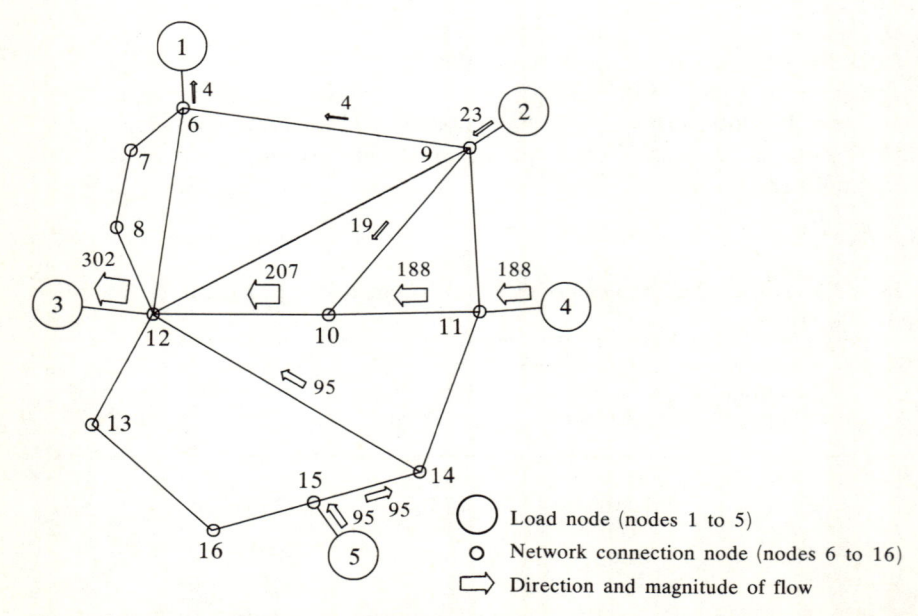

Figure 3.4. Archerville: the trip flows resulting from the minimum-cost flow assignment algorithm with constant link costs (no constraints).

If it were desired that some of the network links have a maximum allowable flow, then some constraints of the form of equations (3.11) would have to be added. Suppose, for example, that it was necessary to limit link $x_{11\ 10}$ to a maximum flow of 100 workers, and to limit link $x_{10\ 12}$ as well. These limits would be added to the problem by adding the constraints

$$x_{11\ 10} \leq 100\ , \qquad (3.11.1)$$

$$x_{10\ 12} \leq 100\ . \qquad (3.11.2)$$

When this is done and the simplex algorithm applied to the problem the resulting flows are as shown in figure 3.5. It is easy to see where 88 trips have been directed from the constrained links to the next-shortest path, along links $x_{11\ 14}$ and $x_{14\ 12}$. Note that the minimum systemwide cost in the unconstrained problem is 2085 units, whereas in the case with the two constrained links the systemwide cost increases to 2701 units because some trips are forced along a longer path.

There are many other illustrations which could have been given here. In particular the use of linear programming for the allocation of land-using activities has not been shown. This will be explored in some detail in chapters 4 and 5. The remainder of this chapter will be devoted to explaining the consequences of the objective function, the constraints, or both, being nonlinear.

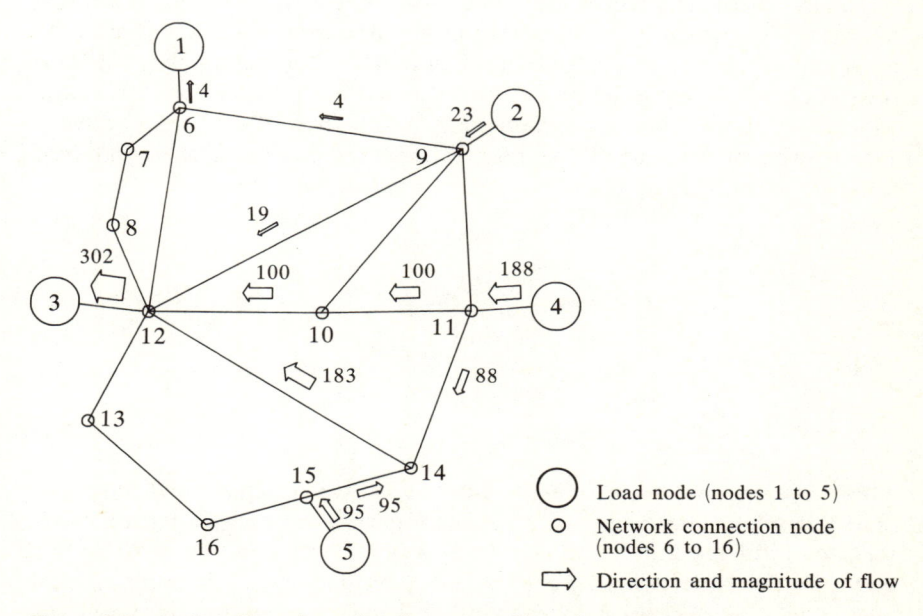

Figure 3.5. Archerville: the trip flows resulting from the minimum-cost flow assignment algorithm with constant link costs (with constraints).

3.4 Nonlinear programming: unconstrained
3.4.1 Algebraic methods
As mentioned above, a linear objective function without constraints is not an interesting problem. Such situations are unbounded and are not really programming or optimization problems at all. Only when constraints enter a problem does the linear objective function have any useful purpose. In the case of a nonlinear objective function the situation can change dramatically. Many nonlinear functions, with one, several, or many variables, can present quite challenging optimization problems.

Before continuing, the reader is again reminded that these few pages are by no means an adequate substitute for a comprehensive survey of nonlinear programming. Several excellent texts exist to which the reader is referred for further study (Avriel, 1976; Fiacco, 1983; McCormick, 1983; Reklaitis et al, 1983). The purpose here is to describe some of the specific techniques used in transport and location model work, again, so that the reader will have the necessary background to continue on in this book.

The simplest nonlinear optimization problems involve straightforward functions whose optima can be easily obtained by algebra and calculus. One such function might be

$$f(X) = 2X^3 + 3X^2 - X + 5 \;. \tag{3.12}$$

Normally, one would expect the values of X to be confined to the set of real numbers. This would be signified by putting the expression $\forall X \in \mathbb{R}$ after the equation, which means it is valid for all X in the set of real numbers, $\mathbb{R}$. This restriction will apply to all problems discussed in this book. For a function which is continuously differentiable, any point where the first derivative is zero *may* be an optimum. To be more specific, there are *necessary* conditions for some point X^* to be a local minimum or maximum. These are

$$\left. \frac{df}{dX} \right|_{X=X^*} = 0 \;, \tag{3.13}$$

$$\left. \frac{d^2f}{dX^2} \right|_{X=X^*} \geq 0, \quad \text{local minimum} \;, \tag{3.14.1}$$

$$\left. \frac{d^2f}{dX^2} \right|_{X=X^*} \leq 0, \quad \text{local maximum} \;. \tag{3.14.2}$$

The necessary condition that the first derivative is equal to 0 [equation (3.13)] defines a stationary point. A stationary point may be a local minimum or local maximum, but it may also be an inflection point or saddle point. To determine whether a stationary point is a local minimum or local maximum it is necessary to examine the *sufficient* conditions. From the necessary conditions [equation (3.13)] it is known that the first derivative(s) are zero. Denote the first nonzero higher order derivative as n. The sufficient

Optimization and optimizing models

conditions are:
if n is odd, then X^* is not a optimum, it is an inflection point or saddle point;
if n is even, the X^* is a local optimum,
and if

$$\left.\frac{d^n f}{dX^n}\right|_{X=X^*} > 0, \quad X^* \text{ is a local minimum}, \tag{3.15.1}$$

$$\left.\frac{d^n f}{dX^n}\right|_{X=X^*} < 0, \quad X^* \text{ is a local maximum}. \tag{3.15.2}$$

As an example, consider the algebraic function

$$f(X) = \frac{25}{6}X^6 - 45X^5 + \frac{325}{2}X^4 - 200X^3 + 189. \tag{3.16}$$

The first derivative of this function is

$$\frac{df}{dX} = 25X^5 - 225X^4 + 650X^3 - 600X^2, \tag{3.17.1}$$

which can be factored to

$$\frac{df}{dX} = 25X^2(X-2)(X-3)(X-4). \tag{3.17.2}$$

It can easily be seen that the first derivative equals 0 when $X = 0, 2, 3,$ or 4. The second derivative of equation (3.16) is

$$\frac{d^2 f}{dX^2} = 125X^4 - 900X^3 + 1950X^2 - 1200X. \tag{3.18}$$

Evaluating the second derivative at the four points where the first derivative is equal to 0, we find

X	0	2	3	4
$\dfrac{d^2 f}{dX^2}$	0	200	-225	800

With use of this further information and the above description of necessary and sufficient conditions. The points $X = 2$ and $X = 4$ are local minima and that where $X = 3$ there is a local maximum. To further evaluate the point where X equals 0 it is necessary to evaluate the third derivative:

$$\frac{d^3 f}{dX^3} = 500X^3 - 2700X^2 + 3900X - 1200. \tag{3.19}$$

At X equals 0, the third derivative equals -1200. Since n is odd, at X equals 0 there is neither a minimum nor a maximum, but an inflection point. A plot of the function of equation (3.16) is given in figure 3.6.

Another example of this sort of optimization analysis is that of determining the parameters of a linear regression. Consider the function

$$\hat{Y}_i = bX_i + a , \tag{3.20}$$

where
$\hat{Y}_i$ is an estimated value of Y_i,
X_i, Y_i are observed variables,
a, b are coefficients to be estimated.

The problem is to determine a and b so as to minimize the difference between the observed values, Y_i, and the estimated values, $\hat{Y}_i$. It will be convenient to transform the X_i to deviations around their mean, such that

$$\mathcal{X}_i = X_i - \overline{X} , \tag{3.21}$$

where

$$\overline{X} = \frac{1}{n} \sum_i X_i . \tag{3.22}$$

The customary criterion for 'goodness-of-fit' of the observed values, Y_i, to the estimated values, $\hat{Y}_i$, is the sum of the squared deviations. Thus the problem is to

$$\text{minimize } Z = \sum_i (Y_i - \hat{Y}_i)^2 . \tag{3.23}$$

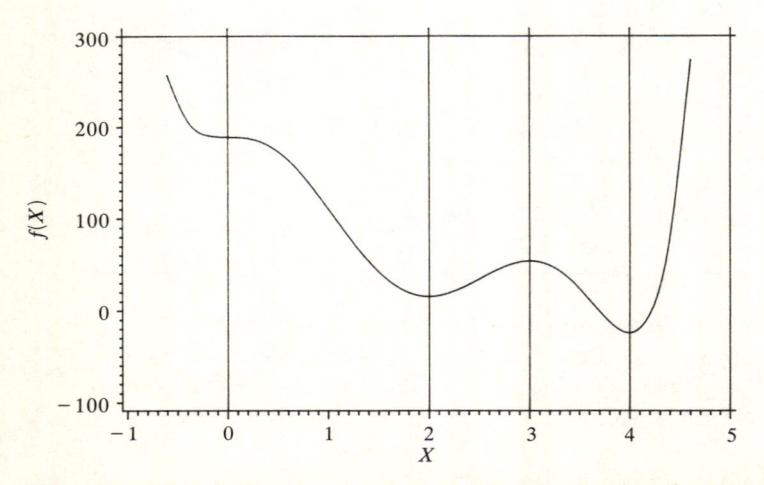

Figure 3.6. Plot of the function expressed by equation (3.16), showing stationary points at $X = 0, 2, 3, 4$.

Optimization and optimizing models

Substituting from equations (3.20) and (3.21), we get

$$\text{minimize } Z(a, b) = \sum_i (Y_i - b\mathcal{X}_i - a)^2 \ ; \tag{3.24}$$

and taking partial derivatives we obtain

$$\frac{\partial Z(a, b)}{\partial a} = -2\left(\sum_i Y_i - \sum_i a - \sum_i b\mathcal{X}_i\right). \tag{3.25}$$

If this partial derivative is set equal to 0, and given that the summation over $i\ (= 1, \ldots, n)$ of a is equal to na and that of $\mathcal{X}_i$ is equal to 0,

$$a = \frac{1}{n}\sum_i Y_i \ ; \tag{3.26}$$

thus a equals the mean of the Y_i. Similarly

$$\frac{\partial Z(a, b)}{\partial b} = -2\sum_i \mathcal{X}_i(Y_i - a - b\mathcal{X}_i) \ , \tag{3.27}$$

again, as the summation of i of $\mathcal{X}_i$ is equal to 0, when this partial derivative [equation (3.27)] is set equal to 0 the result is

$$b = \sum_i \mathcal{X}_i Y_i \bigg/ \sum_i \mathcal{X}_i^2 \ . \tag{3.28}$$

Thus, to find the values of a and b which give the best linear-regression fit to a given set of data points it is only necessary to substitute those data into equations (3.26) and (3.28).

3.4.2 Search methods: estimation of parameters of location models

By a similar approach it is possible to estimate parameters for location models such as those used for the numerical examples in chapter 2. Standard methods for their calibration (parameter estimation) use a nonlinear optimization procedure known as gradient search. The gradient of a function is the vector of its partial derivatives. Gradient search is an efficient procedure for varying the parameters of a function in order to maximize or minimize that function. The numerical evaluation of the gradient provides information on search direction and the size of steps to be taken in that direction.

The residential location model used in the numerical example of chapter 2 was of the form

$$N_i = \sum_j E_j B_j W_i \mathrm{f}(c_{ij}) \ , \tag{3.29}$$

where
c_{ij} is the travel cost from zone i to zone j,
E_j is the number of employees at place of work, in zone j,
N_i is the number of households at place of residence, in zone i,

B_j is a 'balancing' factor of the form

$$B_j = \left[\sum_k W_k \, \mathrm{f}(c_{kj}) \right]^{-1} \tag{3.30}$$

and

$$W_i = L_{i,\mathrm{R}}^{\gamma}(1 + n_{i,1})^{\sigma}(1 + n_{i,2})^{\rho} \tag{3.31}$$

where
$L_{i,\mathrm{R}}$ is the amount of residential land in zone i, and

$$n_{i,1} = \frac{N_{i,1}}{\sum_h N_{i,h}} = \frac{N_{i,1}}{N_{i,1} + N_{i,2}}. \tag{3.32}$$

Gathering all the terms, the equation form for each household type h is

$$N_{i,h} = \sum_j E_{j\mathrm{T}} [L_{i,\mathrm{R}}^{\gamma_h}(1 + n_{i,1})^{\sigma_h}(1 + n_{i,2})^{\rho_h} \exp(-\beta_h c_{ij})]$$

$$\left[\sum_k L_{k,\mathrm{R}}^{\gamma_h}(1 + n_{k,1})^{\sigma_h}(1 + n_{k,2})^{\rho_h} \exp(-\beta_h c_{kj}) \right]^{-1}. \tag{3.33}$$

In that example arbitrary numerical values were assumed for the parameters γ, σ, ρ, and β. If instead only the 'data' had been available, it would have been necessary to estimate the parameter values which gave the best fit of the observed values, $N_{i,h}$, to the estimated values, $\hat{N}_{i,h}$. In the case of a nonlinear function such as this, the appropriate goodness-of-fit criterion is known as the 'likelihood criterion'. The problem then becomes one of finding the parameter values which maximize the likelihood criterion.

There are several possible ways of constructing a precise form of likelihood function for these purposes, with one form (from Putman, 1983a) being

$$\mathfrak{L} = \sum_i N_{i,h} \ln \hat{N}_{i,h} - \sum_i N_{i,h} \ln N_{i,h}. \tag{3.34}$$

This function will have its maximum where $N_{i,h}$ and $\hat{N}_{i,h}$ are equal and where $\mathfrak{L}$ is zero. As an equation is given for the estimation of $\hat{N}_{i,h}$, it is clear that one may think of $\mathfrak{L}$ as a function of observed data (given) and of the parameters γ, σ, ρ, and β. As before, the function will have its maximum where each of its partial derivatives (with respect to the parameters) is equal to zero.

The algebra for these partial derivatives is somewhat tedious and has been given elsewhere (Putman, 1983a). Thus the equations are stated here without their derivation. Note that the minus sign in front of β has been dropped here, the search procedure will find β to be negative.

$$\frac{\partial L}{\partial \beta_h} = \left[\frac{\partial \mathfrak{L}}{\partial \hat{N}_{i,h}} \right] \left[\frac{\partial \hat{N}_{i,h}}{\partial \beta_h} \right], \tag{3.35}$$

$$= \sum_i \left[\frac{N_{i,h}}{\hat{N}_{i,h}} \right] \left[\sum_j T_{ij} \left\{ \ln c_{ij} - B_j^h \sum_k c_{ik} \exp(\beta_h c_{ik}) \right\} \right], \tag{3.36}$$

where

$$T_{ij} = E_j B_j^h W_i^h \exp(\beta_h c_{ij}), \tag{3.37}$$

$$W_i^h = L_{i,R}^{\gamma_h}(1+n_{i,1})^{\sigma_h}(1+n_{i,2})^{\rho_h}, \tag{3.38.1}$$

$$B_j^h = \left[\sum_k W_k^h \exp(\beta_h c_{kj})\right]^{-1}. \tag{3.38.2}$$

Next,

$$\frac{\partial \mathfrak{L}}{\partial \gamma_h} = \sum_i \left[\frac{N_{i,h}}{\hat{N}_{i,h}}\right]\left[\sum_j T_{ij}\left\{\ln L_{j,R} - B_j^h \ln L_{j,R}\sum_k W_k^h \exp(\beta_h c_{ik})\right\}\right]; \tag{3.39}$$

then,

$$\frac{\partial \mathfrak{L}}{\partial \sigma_h} = \sum_i \left[\frac{N_{i,h}}{\hat{N}_{i,h}}\right]\left[\sum_j T_{ij}\left\{\ln(1+n_{j,1}) - B_j^h \ln(1+n_{j,1})\sum_k W_k^h \exp(\beta_h c_{ik})\right\}\right]; \tag{3.40}$$

and, finally,

$$\frac{\partial \mathfrak{L}}{\partial \rho_h} = \sum_i \left[\frac{N_{i,h}}{\hat{N}_{i,h}}\right]\left[\sum_j T_{ij}\left\{\ln(1+n_{j,2}) - B_j^h \ln(1+n_{j,2})\sum_k W_k^h \exp(\beta_h c_{ik})\right\}\right]. \tag{3.41}$$

Once these equations are prepared the numerical values of the variables may be substituted into them and the values of $\mathfrak{L}$ and its partial derivatives calculated. A positive value of a partial derivative indicates that the parameter should be increased in order to move towards the maximum value of $\mathfrak{L}$. A negative value indicates that the parameter should be decreased. A zero value indicates that the parameter is at its optimal value with respect to $\mathfrak{L}$.

To illustrate the operation of the procedure the Archerville data were used. The values assumed in chapter 2 for the parameters were used as initial values in the search procedure. Fifteen iterations of the search procedure for each of the two household types are shown in tables 3.2(a) and 3.2(b). Note how the derivatives all approach zero as does the criterion $\mathfrak{L}$. Note also that the trajectories of the parameters as they converge towards their final values show some fluctuation rather than being a smooth progression. This is because the gradient search procedure is, in fact, a search process, and does tend to move in one direction, pass the optimum, move back, and so on in a zigzag fashion. This pattern is characteristic of such algorithms, and can sometimes result in unacceptably slow convergence. A variety of other algorithms have been developed in an attempt to circumvent such problems, and these are described in various texts on the topic such as those by Himmelblau (1972) and Beveridge and Schecter (1970) as well as others. Thus we note that the search procedure was not forced to reach the point where all the partial derivatives were equal to zero, as this

Table 3.2. Archerville: results from 15 iterations of the gradient-search procedure for the estimation of parameters in the residential location model for (a) low-income households, and (b) high-income households. Note: see the text for definitions of the variables.

Iteration	γ	$\partial\Omega/\partial\gamma$	σ	$\partial\Omega/\partial\sigma$	ρ	$\partial\Omega/\partial\rho$	β	$\partial\Omega/\partial\beta$	Ω
(a) *Low-income households*									
1	0.900	−24.068	2.000	31.040	0.900	−32.269	−0.200	−20.198	−33.332
2	0.444	20.347	2.588	7.780	0.289	−8.576	−0.583	−3.382	−3.490
3	0.660	−8.221	2.670	3.383	0.198	−3.507	−0.618	−3.937	−1.084
4	0.576	7.495	2.705	6.618	0.162	−7.168	−0.658	−4.295	−0.492
5	0.656	−8.205	2.775	−0.527	0.086	0.625	−0.704	−2.264	−0.484
6	0.602	−1.152	2.771	−0.162	0.091	0.126	−0.719	−1.429	−0.199
7	0.581	1.781	2.768	−0.073	0.093	−0.018	−0.744	−0.953	−0.175
8	0.607	−1.320	2.767	−0.242	0.093	0.206	−0.758	−1.198	−0.155
9	0.583	2.079	2.763	−0.072	0.096	−0.031	−0.780	−0.701	−0.141
10	0.609	0.060	2.762	−0.185	0.096	0.117	−0.788	−1.064	−0.121
11	0.619	0.673	2.733	0.011	0.115	−0.126	−0.957	−0.008	−0.036
12	0.623	0.098	2.733	−0.021	0.114	−0.083	−0.957	−0.084	−0.034
13	0.627	−0.347	2.732	−0.084	0.110	−0.008	−0.961	−0.123	−0.034
14	0.623	0.173	2.731	−0.060	0.110	−0.042	−0.962	−0.047	−0.033
15	0.626	−0.097	2.730	−0.067	0.109	−0.031	−0.963	−0.082	−0.033
(b) *High-income households*									
1	3.000	−103.426	0.200	−36.714	3.000	39.221	−0.100	33.704	−152.027
2	1.184	−5.353	−0.445	6.107	3.689	−5.643	0.492	−22.893	−19.574
3	1.016	−16.991	−0.252	−3.107	3.511	3.671	−0.229	−3.673	−7.507
4	0.623	−8.774	−0.324	1.886	3.596	−1.454	−0.314	−12.605	−4.142
5	0.523	7.729	−0.302	−1.729	3.579	2.057	−0.458	−9.392	−3.270
6	0.661	−4.402	−0.333	−2.678	3.616	3.041	−0.627	−2.283	−1.913
7	0.455	−0.974	−0.458	−2.263	3.758	2.595	−0.733	−3.987	−1.293
8	0.321	2.479	−0.770	2.552	4.115	−2.528	−1.282	−2.068	−0.584
9	0.383	1.656	−0.706	0.361	4.052	−0.296	−1.334	2.117	−0.379
10	0.358	0.391	−0.701	0.568	4.047	−0.507	−1.302	0.936	−0.309
11	0.428	−0.359	−0.599	0.269	3.957	−0.145	−1.134	−0.014	−0.155
12	0.421	0.056	−0.594	0.143	3.954	−0.019	−1.135	−0.030	−0.153
13	0.424	−0.123	−0.587	0.016	3.953	0.111	−1.136	0.166	−0.152
14	0.421	0.108	−0.587	0.088	3.956	0.037	−1.132	−0.015	−0.152
15	0.423	−0.011	−0.585	0.076	3.956	0.050	−1.133	0.039	−0.152

might well have taken a good many additional iterations. The results given here, achieved after fifteen iterations, illustrate the point adequately; and the mean absolute percent deviations of the estimated $\hat{N}_{i,h}$ from the observed $N_{i,h}$ were 0.6% and 2.3% for the two household types.

The estimation of parameters for the residential location model could have been done by setting all the partial derivative equations equal to 0 and then by solving that set of equations, simultaneously, for the values of the parameters. The solution of simultaneous nonlinear equations is, however, a rather problematic undertaking and is best avoided whenever possible. Other search routines might also have been used, involving, perhaps, second-order partial derivatives and a procedure known as Newton's method. The point here was to illustrate a use of unconstrained nonlinear optimization, and this has been done. Next, the consideration of adding constraints to the problem is discussed.

3.5 Nonlinear programming: constrained
3.5.1. The nonlinear minimum-cost flow problem

The addition of constraints to a nonlinear optimization problem not only alters the problem, but alters the selection of possible solution methods as well. It is again impossible to attempt a complete coverage of the topic, and readers are referred to any of the general texts cited above (subsection 3.4.1). The constraints in such problems may be linear, or they may themselves be nonlinear. If there are nonlinear constraints an optimization problem falls into the category of nonlinear programming even if the objective function is linear. The interesting point is how much more complex the problems can become as a result of what may, initially, seem to be rather modest changes in the formulation of an objective function and/or constraint.

To illustrate some aspects of the nonlinear programming formulation, the minimum-cost flow problem described above can be reconsidered. In the previous formulation the linear objective function [equation (3.8)] was simply the sum of the trips (flows) on each network link multiplied by the travel cost of the link. The link costs were fixed, remaining constant regardless of link flows (though it was shown by use of additional constraint equations how the link flows could themselves be restricted). Suppose the more realistic view was taken that link costs depended upon link flows. This proposition will be discussed in considerably greater detail in chapter 6. For sake of illustration, consider the following function

$$c_{ij} = c_{ij}^{(0)}(1.0 + \delta x_{ij}^2) , \qquad (3.42)$$

where
c_{ij} is the 'congested', or flow-related, link travel cost,
$c_{ij}^{(0)}$ is the free-flow link travel cost,
x_{ij} is the link flow volume (number of trips),
δ is a parameter.

With this function the link travel cost is equal to the free-flow cost when the link flow volume is zero. As link flow volume increases, link travel cost increases too.

In the linear version of this problem the solution involved only the finding of the minimum paths and the subsequent routing of trips along those paths. If there were specified link flow volume constraints then the excess trips were simply routed to the second-shortest path. When link flows determine link costs the essential nature of the problem changes. The solution becomes a matter of adjusting the volumes, observing the resulting costs, and then adjusting the volumes again. Thus in a very real sense, even for the small-sized problem of the Archerville data the complexity of the problem begins to defy solution by inspection. This was also the case for the parameter-estimation procedure described above. The introduction of simultaneity and/or nonlinearity to a problem often transforms the problem into one which lies beyond an intuitive solution.

By incorporating equation (3.42) into the original objective function of equation (3.8), we obtain

$$\text{minimize } Z = \sum_{i,j} \left[c_{ij}^{(0)}(1.0 + \delta x_{ij}^2) x_{ij} \right], \tag{3.43}$$

and, thus, the objective function becomes a cubic equation. With the same linear constraints as before [equation (3.9)] the flow-balance relationships still hold true, as do the nonnegativity constraints of equations (3.10). This new set of equations is a nonlinear programming problem with a nonlinear objective function and linear constraints. There are many computer programs for nonlinear optimization and/or nonlinear programming problems. The gradient-search procedure used in the illustration of parameter estimation was CALIB (Putman, 1985). The computer program used to solve this nonlinear minimum-cost flow example was MINOS-5.0 (Murtagh and Saunders, 1983) which was embedded in the GAMS package (Kendrick and Meeraus, 1985). For problems with nonlinear objective functions and linear constraints MINOS-5.0 uses a reduced-gradient algorithm (Wolfe, 1962) in conjunction with a quasi-Newton algorithm (Davidon, 1959).

It was necessary to set a value for the parameter δ. A value of 0.0002 was selected so that link flow volumes in the order of 100 trips would result in a tripling of link cost. At this scale, link costs increase significantly, but not astronomically, with link flows on the order of those observed in the linear form of the problem described in subsection 3.3.4. The flows on the network which result from this new nonlinear problem are shown in figure 3.7, and the link volumes, free-flow link costs, and the congested link costs are shown in table 3.3 (only those links which have flows are included). The results hold no great surprises, but do show a clear response to the reformulation of the problem so that link costs are a function of link flows.

It is interesting to compare the results in figure 3.7, obtained from the nonlinear programming (NLP) solution, with those in figure 3.4, obtained

Optimization and optimizing models

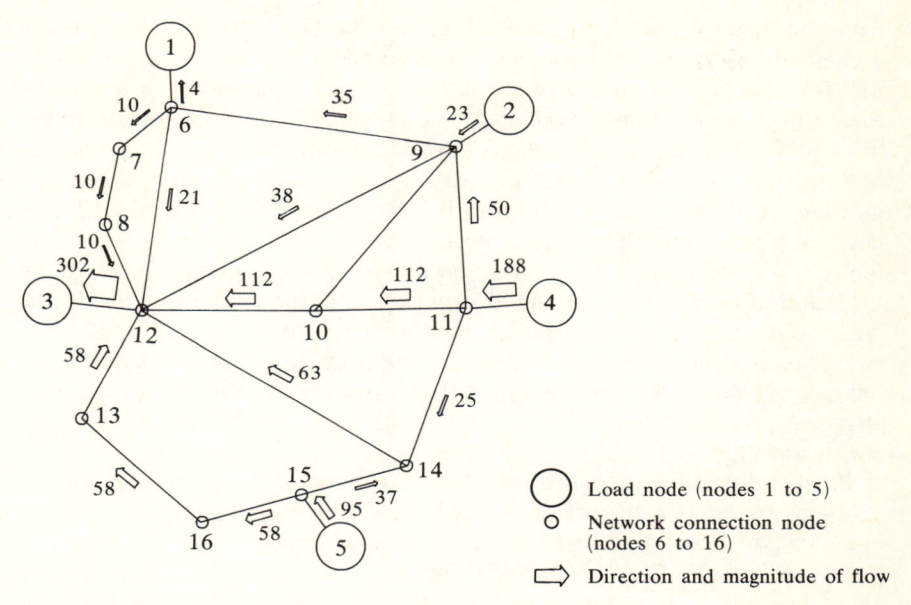

Figure 3.7. Archerville: the trip flows resulting from the minimum-cost flow assignment algorithm with variable link costs (no constraints).

Table 3.3. Archerville: comparison of free-flow and congested link costs for the nonlinear objective function.

Link (node–node)	Volume	Free-flow cost	Congested cost
2–9	23	1.00	1.11
4–11	188	1.00	8.07
5–15	95	1.00	2.81
6–1	4	1.00	1.00
6–7	10	2.00	2.04
6–12	21	5.00	5.44
7–8	10	2.00	2.04
8–12	10	2.00	2.04
9–6	35	4.00	4.96
9–12	38	7.00	9.07
10–12	112	2.00	7.06
11–9	50	5.00	7.52
11–10	112	1.00	3.53
11–14	25	4.00	4.52
12–3	302	1.00	19.24
13–12	58	2.00	3.33
14–12	63	6.00	10.74
15–14	37	2.00	2.56
15–16	58	2.00	3.33
16–13	58	4.00	6.65

from the linear programming (LP) solution. The NLP solution shows much greater utilization of network links because of the effects of congestion on link travel costs. For the LP solution only 11 links were used, whereas the the NLP solution 20 links were used. As a result, the number of trips on the links between nodes 11 and 10 and between nodes 10 and 12, which was 188 in the LP solution, is only 112 in the NLP solution. When the maximum-flow constraint of 100 trips is imposed on those links, only 12 trips need to be distributed to other links. Again the NLP solution, shown in figure 3.8, differs from the LP solution by virtue of using more links. In the LP solution all 88 of the rerouted trips went on one route, whereas in the NLP solution the 12 trips to be rerouted were split amongst several alternative routes. Again, the imposition of link-flow constraints results in an increase of the objective function value [equation (3.43)]. For the unconstrained problem (figure 3.7) $Z = 11536.5$, and with the addition of constraints (figure 3.8), $Z = 11583.1$.

Precise descriptions of the solution algorithms used to solve the above problem are beyond the scope of this text. So, for that matter, are the dozens of algorithms which have been developed—some for whole groups of problems, others for one specific formulation. There is, however, one algorithm which, by virtue of its regular use for the types of problem with which we are concerned, should be discussed here.

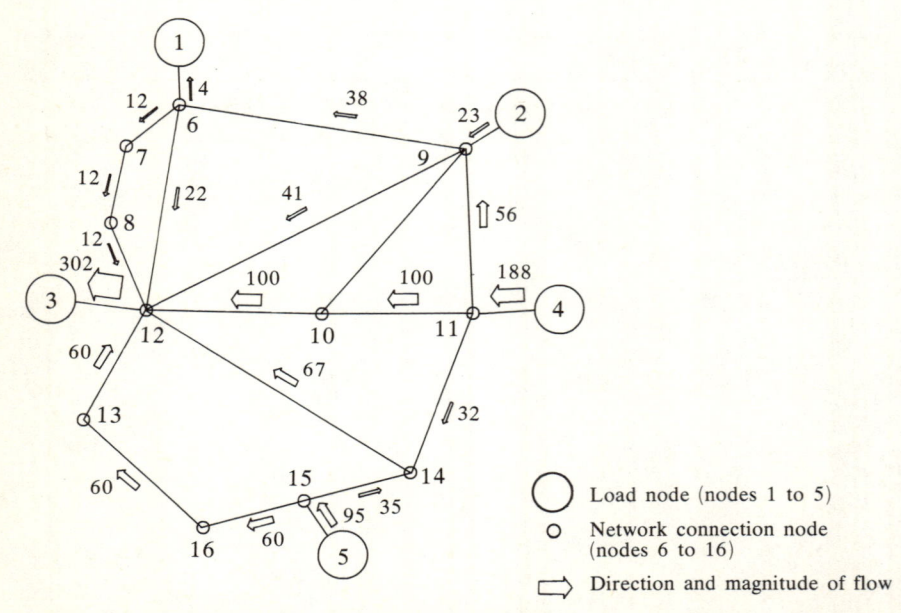

Figure 3.8. Archerville: the trip flows resulting from the minimum-cost flow assignment algorithm with variable link costs (with constraints).

3.5.2 The Frank–Wolfe algorithm

This algorithm is a procedure for solving nonlinear programming problems where there is a nonlinear objective function and linear constraints. The procedure is essentially one of successive linear approximations to the nonlinear objective function. Consider the general problem

$$\text{minimize } Z = f(X_1, X_2, \ldots, X_n), \tag{3.44}$$

subject to

$$\sum_i a_{ij} X_i \geq b_j, \quad \forall j, \tag{3.45}$$

and

$$X_i \geq 0, \quad \forall i. \tag{3.46}$$

The Frank–Wolfe algorithm proceeds as follows.

Step 0 Find a numerical feasible starting point or 'initial feasible solution' to the NLP problem.

Step 1.1 Take (algebraically) the partial derivatives of the original objective function and evaluate them (numerically) at the initial feasible solution.

Step 1.2 The evaluated partial derivatives will be the coefficients in a linear equation. Solve for the minimum of this equation subject to the same constraints as in the original NLP problem. The line connecting the initial feasible solution to this newly calculated minimum gives the direction of search.

Step 2 Find the minimum of the original NLP objective function along the line determined in step 1.2. This minimum is the new solution to the NLP problem. If some prespecified convergence tolerance is met, the algorithm is completed, if not, the new solution is substituted for the previous solution (or the initial feasible solution) and the procedure is repeated from step 1.1.

To illustrate the operation of the Frank–Wolfe algorithm let us consider the following problem:

$$\text{minimize } Z = (X_1 - 2)^2 + (X_2 - 2)^2, \tag{3.47}$$

subject to

$$-X_1 - X_2 + 3 \geq 0, \tag{3.48}$$

$$10X_1 - X_2 - 2 \geq 0, \tag{3.49}$$

and

$$X_1, X_2 \geq 0. \tag{3.50}$$

First an initial feasible solution must be found. The traditional first guess is $X_1 = 0$, $X_2 = 0$, but this point violates constraint (3.49). A closer look at the constraints suggests that $X_1^{(0)} = 0.2$, $X_2^{(0)} = 0$ will be a feasible point (where the superscript in brackets is the iteration number).

Next the partial derivatives of the objective function must be derived:

$$\frac{\partial Z}{\partial X_1} = 2X_1 - 4 , \qquad (3.51)$$

$$\frac{\partial Z}{\partial X_2} = 2X_2 - 4 . \qquad (3.52)$$

Evaluating these partial derivatives at the initial feasible solution, yields

$$\frac{\partial Z}{\partial X_1} = -3.6, \quad \frac{\partial Z}{\partial X_2} = -4 .$$

These values become the coefficients in a linear objective function:

$$\text{minimize } Q = -3.6 Y_1 - 4 Y_2 , \qquad (3.53)$$

which is subject to the same constraints as the original problem [constraints (3.48), (3.49), and (3.50)]. This is a standard LP problem which can be solved with use of the simplex method, and which yields $Y_1^{(1)} = 0.455$, $Y_2^{(1)} = 2.545$. The line connecting the initial feasible solution to this new point is the direction of steepest descent, and is the 'search' direction.

The next step is to find the minimum value of the NLP objective function given by equation (3.47) along the line between the points (0.200, 0.000) and (0.455, 2.545). This minimum can be found by any of several line-search techniques. As the minimum is along the line between the two points, it will be some proportion α away from the initial feasible solution. If the new point is denoted by $X_1^{(1)}$, $X_2^{(1)}$, then

$$(X_1^{(1)}, X_2^{(1)}) = (X_1^{(0)}, X_2^{(0)}) - \alpha [(Y_1^{(1)}, Y_2^{(1)}) - (X_1^{(0)}, X_2^{(0)})] . \qquad (3.54)$$

By substituting the numerical values of the two known points, we get

$$X_1^{(1)} = 0.2 + 0.255\alpha ,$$

$$X_2^{(1)} = 2.545\alpha .$$

Substitution of these values into the original objective function [equation (3.47)] gives

$$\text{minimize } Z = (0.255\alpha - 1.8)^2 + (2.545\alpha - 2)^2 . \qquad (3.55)$$

The minimum value of Z is found (by line search) where $\alpha = 0.848$, giving, by substitution into the previous equations, $X_1^{(1)} = 0.416$ and $X_2^{(1)} = 2.158$.

The entire set of steps is now repeated, beginning with a substitution of these new values into the partial derivative equations (3.51) and (3.52). Now

$$\frac{\partial Z}{\partial X_1} = -3.168, \quad \frac{\partial Z}{\partial X_2} = 0.317 .$$

These in turn give a new linear objective function:

$$\text{minimize } Q = -3.168 Y_1 + 0.317 Y_2, \qquad (3.56)$$

which is again subject to the original constraints. The solution to this LP gives $Y_1^{(2)} = 3.000$ and $Y_2^{(2)} = 0.000$. The next step will be to find the minimum of the original objective function [equation (3.47)] along the line between the points $(X_1^{(1)}, X_2^{(1)})$ and $(Y_1^{(2)}, Y_2^{(2)})$. Following the same procedure as before, we find $\alpha = 0.391$, $X_1^{(2)} = 1.427$, and $X_2^{(2)} = 1.314$. Substitution of these values into the partial derivative equations and use of the results to form a new linear objective function yields

$$\text{minimize } Q = -1.146 Y_1 - 1.372 Y_2, \qquad (3.57)$$

and so forth. The minimum of the NLP problem is at the point where $X_1^* = 1.500$ and $X_2^* = 1.500$. Note that if the objective function had been unconstrained the minimum would be at $(X_1 = 2.00, X_2 = 2.000)$, a solution not permitted by constraint equation (3.48).

The operation of this algorithm may be gained in the graphical sense from figure 3.9. First note that the objective function [equation (3.47)] describes an hemispherical surface with its center at the point where $X_1 = 2$ and $X_2 = 2$. As this is a concave surface its minimum will be at the same point as shown by the circular contour lines. Constraint (3.48) is shown as the line from point A on the X_2 axis to point B on the X_1 axis. Constraint (3.49) is shown as the line from point C on the X_1 axis continuing through the intersection with the constraint (3.48) line. The feasible region is therefore the triangle enclosed by the two constraint lines and the X_1 axis. By inspection it can be seen that the NLP minimum will be at a point on the line drawn from constraint (3.48), from which a perpendicular line will go through the minimum of the unconstrained objective function at $X_1 = 2$ and $X_2 = 2$.

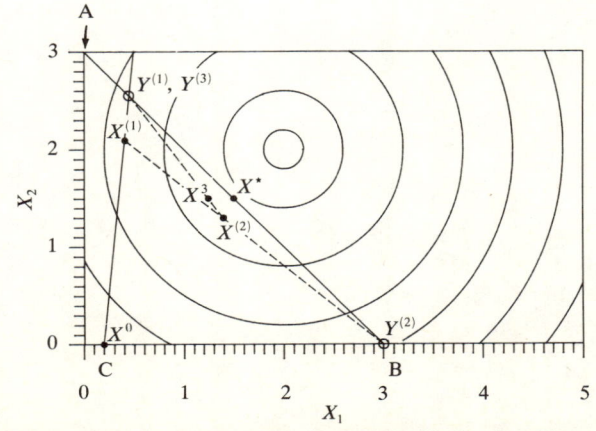

Figure 3.9. An example of the Frank–Wolfe algorithm.

The initial feasible solution $X_1^{(0)} = 0.2$ and $X_2^{(0)} = 0$ is shown on the X_1 axis at the point marked $X^{(0)}$. The first LP solution $Y_1^{(1)} = 0.455$ and $Y_2^{(1)} = 2.545$ is shown at the point marked $Y^{(1)}$ where the two constraint lines intersect. The minimum of the objective function along the line connecting $X^{(0)}$ and $Y^{(1)}$ is shown labelled as $X^{(1)}$. The second LP solution yields the point marked $Y^{(2)}$ on the $X^{(2)}$ axis, where the line drawn from constraint (3.48) intersects the $X^{(2)}$ axis. The minimum of the objective function along the line connecting $X^{(1)}$ and $Y^{(2)}$ is shown labelled as $X^{(2)}$. The third LP solution yields the same answer as the first, and is shown as the point $Y^{(3)}$ where the two constraints intersect. The minimum of the objective function along the line connecting $X^{(2)}$ and $Y^{(3)}$ is shown as $X^{(3)}$. The algorithm would continue this pattern of zigzag solutions, each more closely approaching the optimum, at X^*, than the last. A problem with this algorithm, as will be discussed in later chapters where it is applied, is that convergence, though assured, can be quite slow.

Other optimization techniques will be used in this book as well. Rather than overextending this introductory chapter, these other techniques will be introduced where needed. For the moment the introductory material on optimization can be ended, and attention may next be focused on the applications of linear and nonlinear programming approaches to the problems of modelling residential location, which follow in chapters 4 and 5.

Location models in optimizing frameworks: 1

4.1 Introduction
Chapters 2 and 3 introduced two of the principal elements on which this book is focused. These were, first, the subject (location and transportation models) and, second, the optimization approach. In this chapter the question of activity location will be discussed in the context of optimization and mathematical programming formulations.

In this and the next chapter several mathematical programming model formulations for activity location will be discussed. The models discussed in this chapter will include the Schlager model and the Herbert–Stevens model, and several variations. The next chapter will be focused on Brotchie and Sharpe's TOPAZ model, and other models will also be discussed. There will be an introductory discussion in chapter 5 of the relationship of these mathematical programming models to the entropy-based spatial interaction models. Several numerical examples will be given in both chapters, making use of the Archerville data again.

4.2 A land-use plan design model
Although not the first in which mathematical programming was used, this model, formulated by Schlager for the Southeastern Wisconsin Regional Planning Commission (SEWRPC), is one of the most straightforward and provides a convenient starting place (Schlager, 1965; SEWRPC, 1966). A very direct approach was taken, beginning with the following statement of the problem (Schlager, 1965).

1 The given design requirements are:
  (a) a set of design standards in terms of restrictions on land-use relationships that may exist in the plan;
  (b) a set of needs or demands for each type of land use, based on forecasts of future urban activity.

2 A land-use plan design must be synthesized that satisfies the land-use demands and design standards (with consideration being given to the current state both of natural and of human-made land characteristics) at a minimal combination of public and private costs.

A standard linear programming formulation was postulated. The objective function was simply to minimize the cost of land development in the region while meeting the development demands and other constraints. The original statement of the model had some rather confusing omissions of subscripts. In equation form (corrected) the model was stated as follows:

$$\text{minimize } C = \sum_{m,k,j} c_{kj} L_{mkj} \,, \tag{4.1}$$

subject to

$$\sum_{m,j} d_k X_{mkj} = L_k , \qquad (4.2)$$

$$\sum_j X_{mkj} \leq L_{mk}^{\max} , \qquad (4.3)$$

$$G_{ab} \sum_{m,j} X_{maj} \leq \sum_{m,j} X_{mbj} , \qquad (4.4)$$

$$g_{mab} \sum_j X_{maj} \leq \sum_j X_{mbj} , \qquad (4.5)$$

$$X_{mkj} \geq 0 , \qquad \forall\, m, k, j , \qquad (4.6)$$

where
- C is the total regional cost of land development,
- c_{kj} is the cost of land development of type k on soil of type j,
- L_{mkj} is the land use in zone m of type k on soil of type j,
- L_k is the exogenously estimated regional total amount of land use of type k to be allocated to the region's zones,
- d_k is the coefficient of the service ratio, which inflates land use of type k to account for ancillary land requirements for supporting services for land use k,
- $L_{mk}^{\max}$ is the maximum allowable amount of land use of type k in zone m,
- G_{ab} is the regional constraint on the ratio of total land use of type a to total land use of type b,
- g_{mab} is the zonal constraint on the ratio of total land use of type a in zone m to total land use of type b in zone m.

The first set of constraints, equation (4.2), is to ensure that the exogenously determined regional demand for land use of type k is met. The second set of constraints, equation (4.3), ensures that zonal constraints on land use of type k are met. The third set of constraints, equation (4.4), keeps the regional ratio of each land-use type to each other land-use type less than some prespecified amount, and equation (4.5) serves a similar purpose for intra-zonal land-use ratio constraints. Last, there are the customary nonnegativity constraints of equation (4.6).

Taken together, this model formulation provides a convenient framework for the analysis of a rarely encountered situation. Planning agencies would first need to obtain the cost of land-development data, which would have to include site costs and construction costs. The development of data on ancillary land-use requirements would also be somewhat problematic. Policy definitions might then be used to determine the various absolute and ratio constraints on land use. This all being done, the model could then be used to examine least-cost land-use allocations. The effects of increasing the costs, say by requiring developer-financed remedies to undesirable ecological consequences to be made, could also be examined. By repeated reformulations and solutions of the model the planners could develop a good sense of the effects of such costs and constraints. By extending the cost definitions in the objective function to include measures of 'social' cost, a more elaborate procedure of policy review could be developed. And yet, the whole structure is somewhat contrived. The linearities may well be far from realistic, as well as the notions of the constraint formulations. It does, however, provide a convenient way of introducing such formulations as frameworks for land-use (and, by inference, activity) allocations. The next formulations discussed, actually proposed somewhat prior to this one, attempt a more realistic and conceptually more complete structure.

Before moving to the next model a comment from the Schlager article (1965) is worth noting, as it points out one reason why the techniques being discussed in this book have practical value now, but did not twenty years ago even though some of them were known at that time. Schlager says, "For a region subdivided into about 30 zones, the size of a typical linear program for a land use plan design is about 60 equality and inequality constraints and 400 variables. Computer time on an IBM 1620 computer is about three hours. On larger systems such as the IBM 7090, it would take less than 30 minutes" (page 109). As this book is being written such problems can be solved on an IBM PC/XT microcomputer in just a few minutes and on an IBM 3081 mainframe computer in just a few seconds. Now, there are much faster mainframe computers and even faster supercomputers. Mathematical programs that would have been prohibitively expensive to solve in 1965 are almost trivial today, and there is every reason to believe that this trend will continue.

4.3 The Herbert–Stevens model: original forms
The mathematical programming model formulation which has stimulated the most subsequent research is the Herbert–Stevens model, originally proposed as a residential location model in 1960 (Herbert and Stevens, 1960b). The theoretical basis of the model is an extension of a structure developed by Alonso (1960; 1964) whereby workers are viewed as competing for residential locations in a bidding process.

Before describing the basis of this process a few definitions are necessary. A *residential bundle* is a unique combination of a house, an amenity level, a trip set, and a site of a specific size. The *amenity level* is what it sounds like and will be defined more specifically below. A *trip set* consists of the numbers of trips of all types taken (per year) by a household. A *market basket* is a particular combination of a residential bundle and all the other commodities consumed (per year) by a household.

The underlying process of the model is defined by first assuming that a household considers its total budget, a set of market baskets, and the relevant costs. The set of market baskets is defined as that set to which the particular household type is equally indifferent. Despite the connotation of the word *indifferent* in the English language, the meaning of it here is 'equal preference'. All market baskets in the set offer equal 'utility' or satisfaction to the locating household. The appropriate level of satisfaction for each household type is determined by observing their previous locational decisions.

It is next assumed that each locating household optimizes its location by selecting, from the 'equal satisfaction' set of market baskets, that market basket which maximizes the household's savings. These savings equal the amount of budget remaining after the costs of all other commodities except residential location have been subtracted from the household's total fixed budget. The amount left as a residual is available for residential location, and thus is the residential budget. The costs of 'other commodities' and the residential location costs, or residential budget, will vary from market basket to market basket. The difference between the residential budget and the sum of the house cost, the amenity-level cost, and the trip-set cost is the amount the household can afford to pay *for the site* on which the house is located (in the zone which has the level, or amount, of amenity and the trip set). This difference is called the household's maximum rent-paying ability, or the bid rent *for the site*.

An example adapted from the original paper by Herbert and Stevens (1960a; note that the 1960b paper is the 1960a paper abridged) will serve to illustrate some of these concepts. Assume a housing market with three houses on unit-sized sites, called A, B, and C. All houses are owned by one landlord. Assume there are three households who must find a place of residence, given existing places of work. The households are called H_1, H_2, and H_3. All three houses are equally satisfactory to all three households, but the households have different residential budgets for each of the houses. Each household will have a different residential budget at each location—recall that these budgets are a residual which remains after subtracting other costs from the household's total budget.

For H_1 the maximum savings resulting from choice of location are $10 at A, $7 at B, and, $2 at C. Of course, these could only be achieved at $0 rent for the site. H_1 will try to obtain one of the sites at the lowest possible cost and thus at the highest possible savings. Assume therefore that H_1 will begin by bidding $1 for site A. This bid if accepted would yield savings of $9, the maximum possible for a $1 bid from H_1 (we further assume that in this example bids must be integers). In order to keep the example simple we further assume that neither we nor H_1 know the savings available to H_2 and H_3 at the three sites—they could be the same as those for H_1 or, more likely, they could be different. Without going bid-by-bid, let us say that H_2 obtains C with a bid of $3. Note that at $3, H_1 does not want C, and indeed cannot afford it, as the net savings would be $-$1$, requiring some form of subsidy. Thus H_1 and H_3 are left competing for A and B.

Suppose that given its budget, and so on, H_3 can achieve savings of $9 for A and $5 for B. H_1 would be indifferent to a $9 bid for A and a $6 bid for B, as in either case H_1 would achieve a saving of $1. Thus it would appear that H_1 will locate at B for a bid of $6, and thus H_3 would have to locate at A for a bid of $9.

But enter the 'omniscient' landlord who knows that H_1 can afford to pay $7 for B. If the landlord permits H_1 to locate at B for $6 and H_3 to locate at A for $9, the total rent received will be $15. Similarly if the landlord permits H_3 to locate at B for $5 and then H_1 is forced to locate at A, even at H_1's maximum rent for A of $10, the landlord receives a total of $15. If, on the other hand, the landlord can force H_1 to locate at B for $7 and then allows H_3 to locate at A for $9, then the landlord will receive $16 from the two sites.

For reasons which will become more clear in the discussion below, Herbert and Stevens argued that maximization of total rents payable by *all* households is equivalent to maximizing their savings, or the consumer surplus of the system. Based on this premise, the model can be formulated as a linear programming (LP) problem:

$$\text{maximize } Z = \sum_{w, i, k} X_{ik}^w (b_{ik}^w - C_{ik}) ; \qquad (4.7)$$

subject to

$$\sum_{w, k} q_k^w X_{ik}^w \leq L_i ; \qquad (4.8)$$

$$\sum_{i, k} -X_{ik}^w = -N^w ; \qquad (4.9)$$

and

$$X_{ik}^w \geq 0, \quad \forall i, k, w ; \qquad (4.10)$$

where
X_{ik}^w is the number of households of type w that are assigned to houses of type k in zone i,
b_{ik}^w is the residential budget of households of type w that is available to spend on houses of type k in zone i,
C_{ik} is the cost of a house of type k in zone i, exclusive of site cost,
q_k^w is the size of site required by a household of type w for a house of type k,
L_i is the amount of residential land available in zone i,
N^w is the number of households of type w to be located.

The difference term $(b_{ik}^w - C_{ik})$ in the objective function is the amount that a household of type w in zone i can afford to pay for the site on which a house of type k can be located. In the original Herbert and Stevens formulation this is the precise definition of the bid rent. It should be noted, in passing, that different authors have given different explicit definitions to this term and to the two variables of which it is composed. These interpretations will be discussed below. Thus the objective function is the sum over all household types, zones, and house types of the number of households multiplied by the amount they can afford to pay (and therefore *do* pay) for site location in a specific house type and zone. The objective is to maximize the total bid rent for the entire region.

The constraint equation (4.8) prevents the consumption of land in each zone from exceeding that which is available. The constraint equation (4.9) forces all households in the region to be located to houses in various zones. Without this constraint it is both possible and likely that some households would not be allocated. The importance of this constraint and of the minus signs in it will become clear in the discussion of the dual of this LP problem. Last, the constraint (4.10) is the standard nonnegativity constraint.

Before moving to a discussion of the dual, we should note the fact that, as defined above, all variables except X_{ik}^w are exogenously determined. The version of the model given above is sometimes referred to as the land or land-consumption version. Because of the exogenous origin of the variables used to calculate the bid rents, this model formulation cannot achieve a maximization of the utilities of the households. In effect, the utility levels are given, by virtue of the exogenously calculated bid rents, for given levels of utility or preference. In this programming formulation, the total bid rents for fixed levels of utility are maximized, but in a proper formulation, utility-maximizing behavior by individual households could be assumed. Such a formulation would require endogenous (to the model) calculation both of utility and of bid rents.

In the discussion and interpretation of the dual of the Herbert–Stevens model some interesting properties appear. The formulation of the dual is

$$\text{minimize } Z' = \sum_i r_i L_i + \sum_w v^w(-N^w) ; \tag{4.11}$$

subject to

$$q_k^w r_i - v^w \geq b_{ik}^w - C_{ik} ; \tag{4.12}$$

and

$$r_i \geq 0 ; \tag{4.13}$$

where the variables from the primal formulation have the same definitions as before, and the new variables are

r_i is the annual per-unit rent for land in zone i;
v^w is the consumer subsidy (or surplus, if negative) for all households of type w.

The first term in the objective function of the dual [equation (4.11)] represents the total site rent for the region. Thus the dual involves minimizing the total rents paid by households. But, in the constraint equation (4.12), if for the moment the variable v^w on the left is ignored, the site rent paid must be greater than or equal to the rent-paying ability of any household which might locate on the site. In essence, the dual describes the situation where suppliers of housing, that is, the landlords, are competing with each other for prospective tenants. The variable v^w plays the role of surplus or subsidy to consumers whose presence in a particular house and zone results from the imposition of constraints (4.12), ensuring that all households will be located even if some cannot 'afford' the available housing. Note that the value of v^w is not constrained. An inspection of the objective function [equation (4.11)] shows clearly the desirability of having v^w positive. At the optimal solution of the dual, that is, at a fixed value of Z', a negative value of v^w (that is, a surplus) will increase the objective function (which is supposed to be minimized) unless it is offset by a corresponding reduction in rents. It should again be noted that the presence of these subsidy–surplus terms in the dual formulation are the result of constraint (4.9) in the primal, which forces the location of all households somewhere in the region.

It is important to realize that in the model formulation it is not necessarily the household with the highest income which can afford the highest bid rent for a particular parcel. The bid rent will depend on the residential budget of the locating household. This residential budget is the residual after all other costs are subtracted from the household's total budget. Thus, for example, a household with a lower total budget may be able to afford a higher rent for a particular site because its 'other' costs are lower than a household with a greater total budget, but greater 'other' costs as well. Owing, however, to the requirement from constraint (4.9) that all households be located, it is possible that high-income households will receive locational subsidies when low-income households do not.

The operation of this model can result in household location leading to four possible situations. (1) All the land in a zone may be used by a single locating household type, if that household type can yield the highest unit rent for the land and if there are enough of those households to fill the zone. (2) The land in a zone may not be entirely consumed if there are not enough households for whom land in that zone yields the maximum locational advantage (payable rents). (3) No land will be consumed in zones for which there are other zones that, for all household types, yield higher locational advantage and can accommodate all locators. (4) Land available in a zone may be consumed by more than one type of household, if there are not enough households of the type with the highest rent paying ability in the zone to use all the land in that zone. Other households with equal rent-paying ability in that zone, or with lower rent-paying ability in that zone, but for whom that zone is the best available may then locate in the zone and consume the remaining land.

To return now to a direct comparison of the primal and dual formulations, it is convenient to think of these formulations as the two sides of the housing market. The primal problem has the locating households being sited in such a way as to maximize the rent-paying ability for the whole system. The dual problem has the landlords providing locations in such a way as to maximize the total rents received from all locators in the system. As a side point to which we will return, it should be noted that in, for example, the primal problem, it is not any individual household's rent-paying ability that is being maximized; rather it is the total over all zones and household types. This criterion should result in what is known as a Pareto optimum (after Vilfredo Pareto a late-nineteenth, early-twentieth century economist). At such an optimum, no household can move to increase its rent-paying ability without decreasing the rent-paying ability of some other household by at least an equal amount, and thus at best leaving unchanged, and more likely decreasing, the aggregate rent-paying ability of the system. It would be possible to define an alternative equilibrium condition where the concern would be with the maximization of each household's rent-paying ability. At such an equilibrium no household could move to increase its rent-paying ability, regardless of the effect of such a move on the rent-paying ability of any other household.

One of the properties of a linear programming solution to problems such as the one defined above has to do with the number of nonzero values of X_{ik}^w to be expected in the optimum solution. In this case, the optimal solution will normally contain a number of nonzero values equal to the total number of constraints in the problem. The number of constraints represented by inequality (4.8) is equal to the number of zones in the system, and the number of constraints represented by inequality (4.9) is equal to the number of household types. The immediate consequence of this is that *if at least one household locates in each zone, then there will only be a few zones (the number of household types) with a second household type locating as well*. The equilibrium, or optimal, solution produced by the model may well turn out to produce what might be termed 'lumpy' allocations of households and thereby miss out on the complex blending of household types in zones that can be observed in virtually all metropolitan areas.

4.4 The Herbert-Stevens model: application issues

Not too much attention was paid in the above discussion to what might be involved in actually attempting to apply the Herbert-Stevens model. Despite the fact that the model equations are themselves rather simple, the definitions of the variables can lead to rather difficult problems of application and interpretation.

Consider, to begin, the variable which represents the rate of consumption of land, q_k^w. The original definition was; 'the amount of land used by a household of type w for a house of type k'. Note that this definition does not vary over the zones of the region. Wheaton and Harris (1971) use a definition which varies by house type and zone, q_{ik}, and which is independent of household type. Not long thereafter Wheaton (1974) again redefined the variable, making it vary only by house type, q_k, and making it independent of both zone and household type. This latter definition was used by Senior and Wilson (1974a) as well. From the standpoint of the model equations, it does not matter which of these definitions is used. The land-consumption variable appears only in the constraint equation (4.8). What is important is what the model developer and/or user believes is the most realistic interpretation of land consumption. Does it vary by type of house, by zone, by both? Does it vary with type of household occupying the house? The determination of which definition to use will depend upon the application intended.

A considerably more difficult set of questions is associated with the rent functions, budget functions, and the definition of preferences. In the original formulation, households were assumed to be indifferent to a set of market baskets. These market baskets were defined to be a residential bundle and all the other commodities consumed by the household. The residential bundle was a unique combination of a house, an amenity level, a trip set, and a site. To apply this model it is necessary to develop precise definitions for these and, thus, for the set of market baskets to which the household type is indifferent. To borrow a bit from Shakespeare, "ay, there's the rub". The definition of these preference functions has proved to be a very difficult problem. Herbert and Stevens ignored the problem entirely.

The first serious attempt to estimate these functions was made under Harris's direction (Harris et al, 1966). A summary of the work was also given in Harris (1966). To understand the role of preference function analysis in the Herbert-Stevens model, it is useful to begin with a somewhat more general description of utility functions in such models. To begin, the general concept of utility is one of the degree-of-satisfaction which accrues to a decisionmaking unit by virtue of actions taken. In analyses of the sort to be described here, utility may not be directly measured, but has to be inferred from some external structure or hypotheses. In general one may consider a function of the form

$$U = f(Q_1, Q_2, Q_3, ..., Q_n) \,, \tag{4.14}$$

where U is utility and the Q_n are attributes of the situation producing, or the individuals achieving, that level of utility. The term 'indifferent' was used at the beginning of the description of the Herbert-Stevens model, and can now be better described. Take a specific value of utility, say, U_1. It should be clear that there will be many possible combinations of the Q_n that, when substituted into the function in equation (4.14), will yield exactly U_1. All these combinations of the Q_n define a mathematical surface in n-space. At any point on this surface the value of the function in equation (4.14) will be exactly U_1. This surface is an 'indifference surface', or 'preference function'. As each value of U implies such a surface, the set of all surfaces for all different values of utility is the definition of the 'preference structure' of the decisionmaking unit.

The traditional assumption of economic analyses of this sort is that consumers are attempting, individually, to maximize their received or achieved utility. It is customary to assume that for each decisionmaker, in this case, each locating household, there is a budget equation and a utility function. The budget equation is of the form

$$Y = M + R + T \,, \tag{4.15}$$

where
Y   is the household's total expendible income,
R   is the expenditure on house rent,

T is the expenditure on transportation,
M is the expenditure on all other goods.
The utility function is of the form

$$U = f_1(M) + f_2(Q) + f_3(A) , \qquad (4.16)$$

where
U is the utility achieved,
Q is the expenditure on a particular quantity of housing (including land or site cost),
A is the expenditure on accessibility to activities in the region,
M is the expenditure on all other goods, as before.

In the work by Harris on preference functions it was generally assumed, from the work by Alonso, that R and T are jointly determined, as the choice of housing type and location are jointly determined (Harris, 1966). It is then assumed that the specific functional form of the utility function [equation (4.16)] is

$$U = \ln M + k_1 \ln Q + k_2 \ln A , \qquad (4.17)$$

and, transposition of the terms in equation (4.15) gives

$$M = Y - R - T , \qquad (4.18)$$

which can be substituted into equation (4.17) to give

$$U = \ln(Y - R - T) + k_1 \ln Q + k_2 \ln A . \qquad (4.19)$$

In this equation all the variables on the right-hand side are, at least in theory, measurable quantities.

A key assumption in the Alonso theory is that at equilibrium all members of any given household type have achieved the same level of utility. Thus for a particular household type, U may be considered to be constant. As a consequence, if data for the variables on the right-hand side of equation (4.19) were available for a large number of households of a given type, it would be possible to use regression analysis to estimate the parameters in the equation. In the analyses by Harris the available data were by census tract rather than by individual households (as would be more desirable). As a result it was clearly not safe to assume that they were dealing with specific, clearly defined household types. Thus a further adjustment was made to account for income disparities within groups of household types. The definition of U was

$$U = f_4(S) + \ln Y , \qquad (4.20)$$

where S was defined to be a set of socioeconomic variables. Equation (4.20) was then substituted into equation (4.19), and the terms were rearranged to yield

$$\ln\left(\frac{Y - R - T}{Y}\right) = f_4(S) - k_1 \ln Q - k_2 \ln A . \qquad (4.21)$$

In Harris's analysis of the Hartford (CT) data the population was stratified by size of household, by income, by race, and by housing tenure. The coefficients f_4, k_1, and k_2 in equation (4.21) were then estimated for each population (household) type.

Of course, in order to estimate the parameters of equation (4.21), rather more specific definitions of the variables were necessary. In the analysis of the Hartford data a series of somewhat convoluted transformations was made, but in general the following three types of variable were used.

Socioeconomic variables: median family income, median number of persons per household, percentage white population, percentage owner-occupied housing units;

Housing variables: number of rooms per (housing) unit, amount of residential land per unit, median age of structure, mean structure type, percentage units 'sound with all plumbing';

Accessibility variables: twenty different accessibility measures were constructed by using highway and transit networks and measuring travel time and/or cost to the spatial distributions of five activity types, four employment types, and total population. These measures were then transformed into four orthogonal components via principal component analysis. These components were used to create the actual variables used in the regression.

All told, with the addition of a number of additional 'cross-product' variables, a total of forty-nine independent variables were calculated. With the inclusion of all of these variables an R^2 of 0.916 was obtained. Subsets of the variables were tested, with the smallest (containing ten independent variables) yielding a value of 0.795 for R^2. These results were then used in running modified versions of the Herbert–Stevens model for an artificial data set. No attempt was made to test the results of the preference function analysis and the model against the Hartford data.

Before further discussion of this approach and, perhaps more importantly, its connection to other approaches is proceeded with, it will be informative to do a numerical example.

4.5 The Herbert–Stevens model: numerical examples

In order to give the reader a clearer sense of the workings of this model, use will be made of the Archerville data to do a small numerical example. To begin, it should be noted that several different versions of the model have been proposed by different authors. These differences result from alternate assumptions about the preference functions and, more importantly, about the precise definitions of the equilibrium solution being sought. Some of these issues began to be discussed rather soon after the original model was published (Harris, 1962). First off, the original formulation specifically separates site cost from the cost of a house. From the standpoint of practical application this creates a difficult problem with the data, and has no real benefit in the model formulation. Second, the version of the model given above may be known as the 'land' version, as the supply constraint of equation (4.8) is in terms of available residential land per zone. An alternative formulation, the 'stock' version, reformulates the supply constraint in terms of an exogenously specified quantity of housing stock.

Quite a lot of data preparation was necessary to augment the Archerville data for use in the Herbert–Stevens model. The major matter was to specify a preference function and the data necessary to produce the required numerical values. To begin with, two household types were defined in the original Archerville data given in chapter 2. The low-income (LI) households are assumed here to have an annual income of $25 000. The high-income (HI) households are assumed to have annual incomes of $60 000.

The first form of preference function consists of two access terms; a form which derived in a very general way from some of the empirical work on preference functions which has been discussed above. The first of these was access of households of type w in zone i to total employment, A_{iw}^E of the form

$$A_{iw}^E = \sum_j E_j^T \exp(-\beta_w c_{ij}) , \qquad (4.22)$$

where
E_j^T is the total employment in zone j (see table 2.1),
c_{ij} is the cost of travelling from zone i to zone j on the highway network (see table 2.6),
β_w is the 'empirically' estimated trip function parameter (taken from chapter 2 to be 0.2 for β_1 and 0.1 for β_2).

The second accessibility measure was the accessibility to 'shopping', calculated by substituting commercial employment (from table 2.1) into equation (4.22) to give

$$A_{iw}^S = \sum_j E_j^C \exp(-\beta_w c_{ij}) , \qquad (4.23)$$

where
E_j^C is the commercial employment in zone j.

In this example, the household's residential budget, b_{ik}^w, was calculated as

$$b_{ik}^w = kY_w - \alpha_w^E A_{iw}^E - \alpha_w^S A_{iw}^S , \qquad (4.24)$$

where
b_{ik}^w is as above, the residential budget of type w households for type k houses in zone i;
Y_w is the annual income of type w households
k is the proportion of annual income remaining after subtracting expenditures on 'all other goods' [M in equation (4.15)], and in this example $k = 0.6$;
α_w^E, α_w^S are empirically estimated coefficients in the preference functions.
For this example the values of α were
$\alpha_1^E = 12.5, \qquad \alpha_1^S = 6.5,$
$\alpha_2^E = 20.5, \qquad \alpha_2^S = 15.0.$

These values were selected by trial and error to scale the accessibility values into annualized dollar terms (to correspond to the income figures), to give more importance to access-to-employment than to access-to-shopping, and to have HI households 'pay' more than LI households for their access.

Last, it was necessary to generate housing data for Archerville. First it was decided that there would be three types of house, small, medium, and large. Each of these would use 0.003889, 0.007778, and 0.015556 units of land, respectively. Again, the values were selected to match the Archerville land-use data in table 2.1 and to allow enough houses for the total households in Archerville: 800 of LI and 650 of HI. The last housing-related data item was that of house prices. The three house types were given basic annualized costs of $4000, $6000, and $10000. A matrix of house cost by type and by zone was then created by arbitrarily (but with some aim at consistency with Archerville) varying the house costs from the basic values. These data are given in table 4.1.

Table 4.1. Archerville: annualized house cost ($) by type and zone.

Zone	House type (size)		
	small	medium	large
1	6000	7000	13000
2	5000	7000	12000
3	5000	8000	12000
4	4000	5000	10000
5	10000	9000	10000

With the data preparation completed, the GAMS package (Kendrick and Meeraus, 1985) was used to solve the original version of the Herbert–Stevens model as given in equations (4.7)–(4.10). The results are shown in table 4.2. The most striking aspect of these figures is the extremely discontinuous distribution of household locations. Of the thirty possible combinations of household by type and by zone, only four are used. This is partly owing to the simple fact that the preference function includes no aspect of house type. As small houses are least expensive the model locates all households in small houses. All available land in zones 4 and 5 was used, with the remaining households locating in zone 2. Note that the HI households took the expensive small houses in zone 5 rather than the cheaper small houses in zone 4, because of the differences in the preference functions and travel functions of the household types.

The obvious next step was to augment the preference function by the inclusion of some aspect of house type. As the only information available in this example is house size, a simple weighting variable was constructed. The preference function of equation (4.24) became

$$b_{ik}^w = kY_w - \alpha_w^E A_{iw}^E - \alpha_w^S A_{iw}^S - \alpha^H Q_k \, , \tag{4.25}$$

where

a^H is an empirically estimated coefficient in the preference function, set in this example to 1700,

Q_k is the 'utility' of housing of each house type k, set in this example to 1.0, 2.5, and 4.5 for each of the three house types, respectively.

Note that a^H could have been made to vary by household type, thus reflecting the preferences of different household types for different quantities of housing. One could argue that LI and HI households have the same preferences for quantities of housing or that they differ owing to 'expectation'. The former argument was taken for the next run of the Herbert–Stevens model. All other variables and parameters were unchanged, but the augmented preference function of equation (4.25) was used to calculate the bid rents. The results of this second test of the model are given in table 4.3.

Table 4.2. Archerville: results from test 1 of the Herbert–Stevens model.

Zone	House type (size)		
	small	medium	large
Low-income households			
1	0	0	0
2	157	0	0
3	0	0	0
4	643	0	0
5	0	0	0
High-income households			
1	0	0	0
2	264	0	0
3	0	0	0
4	0	0	0
5	386	0	0

Table 4.3. Archerville: results from test 2 of the Herbert–Stevens model.

Zone	House type (size)		
	small	medium	large
Low-income households			
1	0	0	0
2	0	250	0
3	0	0	0
4	457	93	0
5	0	0	0
High-income households			
1	0	321	0
2	0	136	0
3	0	0	0
4	0	0	0
5	0	193	0

The results of the second test run show only a slight increase in dispersion over zones and house types. LI households locate both in small and in medium-sized houses. HI households locate only in medium-sized houses, and no large houses are occupied. These results reflect the interplay of several factors. First, in the preference function the size of house now has value. Thus, were all other things equal, all households would prefer to locate in larger houses. All other things are, of course, not equal and thus households do not all locate in large houses. In fact, expense evidently outweighs utility, as no household locates in a large house. Another factor, though, is that large houses use more land than small or medium-sized houses. All zones but zone 3 have all the available land used by the houses occupied in the model solution. From a look at the marginal values in the model solution (which was done by use of the simplex algorithm) it is clear that the dollar cost of large houses combined with their land requirements considerably outweighs their somewhat higher utility.

A further set of three test runs was made to illustrate this point. In these runs the utility of large houses was set to 5.5, 6.0, and 6.5. The results of the last two of these three additional test runs are shown in tables 4.4, and 4.5, respectively. It is interesting to observe the changes, or lack thereof, in household-to-house allocations. In the original test 2 (table 4.3) there were no occupied large houses. Increasing Q_3 from 4.5 to 5.5 produces no change in household-to-house allocations. A further increase in Q_3 to 6.0 does result in some changes, as seen in table 4.4. Fewer LI households locate in zone 2, but some of those that do so move to large houses. Zone 3 which formerly had no LI households, now has some LI households located in large houses, and zone 4, which formerly had LI households in small and medium-sized houses, now only has them in small houses. The only change

Table 4.4. Archerville: results from test 4 of the Herbert–Stevens model.

Zone	House type (size)		
	small	medium	large
Low-income households			
1	0	0	0
2	0	72	21
3	0	0	64
4	643	0	0
5	0	0	0
High-income households			
1	0	321	0
2	0	0	136
3	0	0	0
4	0	0	0
5	0	193	0

in location of HI households is in zone 2, where all HI households switched from medium-sized to large houses.

The last of these test runs is with Q_3 increased to 6.5, the results of which are given in table 4.5. There was no further change in the location of HI households. The LI households do shift about, with zone 2 now containing LI households in small and large houses, but none in medium-sized houses. LI households in other zones do not change. In part, the nature of the various shifts we can observe results from the land-use constraint as well as from the changes in Q_3. In the original test run zones 1, 2, 4, and 5 are full (in terms of land use), whereas zone 3 has no households of either type. This situation is unchanged for the run where Q_3 increases to 5.5, but changes with subsequent increases. The final two test runs result in the consumption of all available land in all zones, which restricts the ways in which the household locations can change. Nonetheless, the substantial location changes which result from changes in Q_3 in the range 6.0 to 6.5, although in part owing to the nature of the discontinuities inherent in this small example, also point to a deficiency in the model formulation. Even if the utilities of two possible location options are only slightly different, this model locates *all* possible households in the slightly better location.

It is clear that by modifying some of the arbitrary assumptions which were made for these examples some changes could be effected in the solutions. Even so, the major point would not be changed. By virtue of the formulation of the Herbert–Stevens model its solutions will always give unrealistic results as there will be major discontinuities in the household locations both to zone and to house type. In this regard, recall that earlier in this discussion note was made of the fact that one property of an LP solution (called the positivity condition) is that there will be, at most, as many non-

Table 4.5. Archerville: results from test 5 of the Herbert–Stevens model.

Zone	House type (size)		
	small	medium	large
Low-income households			
1	0	0	0
2	48	0	45
3	0	0	64
4	643	0	0
5	0	0	0
High-income households			
1	0	321	0
2	0	0	136
3	0	0	0
4	0	0	0
5	0	193	0

zero values in the solution as there are constraints in the problem. In the case of this numerical example there are five constraints from equation (4.8), one for each zone, and two constraints from equation (4.9), one for each household type. There are thirty possible solution values for X_{ik}^w (five zones multiplied by two household types multiplied by three house types), yet in theory there will be, at most, only seven nonzero values of X_{ik}^w. Even for this small example this does not seem a very realistic solution. For larger problems, with say one hundred or more zones and perhaps four to six household types and as many house types, the solutions will appear even more unrealistic. Another consequence of the model formulation is that for a given house type in a given zone the bid rents must be equal for all households in that house type. This will lead to many zones in which there will be only one household type in any given house type. Again, a 'lumpy' solution will result. The simple numerical tests illustrate these problems amply.

Other work involving analyses of preference functions has been done (Anas, 1975; Galster, 1977; Wheaton, 1977a; 1977b). Anas (1973) enumerates several functional forms which might be suitable for such models as this. Yet, when all is said and done, there are some serious questions left unanswered. The empirical estimation of preference functions is problematic. Anas (1975, page 912) writes that results were "not, however, as good as they would have to be to justify the use of a model such as the Herbert-Stevens because the uniform utility level hypothesis is far from being fulfilled". Wheaton (1977a) obtains reasonably good results, but only by extensive stratification of the sample, thus resulting in 128 different types of household. Of these, only a few were sufficiently well represented in the sample to be of use, and these, in turn were further restricted by his use of only those households whose primary wage earner worked in the central business district (the purpose of all this being to achieve greater uniformity of the indifference surface). Harris (1966, page 29) asserts that, "income alone will explain 96% to 98% of the variance in rents". He says the preference function will explain 50% to 90% of the residual variance (that is, the variance not explained by income). Anas (1982) shows that even though the preference function appeared to give a moderately good fit to data, when the results were substituted back in to the equation to estimate actual rents, the results were not nearly as good. R^2 for the preference function took values from 0.66 to 0.95, whereas R^2 for the rent equation took values from 0.16 to 0.46.

Perhaps even more important than the problems with the preference functions are the problems with the overall Herbert-Stevens formulation. In the economic theory on which the model is based it is assumed that households are attempting to maximize utility. In the Herbert-Stevens formulation the utility levels are exogenously specified—first by the equal utility assumption used in estimating the preference functions and second by the exogenous calculation of the bid rents. In the most frequently discussed form of the model an exogenously specified housing supply is used. Thus, as neither the utility levels nor the housing stock can change, the model cannot actually achieve an equilibrium solution in the economic sense. Further confirmation of this fact is given by the need for the subsidy terms in the solution of the Herbert–Stevens dual. Wheaton and Harris (1971) and later Wheaton (1974) propose a way of dealing with this set of problems by embedding the Herbert–Stevens model in an iterative scheme. In this system the inner model is the Herbert–Stevens, which locates all households, given a particular set of bid rents. The outer model adjusts the bid rents so that the subsidies or surpluses (the v^w of the dual formulation) are zero in the final solution of the iterative system.

Even after all this, Wheaton's (1977b, page 631) conclusion is that "the long-run spatial theory of Alonso, Mills, and Muth empirically contributes little to the explanation of American location–income patterns". Harris (1972, page 79) reporting on statistical results from an analysis of the New York (city) metropolitan area was that, "the model with largely economic variables failed to define properly the behavior of the upper income groups. ... [It was] found that the best results in remedying this situation were produced by adding variables regarding the accessibility of low- and high-income groups, and of non-whites ... It deals a heavy blow to Alonso's belief that market behavior can be explained without reference to externalities." This was precisely the point made in Galster's (1977) analysis.

Anas (1982) rather nicely summarizes the theoretical shortcomings of the Herbert–Stevens, or bid rent, approach to residential location. The first problem stems from the assumption that the demanders of housing (the locating households) and the suppliers of housing (the builders and/or landlords) have perfect information with regard to the housing market. This assumption is clearly incorrect. The second problem stems from the assumption that within household types there are homogenous preference functions. Even with very considerable segmentation (that is, many different household types specified) this assumption is incorrect as well. There will inevitably be variation in preferences within households groups, no matter how finely stratified. A third assumption, obviously incorrect, is that households are assumed to be perfectly mobile, that is, there is no cost of moving from one house to another. As a result of these problems the equilibrium solution to such a model is quite sensitive to rather small variations in preference function coefficients. It is interesting to note that these same problems appear again in the discussion of the algorithms used to assign traffic to transportation networks, given in chapter 6. A sensible way to proceed, here as well as in chapter 6, is to find ways to relax some of these assumptions. In the next section of this chapter, several variations of the Herbert–Stevens formulation are introduced which move in this general direction.

4.6 The Herbert-Stevens model: further developments

There is quite a lot of scope for further modification and elaboration of the Herbert–Stevens model. Some of this takes the form of variations in the specification of the preference function and some in the definitions of the variables of the model, such as the bid rent and land consumption per house. Some of these were discussed above in the text and were illustrated in the elaboration of the numerical examples. Other work has been done with the general model structure itself.

One such effort, although not actually changing the basic construct, does treat explicitly the differences between households locating in existing houses and households locating in newly constructed houses. This resolves an ambiguity in the original formulation where the differences between the two phenomena are not clearly specified. This was the reason that one version of the model was developed as the 'land' version, dealing with land being consumed for newly constructed housing, and an alternate version of the model, the 'stock' formulation, was developed to deal with households being located in an exogenously estimated 'existing' housing stock. The revised model takes the following form (Anas, 1982):

$$\text{maximize } Z = \sum_{w,i,k} \left[X_{ik}^{w\,\text{new}}(b_{ik}^w - C_{ik}) + X_{ik}^{w\,\text{pre}} b_{ik}^w \right] + \sum_i L_i^{\text{alt}} b_i^0 \,, \tag{4.26}$$

subject to

$$\sum_{w,k} q_k X_{ik}^{w\,\text{new}} + L_i^{\text{alt}} \leq L_i \,, \tag{4.27}$$

$$\sum_w X_{ik}^{w\,\text{pre}} \leq S_{ik} \,, \tag{4.28}$$

$$\sum_{i,k} (X_{ik}^{w\,\text{new}} + X_{ik}^{w\,\text{pre}}) \leq N^w \,, \tag{4.29}$$

$$X_{ik}^{w\,\text{new}} \geq 0, \quad X_{ik}^{w\,\text{pre}} \geq 0, \quad L_i^{\text{alt}} \geq 0 \,. \tag{4.30}$$

In these expressions

$X_{ik}^{w\,\text{new}}$ is the number of households of type w allocated to newly constructed houses of type k in zone i;

$X_{ik}^{w\,\text{pre}}$ is the number of households of type w allocated to preexisting dwellings of type k in zone i;

b_{ik}^w is the bid rent or (annualized) budget available of households of type w for residences of type k in zone i;

C_{ik} is the (annualized) cost of constructing a house of type k in zone i;

L_i^{alt} is the number of units of undeveloped land devoted to alternate use such as agriculture, in zone i;

b_i^0 is the bid rent (annualized) for alternative land use in zone i;

q_k is the amount of land required to build a house of type k (note that this could be indexed by i and w if appropriate in a particular application);

L_i is the amount of land available for residential or other use in zone i;
S_{ik} is the total number of preexisting type k houses in zone i;
N^w is the regional total number of type w households to be located.

There are two major differences between this formulation and the original one given in equations (4.7)–(4.10). First, as already mentioned, the locating households are split into those who occupy preexisting houses and those who occupy newly constructed houses. Both of these groups must be separately accounted for in the objective function [equation (4.26)] as well as in the constraints [equations (4.27)–(4.30)]. The second difference is made necessary not only by the split into new and preexisting housing, but is in fact a requirement for the system to achieve a stable-equilibrium solution. The inclusion of an alternate land-using activity, which will use any land not used for residential purposes, in a sense closes or balances off the land-use supply and demand. The need for such an alternate land use was discussed by Herbert and Stevens (1960a) in their original formulation, in terms of the need for an 'other' household group. This prospect was later discussed by Harris (1966; 1972).

Thus it can be seen that the objective function [equation (4.26)] contains locating households, both in preexisting and in newly constructed houses, and the alternate land use. The first constraint [equation (4.27)] insures that land used by newly constructed housing plus land in the alternate use are less than or equal to the total land available. At the optimum, or equilibrium, solution the effect of the alternate use will be to make this equation an equality. The second constraint [equation (4.28)] is simply the limit on the preexisting housing stock. The effect of the third constraint will be to ensure that at equilibrium all households are located. The remaining constraints are the usual nonnegativity conditions.

Even with this elaboration, the model still suffers from the problem that there will be too few nonzero values of X_{ik}^w and thus the allocations of households to zones will be too lumpy. Another way of understanding this problem, besides that of the property of LP solutions giving a predetermined number of nonzero values, is via the distribution of the bid rents. In the above discussion it was emphasized that each household group was assumed to have homogeneous preferences, that is, all households in a particular group are assumed to be on the same indifference surface. For practical purposes this assumption allowed the further assumption that each household in a particular group will have a bid rent surface identical to each other household in the same group. Another way of viewing this situation, as will be discussed again in chapters 6 and 7 where user-equilibrium trip assignment will be compared with stochastic user-equilibrium trip assignment, is to assume that for the purposes of the model that although the bid rents may not be identical, the model is effectively considering only the mean bid rent for each household type. An obvious extension of the model presents itself here in the form of considering that there is a variation of the bid rent surfaces within each household group. This variation might be thought of as resulting from differences in perception on the part of individual households, differences in the housing-market information that is available to individual households, or to actual small differences in their preference functions. An interesting development of the consequences of one of the many possible functional forms which could be adopted to represent bid rent variation has been described by Senior (1974), following on from work by Houghton (1971). Before moving into this discussion, it will be helpful to alter slightly the Herbert–Stevens formulation to make it more consistent both for this discussion as well as for some of the work to be discussed towards the end of this chapter.

The notion of imperfect information about the housing market, combined with some further arguments regarding imperfect functioning of the housing-market, leads (as will be discussed in a later section of this chapter) to suboptimal solutions of the model. To move towards this discussion, and to serve present purposes, it is useful to adopt a formulation of the Herbert–Stevens model put forward by Senior and Wilson (1974a). To work towards what will become a 'stock' version of the model, the formulation proposed begins as the 'land' version, as follows:

$$\text{maximize } Z = \sum_{i,k,w} X_{ik}^w (b_{ik}^w - C_{ik}^w) , \qquad (4.31)$$

subject to

$$\sum_w X_{ik}^w q_{ik} \leq L_i , \qquad (4.32)$$

$$\sum_{i,k} X_{ik}^w = N^w , \qquad (4.33)$$

$$X_{ik}^w \geq 0, \quad \forall i, k, w , \qquad (4.34)$$

where
- X_{ik}^w is the number of households of type w assigned to houses of type k in zone i;
- b_{ik}^w is the residential budget of households of type w available to spend on houses of type k in zone i;
- C_{ik}^w is the cost to a household of type w of a house of type k in zone i, *exclusive* of site cost;
- q_{ik} is the size of site required in zone i for a house of type k;
- L_i is the amount of residential land available in zone i;
- N^w is the number of households of type w to be located.

There are only three differences between this formulation and the original one given in equations (4.7)–(4.10). First, the cost of a house of type k in zone i is postulated to vary as a function of the type of household, w, planning to purchase it. Second, the size of site required for a type k house is posulated to vary with the zone of location only, whereas in the original formulation it was invariant over zones but varied with the type of household purchasing the house. Last, the somewhat peculiar minus signs of the constraint on the total number of households [equation (4.9)] are removed in the equivalent constraint here [equation (4.33)]. These minus signs were originally used by Herbert and Stevens in order to control the signs of the surplus variable in the dual, but were later found to be unnecessary.

The formulation is next converted to the 'stock' version, with the site costs that were included in house costs and transport costs being separated out into an explicit variable, to yield:

$$\text{maximize } Z = \sum_{i,k,w} X_{ik}^{w}(b_{ik}^{w\,\text{inc}} - c_i^{w}), \tag{4.35}$$

subject to

$$\sum_{w} X_{ik}^{w} \leq S_{ik}, \tag{4.36}$$

$$\sum_{i,k} X_{ik}^{w} = N^{w}, \tag{4.37}$$

$$X_{ik}^{w} \geq 0, \quad \forall i, k, w, \tag{4.38}$$

where

$b_{ik}^{w\,\text{inc}}$ is the residential budget of households of type w that is available to spend on a house of type k in zone i, *inclusive* of site cost;

c_i^{w} is the average cost of transport for a household of type w residing in zone i;

S_{ik} is the number of preexisting houses of type k in zone i.

It is worth noting here that the explicit inclusion of transport cost as a variable may lead to complications in the definition of the preference functions. In the preference functions employed by Harris, as described above, and in the simplified preference functions utilized in the numerical example, use was made of constructed accessibility variables; these variables included transport costs. The question is open to discussion regarding the extent to which the use of transport costs in both ways in the model is a problem.

A further disaggregation of the model may now be made which is not only interesting in this context but is of considerable importance to the discussion later in this chapter. This disaggregation involves specifying the workplace of the head of the locating household. Note that it is assumed here, as it is in most location models, that each household has only one employee who has, by virtue of his or her journey to work, a considerable effect on the location of the household. Extensions of these models to multiemployee households all of whom jointly determine household location are not difficult in theory, but are very difficult in practice because of the problems of obtaining the necessary data for parameter estimation. In this new version of the model the total number of employed heads of type-w households working in zone j is substituted for the total number of type-w households, yielding the following model form:

$$\text{maximize } Z = \sum_{i,j,k,w} X_{ijk}^{w}(b_{ijk}^{w\,\text{inc}} - c_{ij}^{w}), \tag{4.39}$$

subject to

$$\sum_{j,w} X_{ijk}^w \leq S_{ik} \,, \tag{4.40}$$

$$\sum_{i,k} X_{ijk}^w = E_j^w \,, \tag{4.41}$$

$$X_{ijk}^w \geq 0, \quad \forall\, i, j, k, w \,, \tag{4.42}$$

where
- X_{ijk}^w is the number of heads of type-w households who are employed in zone j and live in a type-k house in zone i;
- $b_{ijk}^{w\,\text{inc}}$ is the residential budget of households of type w, whose heads are employed in zone j, which is available to spend on a house of type k in zone i;
- c_{ij}^w is the average transport (journey-to-work) cost for the head of a type-w household who is working in zone j and living in zone i;
- E_j^w is the number of heads of type w households who work in zone j, to be located.

Now, the interesting aspect of this formulation is that it includes explicitly the spatial patterns both of residences and of workplaces, and also the costs of travelling between these places. In this form, however, as in the case of its predecessors, it is still assumed that all type-w households have the same bid rent surface; but with the inclusion of spatial separation it is possible to discuss Senior's (1974) adaptation of Houghton's (1971) model.

The principal change in this formulation is that the demand for a house of type k located in zone i by an employee from a household of type w, who is working in zone j, is described by a declining exponential function. This demand function may be expressed in equation form as

$$X_{ijk}^w = \theta_{ijk}^w \exp[-\beta(b_{ijk}^{w\,\text{inc}} - c_{ij})] \,, \tag{4.43}$$

where
- θ_{ijk}^w is a parameter;
- β is a parameter, which could be disaggregated but which need not be for this exposition;
- c_{ij} is the journey-to-work cost between zone i and zone j.

Note that the journey-to-work costs are no longer disaggregated by household type; though this could be done, it may well be unnecessary with the presence of the parameters θ_{ijk}^w and β.

The total value of bids made by any given type of household is obtained by integration:

$$\text{total value of bids} = \int_0^{E_j^w} (b_{ijk}^{w\,\text{inc}} - c_{ij}) \, \mathrm{d} X_{ijk}^w \,. \tag{4.44}$$

The mean bid for the group is simply the total value of bids divided by the number of households in the group, which is, after integration and division, given by:

$$\text{mean bid} = \bar{b}_{ijk}^{w\,\text{inc}} - \bar{c}_{ij}. \tag{4.45}$$

The mean bid of c_{ij} is c_{ij}, as the c_{ij} do not vary by household type. The demand function [equation (4.43)] can be rearranged to give

$$b_{ijk}^{w\,\text{inc}} - c_{ij} = -\frac{1}{\beta} \ln \frac{X_{ijk}^w}{\theta_{ijk}^w}. \tag{4.46}$$

If we substitute this result into the integration of equation (4.43), calculate the mean again, and rearrange terms, we set

$$\ln \theta_{ij}^{kw} = \beta(\bar{b}_{ijk}^{w\,\text{inc}} - c_{ij}) - 1 + \ln E_j^w. \tag{4.47}$$

Equation (4.47) may be substituted into equation (4.46), which gives an expression for bid rents:

$$b_{ijk}^{w\,\text{inc}} - c_{ij} = -\frac{1}{\beta} \ln \frac{X_{ijk}^w}{E_j^w} - \frac{1}{\beta} + \bar{b}_{ijk}^{w\,\text{inc}} - c_{ij}. \tag{4.48}$$

If this expression is substituted into equation (4.43) and the result integrated, a new expression for total value of bids is produced. Recall that the objective function for the forms of the Herbert–Stevens model is to maximize the total value of bid rents; then, after the just-described manipulations, the following objective function results:

$$\text{maximize } Z = \sum_{i,j,k,w} X_{ijk}^w \left(-\frac{1}{\beta} \ln \frac{X_{ijk}^w}{E_j^w} + \bar{b}_{ijk}^{w\,\text{inc}} - c_{ij} \right), \tag{4.49}$$

subject to constraint equations (4.40), (4.41), and (4.42). In the objective function [equation (4.49)] the operand of the logarithmic term is the percentage of type-w heads of households (recall the assumption that there will be one employee per household, to give $N_j^w = E_j^w$) who choose a house of type k at a particular bid rent. The maximum value of this operand is 1.0 when all type-w heads of households choose a house of type k. In this circumstance the logarithmic term becomes zero and the bid rent is equal to the mean bid rent. In this special case, equation (4.49) becomes identical to equation (4.39) and thus reverts to the earlier version of the model with constant bid rents. Also note that the ratio in the logarithmic term moves counter to bid rents. The higher the bid rents the smaller the percentage of type-w heads of households able to bid. The parameter β operates as a dispersion parameter. For a given value of the ratio in the logarithmic term, a large value of β will imply a small deviation from the average bid rent, and a very small value of β will imply a large deviation from the average bid rent.

4.7 Numerical examples of the modified forms of the Herbert–Stevens model

In order to better appreciate the consequences of the modifications to the Herbert–Stevens model which were discussed above, small numerical examples can be examined. Again, the data for Archerville will be used, and the GAMS package will be applied to solve the examples. The first of these modified versions to be tested is the 'stock' version with separate transport costs and with heads of households being located from place of work. This version was given in equations (4.39)–(4.42).

Only a few new data items are required for this example. First it was necessary to convert the data from number of employees at place of work to number of heads of households of type w at place of work. Rather than consider a conversion matrix for transforming employees by employment sector to heads of households by household type, a simple proportion was used. Of the total employment at place of work, 55% are heads of LI households and 45% are heads of HI households (as per the original Archerville data). Second, a scaling factor for the travel costs had to be developed to make the travel costs comparable in absolute value with the other terms in the objective function [equation (4.39)].

The remaining data item to be prepared was the inventory of housing stock. Faced with this task one realizes that there is a certain circularity involved. The intent is to produce 'forecasts' of the location of households in house types by zones. Yet somehow there is to be an exogenously produced set of estimates of the location of the house types in which the households will locate. A careful rereading of some of the abovementioned attempts at application by other authors reveals that those authors faced this quandry as well. Harris et al (1966) used a simple Lowry-like model (Lowry, 1964) to assist in preparing an exogenous set of density and housing-accommodation constraints. For this second set of numerical examples it was decided to begin with a considerable surplus of housing in order to see the result of a largely unrestricted allocation of households. Thus the housing supply was set to 300 houses of each type in each zone. The results of this test are given in table 4.6.

In table 4.6 the place of work of the heads of households is indicated in parentheses beside the numbers of households locating in each house type in each zone. Thus, for example there are 166 LI households locating in medium-sized houses in zone 1, and the heads of all of them are employed in zone 1 as well. In zone 4 there are 300 LI households located in small houses. The heads of 159 of these households are employed in zone 2 and the heads of the remaining 141 are employed in zone 3. The total number of households of each type locating in each house type in each zone is given after the brace covering the disaggregated figures; for example, there are a total of 300 LI households located in medium-sized houses in zone 4.

To turn to the actual results, perhaps their most striking aspect is that, again, there are so few zones in which the households are actually located.

The LI households occupy all the small and medium-sized houses in zone 4, whereas the HI households took all the medium-sized and large houses in zone 5. All the remaining HI households are in small houses in zone 5. Besides a few LI households in large houses in zone 4, all the other LI households not in zone 4 are in medium-sized houses in zone 1, their head's zone of workplace. Thus, despite the fact that the objective function now explicitly includes journey-to-work costs, the location patterns of households which result are still characterized by very few nonzero entries in the matrices of possible locations by house type and zone.

These results, somewhat less constrained by available housing, make it clear that to some degree the model would tend to locate most households in just one or two zones if there were no constraint on the housing stock at all. A second test was done, where the available housing stock was much more limited. Here the total available was limited to a total of 1800 houses, compared with 4500 in the previous test. Admittedly the restriction here is quite arbitrary, but does correspond, say, to a situation of some, but not extreme, housing surplus for a region of 1450 households. A more important issue was that of how the 1800 houses were to be distributed by zone. The actual quantities of housing by type and by zone were set to correspond roughly to the location of households in Archerville as given in the data of table 2.1. From a philosophical point of view the determination of housing stock by an intuitive model is analogous to the use of a mathe-

Table 4.6. Archerville: results from test 1 of the modified Herbert–Stevens model.

Zone	House type (size)		
	small	medium	large
Low-income households			
1	0	166 (1)	0
2	0	0	0
3	0	0	0
4	159 (2) ⎱ 300[a] 141 (3) ⎰	145 (3) ⎱ 110 (4) ⎬ 300[a] 55 (5) ⎰	34 (2)
5	0	0	0
High-income households			
1	0	0	0
2	0	0	0
3	0	0	0
4	0	0	0
5	50 (1)	84 (1) ⎱ 157 (2) ⎬ 300[a] 59 (3) ⎰	165 (3) ⎱ 90 (4) ⎬ 300[a] 45 (5) ⎰

[a] Aggregate.
Note: numbers in parentheses indicate the zone in which the head of household works.

matical model, in that for both cases the exogenously estimated housing stock may have a considerable effect on the resulting distribution of households. The extent to which the estimates of the housing stock are made to conform to certain locational patterns may determine, to a considerable degree, the extent to which the household locations match that pattern as well. The housing stock for Archerville is given in table 4.7.

With the new constraints on housing, given in table 4.7, the model of equations (4.39)–(4.42) was rerun. The results of the second test run of the modified Herbert–Stevens model are given in table 4.8. The effect of the

Table 4.7. Archerville: housing supply.

Zone	House type (size)		
	small	medium	large
1	200	200	100
2	200	200	100
3	100	100	50
4	50	100	100
5	1 [a]	100	200

[a] The use of 1 here, rather than 0, is to prevent numerical problems in the nonlinear programming versions of the model which will be used for subsequent examples.

Table 4.8. Archerville: results from test 2 of the modified Herbert–Stevens model.

Zone	House type (size)		
	small	medium	large
Low-income households			
1	23 (1), 42 (3) } 65 [a]	142 (1)	0
2	193 (2), 7 (3) } 200 [a]	43 (2)	0
3	99 (3)	0	0
4	50 (5)	95 (4), 5 (5) } 100 [a]	85 (3), 16 (4) } 101 [a]
5	0	0	0
High-income households			
1	135 (1)	58 (3)	0
2	0	157 (2)	0
3	0	0	0
4	0	0	0
5	1 (3)	100 (3)	65 (3), 89 (4), 45 (5) } 199 [a]

[a] Aggregate.
Note: numbers in parentheses indicate the zone in which the head of household works.

new constraints on housing is immediately obvious, as there is much more dispersion of locating households. Of the fifteen house-type-by-zone locations possible for each income group, LI households use eight and HI households use six. In both cases this is double the numbers of house-type-by-zone locations used in the first of these tests (table 4.6), when there was much more housing stock 'available'. A careful comparison of the numbers of locating households with the exogenous housing-stock constraints reveals that the constraints are wholly responsible for the apparent increase in the dispersion of household locations. All the household locations are exactly equal to the constraints (subject to rounding errors). This is an important point to keep in mind with regard to the possible use of such models for forecasting purposes. Prior to the use of any such forecast one would wish to be quite sure that such forecasts were not too strictly dependent upon the exogenous constraints or, at least, that the dependence be made known to the users of the forecasts.

The previous two tests of the modified Herbert–Stevens model were made with the housing-utility variable for large houses, Q_3, at its original level of 4.5. For a final test of this model form, the value of Q_3 was increased to 6.5. The results, given in table 4.9 are significant but hardly spectacular. There is a general shift to a greater degree of location in large houses.

Table 4.9. Archerville: results from test 3 of the modified Herbert–Stevens model.

Zone	House type (size)		
	small	medium	large
Low-income households			
1	49 (1)	108 (1)	0
2	9 (1) ⎫ 93 (2) ⎬ 200[a] 98 (3) ⎭	43 (3)	100 (2)
3	0	0	50 (3)
4	50 (3)	35 (3) ⎫ 10 (4) ⎬ 100[a] 55 (5) ⎭	100 (4)
5	0	0	0
High-income households			
1	0	92 (1)	42 (1) ⎫ 100[a] 58 (3) ⎭
2	0	157 (2)	0
3	0	0	0
4	0	0	0
5	1 (3)	100 (3)	65 (3) ⎫ 90 (4) ⎬ 200[a] 45 (5) ⎭

[a] Aggregate.
Note: numbers in parentheses indicate the zone in which the head of household works.

The overall zonal pattern of household location remains unchanged, and once again the exogenously generated housing-stock constraints determine the household allocations in virtually all cases.

Next a series of numerical experiments was done by using the nonlinear version of the Herbert–Stevens model, which incorporates a bid rent dispersion term. This model form [equation (4.49) plus equations (4.40)–(4.42)] required only one additional numerical item. A value of β had to be selected. The mean values of the bid rents were taken to be the values used in the previous numerical examples. The value of Q_3 was left at its previous high value of 6.5, and the housing constraints were left at the values used in the last of the previous tests. A value of 0.1 was arbitrarily chosen for β in this first test of the nonlinear programming (NLP) version of the Herbert–Stevens model. The results of this test, the fourth of this second series and the first of the NLP model, are given in table 4.10.

The dispersion of bid rents results, in this first test run, in only a slight additional dispersion in LI households. There is, however, a significant increase in the dispersion of HI households. One interesting consequence of this increased dispersion is that here, in effect, the two household types are competing directly for a limited supply of housing. In addition to the

Table 4.10. Archerville: results from test 4 of the modified Herbert–Stevens model.

Zone	House type (size)		
	small	medium	large
Low-income households			
1	22 (1)	90 (1)	45 (1)
2	3 (1) ⎫	3 (1) ⎫	2 (1) ⎫
	77 (2) ⎬ 136[a]	77 (2) ⎬ 136[a]	39 (2) ⎬ 69[a]
	56 (3) ⎭	56 (3) ⎭	28 (3) ⎭
3	0	0	50 (3)
4	17 (3) ⎫	34 (3) ⎫	34 (3) ⎫
	22 (4) ⎬ 50[a]	44 (4) ⎬ 100[a]	44 (4) ⎬ 100[a]
	11 (5) ⎭	22 (5) ⎭	22 (5) ⎭
5	0	0	0
High-income households			
1	19 (1) ⎫ 27[a]	77 (1) ⎫ 110[a]	39 (1) ⎫ 56[a]
	8 (3) ⎭	33 (3) ⎭	17 (3) ⎭
2	63 (2)	63 (2)	31 (2)
3	0	0	0
4	0	0	0
5	1 (3)	55 (3) ⎫	111 (3) ⎫
		30 (4) ⎬ 100[a]	60 (4) ⎬ 201[a]
		15 (5) ⎭	30 (5) ⎭

[a] Aggregate.
Note: numbers in parentheses indicate the zone in which the head of household works.

increased dispersion of household location overall, there is also an increase in dispersion in terms of the place of work of household heads. For example, in the previous example the heads of all the LI households locating in small houses, in zone 4 all worked in zone 3. In this test run there are heads of households locating in zone 4 who work in zones 3, 4, or 5. Even so, in terms of absolute numbers of households locating in house types, by zone, the results were still rather like those obtained from the LP version of the modified model.

For the next test run the value of β was decreased to 0.01, as it was expected that this would increase the magnitude of the bid rent dispersion portion of the objective function relative to the mean bid rent portion. The results of this run, not tabulated here, showed virtually no difference in the total numbers of locating households by house type and zone. The dis-

Table 4.11. Archerville: results from test 6 of the modified Herbert–Stevens model.

Zone	House type (size)					
	small		medium		large	
Low-income households						
1	12 (1) 10 (2) 15 (3) 5 (4) 2 (5)	44[a]	33 (1) 27 (2) 41 (3) 13 (4) 7 (5)	121[a]	17 (1) 14 (2) 21 (3) 7 (4) 3 (5)	62[a]
2	16 (1) 24 (2) 24 (3) 10 (4) 5 (5)	79[a]	27 (1) 41 (2) 40 (3) 18 (4) 9 (5)	135[a]	14 (1) 20 (2) 20 (3) 9 (4) 4 (5)	67[a]
3	5 (1) 5 (2) 12 (3) 4 (4) 2 (5)	28[a]	3 (1) 3 (2) 8 (3) 2 (4) 1 (5)	17[a]	7 (1) 7 (2) 17 (3) 5 (4) 3 (5)	39[a]
4	6 (1) 8 (2) 14 (3) 7 (4) 3 (5)	38[a]	12 (1) 16 (2) 28 (3) 14 (4) 6 (5)	76[a]	12 (1) 16 (2) 28 (3) 14 (4) 6 (5)	76[a]
5	0		1 (1) 1 (2) 3 (3) 1 (4) 1 (5)	7[a]	2 (1) 3 (2) 5 (3) 2 (4) 2 (5)	14[a]

Location models in optimizing frameworks: 1   93

persion of place of work of household heads did increase somewhat. In the previous test there were 38 nonzero values in the model solution; in this test there were 62 nonzero values. Even so, the results were really not very different.

The final test for this series was done with a value of β of 0.001. The results here were dramatically different from those which preceded them. Houses of all types in all zones are in use (except small houses in zone 5 which are constrained to less than or equal to 1 unit. All the used house types in all zones have heads of households working in at least four zones, and all but two of the twenty-eight available combinations of household by house type by zone (excepting the small houses in zone 5) have heads of households working in all five zones. This represents just about as full a dispersion as can be obtained. There are thirty combinations of household by

Table 4.11. (continued).

Zone	House type (size)					
	small		medium		large	
High-income households						
1	8 (1) 7 (2) 9 (3) 3 (4) 1 (5)	28[a]	23 (1) 19 (2) 26 (3) 8 (4) 3 (5)	79[a]	11 (1) 10 (2) 13 (3) 4 (4) 1 (5)	39[a]
2	8 (1) 13 (2) 11 (3) 5 (4) 2 (5)	39[a]	14 (1) 22 (2) 19 (3) 8 (4) 3 (5)	66[a]	7 (1) 11 (2) 10 (3) 4 (4) 1 (5)	33[a]
3	2 (1) 2 (2) 4 (3) 1 (4)	9[a]	1 (1) 1 (2) 3 (3) 1 (4)	6[a]	2 (1) 3 (2) 6 (3) 2 (4)	13[a]
4	2 (1) 3 (2) 4 (3) 2 (4) 1 (5)	12[a]	4 (1) 5 (2) 9 (3) 4 (4) 1 (5)	23[a]	4 (1) 6 (2) 9 (3) 4 (4) 1 (5)	24[a]
5	0		16 (1) 20 (2) 34 (3) 14 (4) 10 (5)	94[a]	31 (1) 39 (2) 68 (3) 29 (4) 20 (5)	187[a]

[a] Aggregate.
Note: numbers in parentheses indicate the zone in which the head of household works.

house type by zone in this sample problem. Given the formulation each household head can be employed in any of the five zones. Thus, if in the solution a record is also kept of the place of employment of the head of household, there are 150 possible entries in what might be termed the disaggregated solution to the model. It is in this context that the 38 and 62 nonzero values of the solutions from the preceding two tests should be viewed. The results of this last test gave 137 nonzero values. Given that there is only one small house in zone 5, nine possible nonzero values are eliminated by this constraint. Thus the practical maximum on nonzero values is 141, and the test result is obviously very nearly at this maximum. These results are shown in table 4.11. It should be noted that the 'aggregated' results which are shown in table 4.11 are also more dispersed than those of the previous tests.

Although these numerical results are interesting to examine, the major point made here is a theoretical one. A model which does not include a mechanism for representing what might best be called 'preference dispersion' will not be capable of producing realistic representations of reality. The Houghton–Senior version of the Herbert–Stevens model tested above produced much more 'realistic' household-to-housing allocations simply because some variation in preferences was assumed. Other forms of preference distribution might just as easily have been proposed and the resulting NLP version of the Herbert–Stevens model derived. The notion of preference distributions can be easily related to entropy concepts. The development of this NLP version of the Herbert–Stevens model resulted from relaxing the assumption of constant bid rents for a given household group. Relaxation of the assumptions of perfect information and perfect competition of the model leads us, through suboptimal solutions, to models with entropy terms. This concept appears again in the next chapter during the discussion of TOPAZ, and again after that in the discussion of the relationship between mathematical programming models and entropy models.

5
Location models in optimizing frameworks: 2

5.1 Introduction
In chapter 4 mathematical programming models of residential location were introduced. Attention was focused on the Herbert–Stevens model and on variations of that model formulation. In this chapter the main discussion is of the TOPAZ model and its variants. Other formulations of mathematical programming models are also discussed. Numerical examples of TOPAZ and its variants are presented. Some mention is also made of the connection between a version of TOPAZ which incorporates locational dispersion and the combined location and assignment models which will be presented in chapter 8.

5.2 Technique for the optimal placement of activity in zones: TOPAZ
TOPAZ is a mathematical programming technique which was originally proposed in the late 1960s and early 1970s (Brotchie, 1969; Sharpe and Brotchie, 1972). The most complete discussion of the applications of the model is to be found in the book by Brotchie et al (1980).

The model was originally proposed as a method for determining least-cost allocations of activities to zones. In and of itself this was not an unusual goal, being precisely what was intended by Schlager in the land-use plan design model discussed at the beginning of chapter 4. What was different was the inclusion of activity interaction costs in the objective function. The model as first posed by Brotchie (1969) had the following form:

$$\text{maximize } Z = \sum_{i,j} a_{ij} b_{ij} A_i + \sum_{i,j,k,l} a_{ij} a_{kl} b_{ijkl} S_{ik} R_{jl} \,, \tag{5.1}$$

subject to

$$\sum_i a_{ij} A_i \leq X_j \,, \tag{5.2}$$

$$\sum_j a_{ij} = 1.0 \,, \tag{5.3}$$

$$a_{ij} \geq 0 \,, \tag{5.4}$$

where
a_{ij} is the portion of the region's total activity of type i which is allocated to zone j,
A_i is the regional total of activity of type i,
b_{ij} is the sum of the benefits minus the costs (that is, the net benefit) of establishing and operating one unit of activity i in zone j,
S_{ik} is the level of interaction between activities i and k,
R_{jl} is the length of the (minimum) path between zone j and zone l,
b_{ijkl} is the net benefit of one unit of interaction between activities i and k along the minimum path between zones j and l,
X_j is the capacity of zone j.

The objective function, then, is the sum of net location benefits and net interaction benefits resulting from the apportionment of the region's total activity to specific zones. The first set of constraints [equation (5.2)] is to ensure that no zone's capacity is exceeded. The second set of constraints [equation (5.3)] is to ensure that the sum of the portions of activity of type i, over all zones j, equals one, and thereby the regional totals are exactly allocated. The last set of constraints are the usual ones of nonnegativity. Note that this is a quadratic programming problem. The objective function is quadratic because of the a_{ij} a_{kl} product in the second term. The constraints are linear.

As was the case with the first formulation of the Herbert–Stevens model, two groups of questions arise almost instantly. The first are with regard to the formulation per se, that is, is it correct or appropriate to the problem? In the second set of questions the formulation is accepted as a working premise, and precise specification of the variables is asked for. Here it is obvious that there will be questions about several of the variables which appear in the above formulation.

By the time of the Sharpe and Brotchie (1972) paper, the model was named TOPAZ and its formulation slightly revised. Later forms of the model included disaggregation by travel mode and time period (Sharpe et al, 1974). It seems from the various formulations which have been published that two main points were at issue. The first was the joint question of how to deal with new activity levels vis-à-vis existing activity levels, and whether to deal with activity rates or absolute quantities in location (Brotchie and Sharpe, 1974). The second point involved the specification of the interaction variables of the second term of the objective function.

To pursue this latter point further, note first that although the interaction term in equation (5.1) is clearly quadratic with respect to the locating activities, it is far from clear as to how the other variables are to be defined. Dickey (1981) defines the objective function of TOPAZ to be

$$\text{minimize } Z = \sum_{i,j} C_{ij} X_{ij} + \sum_{i,j,k} v_i^P (X_{ij} + e_{ij}) \left[\frac{v_i^a (X_{il} + e_{il}) t_{jl}^{-2}}{v_i^a \sum_l (X_{il} + e_{il}) t_{jl}^{-2}} \right] c_{jk} , \qquad (5.5)$$

subject to

$$\sum_j X_{ij} = A_i , \qquad (5.6)$$

$$\sum_i X_{ij} = L_j , \qquad (5.7)$$

$$\sum_i A_i = \sum_j L_j , \qquad (5.8)$$

$$X_{ij} \geq 0 , \qquad (5.9)$$

where
C_{ij} is the costs minus benefits of establishing one unit of activity i in zone j,
X_{ij} is the amount of activity i allocated to zone j,
e_{ij} is the amount of activity i preexisting in zone j,
v_i^a is the number of vehicular trips attracted to a zone, per unit of activity i in the zone,
v_i^p is the number of vehicular trips produced per unit of activity i in a zone,
t_{jk} is the travel time from zone j to zone k,
c_{jk} is the travel cost between zone j and zone k,
A_i is the total amount of activity-type i in the region,
L_j is the total available land in the zone.

Here the first term of the objective function, equation (5.5), is obviously the cost of locating the activities. Note also that the objective function is to be minimized, whereas Brotchie's original version has an objective function which is to be maximized [equation (5.1)]. There is no great difference, except to reflect, perhaps, different expectations: owing to the sign, the net-benefits term used by Brotchie becomes a net-cost term for Dickey. The second term in the objective function [equation (5.5)] is a more explicit attempt to deal with interaction costs. The fractional portion of the term estimates the proportion, of all trips of activity type i that are attracted to all zones, which is attracted to zone k. Thus the whole expression calculates the sum, over all zone pairs, of the product of the number of trips produced at each origin, the number of trips attracted to each destination, and the travel cost for each trip. The relative attractiveness of each destination results solely from the inverse of the square of the zone-to-zone travel time for each zone pair relative to the sum over all other zone pairs. Although this is a rather rudimentary specification for an interaction term, it is at least explicit. Later versions of the model use more sophisticated interaction terms, as will be discussed below.

The purpose of the various constraints should be mentioned here. The first set of constraints [equation (5.6)] serves to ensure that all of activity-type i is allocated to zones in the region. The second set of constraints [equation (5.7)] ensures that the land used by all activities locating in zone j does not exceed the land available. Two points should be noted here. First, vacant developable land is considered to be a locating activity in order to balance the 'land accounts' within the mathematical programming framework of the model. Second, it is implicit in this equation that locating activities are described in terms of land use. This implies that the regional activity totals must be expressed in terms of land use as well. A further implication is that the land consumption of each locating activity (for example, employment by type or households by type) will have a constant rate over all the zones in the region. This is obviously a matter for further development. The third set of constraints [equation (5.8)] serves to keep the total activities in the region equal to the total capacity (in terms of land) of the region. Last, there are the customary nonnegativity constraints [equation (5.9)].

Brotchie and Sharpe (1974) propose an objective-function formulation similar to that of equation (5.1) but change to an absolute activity formulation from the original proportional formulation. The objective function used had the following form

$$\text{maximize } Z = \sum_{i,j} b_{ij} X_{ij} + \sum_{i,j,k,l} b_{ijkl} X_{ij} X_{kl} \,, \tag{5.10}$$

where
- X_{ij} is the area allocated to activity i in zone j,
- b_{ij} is the benefit less cost per unit of activity i allocated to zone j,
- b_{ijkl} is the benefit less cost of interactions between a unit of activity of type i in zone j and a unit of activity of type k in zone l.

Now in this formulation an equation was given for the calculation of interaction cost. This was

$$b_{ijkl} = w_{ik} d_{ik} P_{ij} Q_{kl} f_{ik} f_{ki} t_{jl}^{-r} \,, \tag{5.11}$$

where
- w_{ik} is a weighting factor,
- d_{ik} is the unit net benefit of travel (benefit minus cost, per trip per unit time) between activity i and activity k,
- f_{ik} is the number of trips to activity k generated per unit of activity i,
- f_{ki} is the number of trips attracted from activity i per unit of activity k,
- t_{jl} is the travel time between zones j and l,
- r is a parameter, empirically determined and expected to be in the range of 1 to 4,

and

$$P_{ij} = \left[\sum_{l} Q_{kl} f_{ki} X_{kl} t_{jl}^{-r} \right]^{-1} \,, \tag{5.12}$$

$$Q_{kl} = \left[\sum_{j} P_{ij} f_{ik} X_{ij} t_{jl}^{-r} \right]^{-1} \,. \tag{5.13}$$

Here, then, a doubly constrained gravity model, as given in equations (5.11)-(5.13), is used to calculate the interaction costs. This is an improvement in the explicit statement of the model, but does leave undefined the rather critical terms f_{ik} and f_{ki}. In a sense, one would expect these terms to result from the gravity model formulation and the b_{ijkl} to be a given or an input variable.

Last, in Brotchie et al (1980) the model is formulated as (with time and mode subscripts omitted):

$$\text{minimize } U = \sum_{i,j,k,l} S_{ijkl} R_{jl} b_{ijkl} + \sum_{i,j} a_{ij} C_{ij} \,, \tag{5.14}$$

subject to

$$\sum_i a_{ij} = L_j ,\quad (5.15)$$

$$\sum_j a_{ij} = A_i ,\quad (5.16)$$

$$\sum_i A_i = \sum_j L_j ,\quad (5.17)$$

$$a_{ij} \geq 0 ,$$

where
S_{ijkl} is the amount of interaction between activity i in zone j and activity k in zone l,
R_{jl} is the length of the travel path between zone j and zone l,
b_{ijkl} is the benefit less the cost of a unit of interaction S_{ijkl} along the path between zones j and l,
a_{ij} is the portion of the region's activity of type i which is allocated to zone j,
C_{ij} is the benefit less the cost of establishing and operating a unit of activity i in zone j,
L_j is the area available in zone j.

In this formulation the interaction term is defined by a doubly constrained gravity model too, but here the definitions are more what one might expect. The formulation is

$$S_{ijkl} = P_{ij} Q_{kl} F_{ij} F_{kl} g(t_{jl}) ,\quad (5.18)$$

where
F_{ij} is the number of trips generated by activity i in zone j,
F_{kl} is the number of trips attracted by activity k in zone l,
$g(t_{jl})$ is a function of the travel time between zone j and zone l, t_{jl}, and

$$P_{ij} = \left[\sum_l Q_{kl} F_{kl} g(t_{jl}) \right]^{-1} ,\quad (5.19)$$

$$Q_{kl} = \left[\sum_j P_{ij} F_{ij} g(t_{jl}) \right]^{-1} .\quad (5.20)$$

It should be noted that this problem cannot be solved directly, as the solution to equations (5.18)–(5.20) must be obtained iteratively. This can be done by setting either the P_{ij} or the Q_{kl} equal to 1.0 and solving for the other, that is, the ones not initially set to 1.0. The process is then repeated until successive changes in the P_{ij} and Q_{kl} reach a predetermined tolerance. For the numerical examples to be discussed next, a simpler formulation will be used.

5.3 TOPAZ: a numerical example

The use of the Archerville data for a simple numerical example of TOPAZ required little further preparation of data, but it did require some modification to the model formulation itself. Sharpe et al (1984) proposed TOPAZ82 in the following form:

$$\text{minimize } Z = \sum_{i,j,k,l} S_{ijkl} c_{ijkl} + \sum_{i,j} b_{ij} X_{ij} + \sum_{i,j} d_{ij} Y_{ij} , \quad (5.21)$$

subject to

$$\sum_{l} S_{ijkl} - s_{ik}(X_{ij} - Y_{ij} + e_{ij}) = 0 , \quad (5.22)$$

$$\sum_{j} S_{ijkl} - r_{ik}(X_{kl} - Y_{kl} + e_{kl}) = 0 , \quad (5.23)$$

$$\sum_{j} (X_{ij} - Y_{ij}) = A_i - \sum_{j} e_{ij} , \quad (5.24)$$

$$\sum_{i} (X_{ij} - Y_{ij}) \leq L_j - \sum_{i} e_{ij} , \quad (5.25)$$

$$X_{ij} \geq 0 , \quad (5.26)$$

$$Y_{ij} \leq e_{ij} , \quad (5.27)$$

$$S_{ijkl} \geq 0 , \quad (5.28)$$

where
- A_i is the regional total of activity i,
- b_{ij} is the unit cost less the benefit of incrementing activity i in zone j,
- d_{ij} is the unit cost less the benefit of decrementing activity i in zone j,
- c_{ijkl} is the unit cost less the benefit of interaction between activity i in zone j and activity k in zone l,
- e_{ij} is the preexisting (base year) level of activity i in zone j,
- s_{ik} is the level of interaction between a unit of activity i per unit of activity k,
- r_{ik} is equal to $s_{ik} A_i/A_k$,
- S_{ijkl} is the level of interaction (trips) between activity i in zone j and activity k in zone l,
- X_{ij} is the amount of activity i to be allocated to zone j,
- Y_{ij} is the amount of activity i to be removed from zone j,
- L_j is the capacity of zone j.

The major point to be noticed about this formulation of TOPAZ is that the interaction term in the objective function is linear as are the trip origin constraints, equation (5.24), and trip destination constraints, equation (5.25). For the actual numerical example the problem was simplified even further, to

$$\text{minimize } Z = \sum_{i,j,k,l} S_{ijkl} c_{ijkl} + \sum_{i,j} b_{ij} X_{ij} , \quad (5.29)$$

subject to

$$\sum_{l} S_{ijkl} - s_{ik} X_{ij} = 0 \, , \tag{5.30}$$

$$\sum_{j} S_{ijkl} - r_{ik} X_{kl} = 0 \, , \tag{5.31}$$

$$\sum_{j} X_{ij} = A_i \, , \tag{5.32}$$

$$\sum_{i} X_{ij} \leq L_j \, . \tag{5.33}$$

This eliminated the base year or 'existing' activities as well as the demolition term in the objective function. Last, in order to make these results compatible with the Herbert–Stevens model experiments, the model was abbreviated to a form for residence location only. In general this involved dropping the k subscript throughout the model equations.

Precise definitions had to be given to some of the variables as well. First, it was assumed, as is customary in these examples, that there is one employee per household, and thus one work trip per household. The Archerville data show there are 0.55 low-income (LI) households per 1.0 employee, and 0.45 high-income (HI) households per 1.0 employee. Thus the X_l in equation (5.36) is employment in zone l, and the formulation of the problem becomes

$$\text{minimize } Z = \sum_{i,j,l} S_{ijl} c_{ijl} + \sum_{i,j} b_{ij} X_{ij} \, , \tag{5.34}$$

subject to

$$\sum_{l} S_{ijl} - s_i X_{ij} = 0 \, , \tag{5.35}$$

$$\sum_{j} S_{ijl} - r_i X_l = 0 \, , \tag{5.36}$$

$$\sum_{j} X_{ij} = A_i \, , \tag{5.37}$$

$$\sum_{i} X_{ij} \leq L_j \, , \tag{5.38}$$

$$S_{ijl} \geq 0, \quad X_{ij} \geq 0 \, . \tag{5.39}$$

Note again that the second term in the objective function is simply a minimum-cost location term. The first term is the linear 'transportation' problem. Taking the location problem first, it is clear that development of the necessary 'data' raises many of the same issues discussed in the development of the numerical examples for the Herbert–Stevens model. The b_{ij} are supposed to be the net values of the costs and benefits of locating a unit of activity type i in zone j. In many TOPAZ applications the virtual impossibility of measuring benefits resulted in the b_{ij} being simply a cost of location, to be minimized. For the Archerville example several possibilities exist. The easiest way would be simply to use average annualized house cost, but then the model would attempt to locate all households in the zone with the lowest house cost. All that would prevent this would be constraints on the amount of activity which could be accommodated in each zone.

Raising the issue of zonal constraints raises, in turn, the issue of converting activity types into land consumed. Here again, there are several possible ways to proceed. These are (1) regional rates of land consumption by activity type, (2) zonal rates of land consumption by activity type, or (3) exogenously developed estimates of housing stock. As a first test of the model a set of regional rates of land consumption was assumed, the values being set so as to allow all households to be accommodated by existing residential land in the region. The rates were 0.00525 land units per household for LI households and 0.00646 land units per household for HI households.

The cost of location by zone was taken to be the weighted average annualized house cost (the housing stock data developed in table 4.1 for the Herbert–Stevens examples were used). This gave locations costs as follows:

zone 1 $7800,   zone 2 $7200,   zone 3 $7600,
zone 4 $6800,   zone 5 $9668.

Then only the location-cost term of the objective function [equation (5.34)] was used to solve for the minimum cost location of Archerville's 800 LI and 650 HI households. As there were no trip costs and no trips there were no trip-end constraints. The resulting location of households is given in table 5.1(a). It is worth noting that for this model solution all residential land is used in zones 2, 3, and 4. Obviously, no residential land is used in zone 5, and zone 1 is partially filled. Here it is quite clear that if there were no land constraint, all households would have located in zone 4 where location was least expensive. Possible solutions to this particular inadequacy of the model lie in the direction of elaborating the description of location costs and benefits.

Table 5.1. Archerville: location of households, based on the location-cost component of the TOPAZ model.

Zone	Households			Zone	Households		
	LI	HI	total		LI	HI	total
(a) Test 1				(b) Test 2			
1	0	294	294	1	0	296	296
2	324	201	525	2	400	59	459
3	0	155	155	3	0	118	118
4	476	0	476	4	400	0	400
5	0	0	0	5	0	177	177

Note: LI, low income; HI, high income.

As a short aside, it should be noted that all residential land available in Archerville was not used by this first location-cost-minimizing solution simply because it was not necessary, given the land-units-per-household (lu/hh) rates used. If these numbers were adjusted so that there were just enough land in Archerville to hold all the households (for example, 0.00625 lu/hh for LI and 0.00846 lu/hh for HI), then a somewhat different allocation pattern would result. These results are shown in table 5.1(b) and should be compared with the results given in table 5.1(a). In this second set of results all available residential land in Archerville is used to accommodate the 1 450 households in the area.

Comparing tables 5.1(a) and 5.1(b) shows the effect of the increased land requirements of the households. Zone 4, with the lowest average cost of housing, formerly accommodated 476 LI households but in the second test could only hold 400. The 'excess' 76 were put in the second-cheapest zone, zone 2. This, in turn, precluded the location of some of the HI households formerly in that zone, which then had to be accommodated elsewhere. Owing to the increased land-per-household requirements, fewer HI households could be accommodated in zone 3 than before, and only 2 more than before could be accommodated in zone 1. The net result was that 177 HI households had to be located in zone 5, the most expensive zone in Archerville. This caused a 4% increase in the objective function, total location cost, from 10 487 867 in the first test to 10 941 883 in the second test. None of this is surprising, but it does, once again, point up the effects of constraints on the solutions of these mathematical programming formulations of location models.

Next the transportation, or interaction, cost term in the objective function [equation (5.34)] was examined. This required the specification of data for the trip-end constraints, equations (5.35) and (5.36), as well. As mentioned above it was assumed that there is only one employee per household and that there are 0.55 LI and 0.45 HI households per employee. Recalling that in this example (a) only home-to-work trips are being dealt with, and (b) there is only one employment type, then both s_i will be equal to one, so that the constraint equation (5.35) will be given by

$$\sum_l S_{ijl} - X_{ij} = 0 \,,$$

or, in words, the sum over all possible destination zones l of all trips of type i leaving (that is, produced in) zone j must equal the trips of type i generated in zone j. In this example the trips of type i (that is, from households of type i) generated in zone j equal the number of households of type i living in zone j.

Following the same reasoning, the r_i will be equal to the ratio of household types to employees, so that the constraint equation (5.36) will be as given above, with $r_1 = 0.55$, $r_2 = 0.45$, and X_l equal to the total employment in zone l. Then, in words, the sum over all possible origin zones j of all trips of type i terminating in zone l must equal the total households of type i

attracted to employment in zone l. The total households of type i attracted to employment in l is simply the conversion rate multiplied by the employment.

In this test run just the interaction or transportation cost term of the objective function, equation (5.34), was used to solve for the minimum-cost location of Archerville's 1 450 households. First, note that with the problem simplified to residence location only, the full allocation of activities constraint, equation (5.37), is made redundant by the pair of trip-end constraints, equations (5.35) and (5.36). The redundant constraint was removed from the problem to eliminate the possibility of encountering infeasible solutions due to computational round-off errors. The solution to this problem is given in table 5.2.

The results in table 5.2 should be compared with those in table 5.1(a), as the lower land-consumption rates were used in both. Note first that the minimum-transport-cost solution gives somewhat more dispersion of households to zones than does the minimum-location-cost solution. That this is largely owing to the exogenously determined location of employment can readily be deduced from the low level of commutation shown in the latter part of table 5.2. Were the employment to be less dispersed then, subject to the land-use constraints, the residential location would be less dispersed as well.

There is another matter which will have consequences for the next set of tests: location and interaction costs are used to determine residential location. The value of the objective function at the minimum-transport-cost solution is 28 600. This is only 0.2% of the value of the objective function for the minimum-location-cost solution. In the first run of the next series, both the location cost and the interaction (transport) cost terms were included in the objective function. In an attempt to make the magnitude of the two components of the objective function more similar, the location cost term was multiplied by 0.01, lowering its value at the optimum to 108 251.

Table 5.2. Archerville: location of households, based on the transport-cost component of the TOPAZ model.

Zone	Households			Zone	Households	
	LI	HI	total		LI	HI
1	166	220	386	1	166 (1)	134 (1)
2	193	157	350			86 (3)
3	190	0	190	2	193 (2)	157 (2)
4	196	228	424	3	190 (3)	0
5	55	45	100	4	85 (3)	138 (3)
					111 (4)	90 (4)
				5	55 (5)	45 (5)

Note: numbers in parentheses indicate the zone in which the head of household works; LI, low income; HI, high income.

Even though this location-cost component was almost four times larger than the transport-cost component, the transport-cost portion of the objective function completely dominated the model solution. In several more test runs the location-cost term in the objective function was multiplied by 0.02, 0.03, and 0.04, respectively, with no effect. It was not until a multiplier of 0.045 was used that a jointly (location plus transport) determined solution began to emerge. This result is given in table 5.3. Note that at the optimum for this test the interaction-cost term in the objective function had the value 38 141 and the location-cost term had a value of 476 309. If the multiplier is moved to 0.05, the results given in table 5.4, and objective-function components (transport cost and location cost) of 40 402 and 526 886, respectively, are produced. A comparison of table 5.3 with table 5.4 shows a slight shifting of

Table 5.3. Archerville: location of households, based on the location-cost and the transport-cost components of the TOPAZ model, for $\lambda_2 = 0.045$.

Zone	Households			Zone	Households	
	LI	HI	total		LI	HI
1	166	135	301	1	166 (1)	135 (1)
2	193	307	500	2	193 (2)	157 (2)
3	190	0	190			150 (3)
4	251	183	434	3	190 (3)	0
5	0	25	25	4	85 (3)	73 (3)
					110 (4)	90 (4)
					56 (5)	20 (5)
				5	0	25 (5)

Note: λ_2, the location-cost multiplier in the objective function; see also table 5.2.

Table 5.4. Archerville: location of households, based on the location-cost and the transport-cost components of the TOPAZ model, for $\lambda_2 = 0.050$.

Zone	Households			Zone	Households	
	LI	HI	total		LI	HI
1	166	160	326	1	166 (1)	135 (1)
2	193	307	500			25 (3)
3	190	0	190	2	193 (2)	157 (2)
4	251	183	434			150 (3)
5	0	0	0	3	190 (3)	0
				4	85 (3)	73 (3)
					110 (4)	90 (4)
					56 (5)	20 (5)
				5	0	0

Note: see table 5.3.

HI households with increase in location cost. These sensitivity tests were continued with increasing values of the location-cost multiplier.

Over the range from 0.045 to 1.00 the location-cost multiplier resulted in a more or less gradual shift from the transport-cost-only solution toward the location-cost-only solution. From a value slightly less than 1.0 to a value of 1.1852 the multiplier causes no change in the model solution. At a value of 1.1855 the multiplier results in a model solution identical to that of the location-cost-only solution. Thus at some critical point where the multiplier of the location-cost term in the objective function is between 1.1852 and 1.1855 there is a sudden shift in the model solution from an apparently stable intermediate solution to the location-cost-only solution. Although it is presented as an aside here, clearly the matter of model sensitivity and solution stability needs further discussion, and will be considered further in section 5.5.

5.4 TOPAZ: a numerical example incorporating dispersion

As was shown to be true of the Herbert–Stevens model, the location patterns produced by the linear programming version of TOPAZ, particularly when examined at the zone-to-zone trip level, are rather lumpy. The addition of a nonlinear dispersion term to the Herbert–Stevens model objective function made a noticeable difference, and, as will shortly be shown, it does with TOPAZ as well. The result is achieved by substituting a constrained gravity model for the linear 'transportation' model portion of the TOPAZ objective function.

The standard doubly (or fully) constrained spatial interaction model has the form

$$S_{jl} = A_j B_l O_j D_l \exp(-\beta c_{jl}) , \qquad (5.40)$$

where

S_{jl} is the number of trips between zone j and zone l,
O_j is the number of trips generated in (originating from) zone j,
D_l is the number of trips attracted to (terminating in) zone l,
c_{jl} is the zone-to-zone travel cost,
β is a parameter,

and, where

$$A_j = \left[\sum_l B_l D_l \exp(-\beta c_{jl}) \right]^{-1} , \qquad (5.41)$$

$$B_l = \left[\sum_j A_j O_j \exp(-\beta c_{jl}) \right]^{-1} . \qquad (5.42)$$

Note that these terms are similar to those included in the version of TOPAZ given in equations (5.14)–(5.20). More importantly, it has been shown by Murchland (1966) and Wilson (1967) that the above model [equations (5.40)–(5.42)] can be derived from an equivalent optimization problem.

The form of the problem is

$$\text{maximize } S = -\sum_{j,l} S_{jl} \ln S_{jl} , \qquad (5.43)$$

subject to

$$\sum_{l} S_{jl} = O_j , \qquad (5.44)$$

$$\sum_{j} S_{jl} = D_l , \qquad (5.45)$$

$$\sum_{j,l} c_{jl} S_{jl} = C , \qquad (5.46)$$

where the only new value is C, taken to represent the total travel cost for the system.

There is a relationship between the β in equations (5.40)–(5.42), and the C in equation (5.46). Recall that the 'transportation' problem of linear programming has the objective function.

$$\text{minimize } Z = \sum_{j,l} c_{jl} S_{jl} , \qquad (5.47)$$

and is subject to the constraints of equations (5.44) and (5.45) and also to nonnegativity constraints. Evans (1973) has shown that the values of S_{jl} which are obtained for the maximization problem of equations (5.43)–(5.46) approach the values of S_{jl} obtained for the minimization problem of equations (5.47), (5.44), (5.45), as $\beta \to \infty$. In other words, the greater the value of β the closer the trip flows from the spatial interaction model come to the trip flows from the minimum-cost flow model. Thus β in the spatial interaction model produces a dispersion of trips away from the optimum or minimum-cost solution. What human behavior might account for this dispersion is unspecified, but presumably includes such factors as variables not in the model, as well as variations in individual perceptions of costs, and differences in individual utility functions.

To introduce the spatial interaction model into TOPAZ in lieu of the minimum-cost 'transportation' model requires that the model of equations (5.43)–(5.46) be substituted for the transport-cost term in the TOPAZ objective function. In the planning and forecasting of situations the value of C from equation (5.46) is usually not known. This can be overcome by using a standard technique for including the constraint [equation (5.46)] in the objective function [equation (5.43)]. The result is a revised objective function:

$$\text{maximize } S = -\frac{1}{\beta} \sum_{j,l} S_{jl} \ln S_{jl} - \sum_{j,l} S_{jl} c_{jl} . \qquad (5.48)$$

It should be noted that this introduces the parameter β into the objective function. This parameter could be obtained by calibration of the spatial interaction model of equations (5.40)–(5.42) against observed data.

Note also that as $\beta \to \infty$ the objective function of equation (5.48) becomes identical with the 'transportation' model objective function. Last, this set of terms can be incorporated into the TOPAZ82 objective function to yield (after changing the signs to allow minimization)

$$\text{minimize } S = \frac{1}{\beta} \sum_{i,j,l} S_{ijl} \ln S_{ijl} + \sum_{i,j,l} S_{ijl} c_{ijl} + \sum_{i,j} b_{ij} X_{ij} , \qquad (5.49)$$

subject to

$$\sum_{l} S_{ijl} - s_i X_{ij} = 0 , \qquad (5.50)$$

$$\sum_{j} S_{ijl} - r_i X_l = 0 , \qquad (5.51)$$

$$\sum_{j} X_{ij} = A_i , \qquad (5.52)$$

$$\sum_{i} X_{ij} \leq L_j , \qquad (5.53)$$

$$S_{ijl} \geq 0, \quad X_{ij} \geq 0 , \qquad (5.54)$$

where
S_{ijl} is the level of interaction (trips) of activity i between zone j and zone l,
b_{ij} is the cost less the benefit of locating activity i in zone j,
c_{ijl} is the travel cost for activity i between zone j and zone l,
s_i is level of interaction (trips generated) per unit of activity i,
r_i is the number of trips attracted by employment, per unit of activity i,
X_{ij} is the amount of activity i to be allocated to zone j,
A_i is the regional total of activity i,
L_i is the area available in zone j.

The Archerville data were again used for tests of the model which, it should be pointed out, is virtually identical (after some algebraic manipulation) to the version of TOPAZ given in equations (5.14)-(5.20). The only new value required was a numerical value of β. This, however, raises an interesting question, which relates directly to the above-described experiments with weightings or multipliers of the terms in the TOPAZ objective function. To simplify the coming discussion it will be convenient to think of the objective function [equation (5.49)] as having three components:

$$\text{minimize } Z = \lambda_3 U_3 + \lambda_1 U_1 + \lambda_2 U_2 , \qquad (5.55)$$

where λ_1, λ_2, and λ_3 are arbitrary weights, and where U_1 is the transport-cost term, U_2 is the location-cost term, and U_3 is the entropy term. Thus the location-cost-only experiments whose results were given in tables 5.1(a) and 5.1(b) were done with λ_1 and λ_3 both set to 0, and λ_2 set to 1. The transport-cost-only tests whose results are given in table 5.2 were done with λ_1 equal to 1 and both λ_2 and λ_3 equal to 0. The set of tests for both transport cost

Table 5.5. Archerville: location of households, based on the location-cost and the transport-cost components of the TOPAZ model, for $\lambda_2 = 0.070$.

Zone	Households			Zone	Households	
	LI	HI	total		LI	HI
1	0	294	294	1	0	135 (1)
2	359	173	532			69 (3)
3	0	155	155			45 (4)
4	441	28	469			45 (5)
5	0	0	0	2	166 (1)	157 (2)
					193 (2)	16 (4)
				3	0	155 (3)
				4	276 (3)	28 (4)
					110 (4)	
					55 (5)	
				5	0	0

Note: numbers in parentheses indicate the zone in which the head of household works; see also tables 5.2 and 5.3.

Table 5.6. Archerville: location of households, based on the location-cost and the transport-cost components of the TOPAZ model, for $\lambda_2 = 0.070$, and incorporating dispersion, with $\beta = 0.05$.

Zone	Households			Zone	Households	
	LI	HI	total		LI	HI
1	45	258	302	1	42 (1)	128 (1)
2	319	205	524		3 (3)	21 (2)
3	20	138	158			84 (3)
4	416	49	465			14 (4)
5	0	0	0			11 (5)
				2	105 (1)	6 (1)
					181 (1)	135 (2)
					25 (3)	17 (3)
					4 (4)	32 (4)
					4 (5)	15 (5)
				3	1 (1)	117 (3)
					19 (3)	12 (4)
						9 (5)
				4	18 (1)	6 (3)
					11 (2)	33 (4)
					230 (3)	10 (5)
					106 (4)	
					51 (5)	
				5	0	0

Note: see also tables 5.2, 5.3, and 5.5.

and location cost taken together were done with λ_1 equal to 1, with varying λ_2, and with λ_3 held constant at 0. In that last set of experiments it was found that a range of 0.045 to 1.00 for λ_2 when λ_1 is 1 (and λ_3 is 0) produces a solution which reflects the effect both of transport cost and of location cost. For the first series of dispersion-term tests the basic, no-dispersion solution will be taken to be the results when λ_1 equals 1, λ_2 equals 0.70, and λ_3 equals 0. These results are shown in table 5.5, and a comparison of them with earlier tables will show them to be towards the location-dominated end of the range of tests, but with significant effects from the transport-cost component.

The value of λ_3 as discussed here is the inverse of the β of equation (5.49) and thus will directly affect the extent to which location is dispersed from the no-dispersion case where $\lambda_3 = 0$ and thus $\beta = \infty$. The first three tests of this series were done with $\lambda_3 = 2$, 5, and 10, respectively ($\beta = 0.5$, 0.2, 0.1) and gave some small degree of dispersion which increased as β decreased. With a λ_3 of 20 and thus with β equal to 0.05, the dispersion became rather more extensive, producing the results shown in table 5.6.

Table 5.7. Archerville: location of households, based on the location-cost and the transport-cost components of the TOPAZ model, for $\lambda_2 = 0.070$, and incorporating dispersion, with $\beta = 0.033$.

Zone	Households			Zone	Households	
	LI	HI	total		LI	HI
1	55	250	305	1	45 (1)	115 (1)
2	311	211	522		3 (2)	31 (2)
3	37	124	161		7 (3)	79 (3)
4	397	65	462			15 (4)
5	0	0	0			10 (5)
				2	89 (1)	16 (1)
					162 (2)	122 (2)
					42 (3)	30 (3)
					11 (4)	29 (4)
					7 (5)	14 (5)
				3	3 (1)	3 (1)
					30 (3)	2 (2)
					2 (4)	98 (3)
					2 (5)	12 (4)
						9 (5)
				4	19 (1)	3 (2)
					17 (2)	17 (3)
					196 (3)	33 (4)
					98 (4)	12 (5)
					67 (5)	
				5	0	0

Note: numbers in parentheses indicate the zone in which the head of household works; see also tables 5.2 and 5.3.

A further increase in λ_3, to 30, giving a value of β of 0.033, yields even greater dispersion, as shown in table 5.7. Numerous other tests were run with different combinations of values for λ_1, λ_2, and λ_3. In all cases, increased values of λ_3 produced increased dispersion of household location. The most dispersion was achieved with values of λ_1 equal to 1, λ_2 in the vicinity of 0.3, and λ_3 equal to 30 or more (though when λ_3 got much above 40 the solutions to the model began to exhibit problems of degeneracy, and nonoptimal values began appearing in the solution vectors). The conclusion here, as before, is simple. The addition of a dispersion term to the objective function of a mathematical programming model of residence location results in a more even distribution of residents to zones and probably represents a more realistic representation of actual human behavior.

5.5 A comment on sensitivity analysis

In the discussion above attention was called to abrupt shifts in the solution of TOPAZ, resulting from gradual changes in objective function coefficients (the value of λ_2 in particular). This phenomenon is the result of a reasonably well known property of linear programming models, and can perhaps best be discussed here in terms of the simplex-algorithm example of the health clinic, as described in chapter 3. The objective function was

$$\text{maximize } e = 3X_H + X_T , \tag{5.56}$$

where
e is the effectiveness of the program,
X_H is the number of house-to-house programs,
X_T is the number of telephone programs.

This function defines a family of parallel lines as was shown in figure 3.3. The slope of these lines is determined by the coefficients, 3 and 1, in the objective function. Now consider the feasible region for the problem as redrawn in figure 5.1. The cut-off corner (upper right) of the feasible region results from the budget constraint of the problem. A straight line parallel to this edge of the feasible region, as shown in figure 5.1, will form an angle A°, with the X_H axis. If the objective function were to be precisely parallel to this line, the problem would have no unique optimum, as any combination of X_T and X_H anywhere along the edge of the feasible region would yield the same value for the objective function. Yet, even the smallest change in the objective function coefficients would alter the situation.

If the slope of the objective function were to be reduced, thus reducing its angle with the X_H axis to less than A°, then the optimal solution would be located at vertex B of the feasible region. This would be true for any values of the objective-function coefficients that resulted in the angle of the objective function with the X_H axis being less than A° but greater than 0°. Similarly, if the slope of the objective function increased so that its angle with the X_H axis was more than A°, the optimum solution would be at vertex C of the feasible region. This would be true for any set of objective-function coefficients which results in the angle of the objective function with the X_H axis being greater than A° but less than 90°.

For this problem it is easy to determine the range of, say, the X_H coefficient and the points at which the optimal solution will jump from vertex B to vertex C, or the reverse. If the coefficients of the budget constraint are divided by 200 then the left-hand side equals $5X_H + X_T$. If this is compared with the objective function, then it is clear that if the coefficient of X_T is held constant at 1, the optimum solution will be at vertex B for any value of X_H greater than 5. For any value of X_H less than 5 the optimum solution will be at vertex C, and if X_H is equal to 5 the problem will be degenerate. If, on the other hand, the coefficient of X_H is kept constant at 3, then for values of the coefficient of X_T between 0 and 0.6 (three fifths), the optimal solution will be at vertex B. The optimal solution will be at vertex C for all values of the coefficient of X_T greater than 0.6.

The same properties described here exist for the linear version of TOPAZ. The TOPAZ example cannot be analyzed graphically as above, because there are too many variables involved. The principal remains the same, however, and accounts for the abrupt changes in the solutions of TOPAZ, discussed above in section 5.3. More discussion of this topic of sensitivity analysis for mathematical programming models can be found in standard texts (Bradley et al, 1977). An advantage of adding the dispersion term to models such as TOPAZ or Herbert–Stevens is that, although still possible, such discontinuities of response are much less likely. A disadvantage is that the sensitivity analysis itself can become much more complex (Fiacco, 1983). Such abrupt solution shifts could still occur, but only when optimal solutions are located at intersections of constraint equations. In linear programs, solutions must be located at such intersections (that is, the vertices of the feasible region). In nonlinear programs, the solutions will

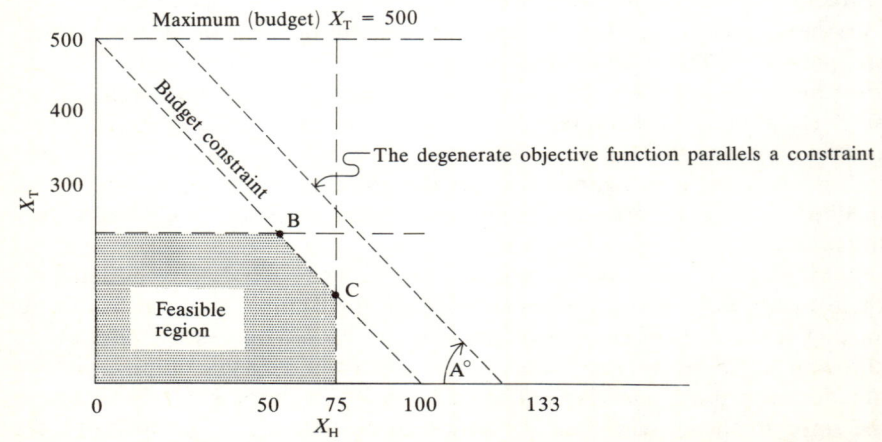

Figure 5.1. Illustration of case of degenerate solution to a linear programming problem.

often, perhaps more often than not, be found at a function minimum or maximum along what will now be nonlinear objective functions or constraints. Of course, for most applications these simple two-dimensional examples must be extrapolated into n-space and n-dimensional feasible regions.

5.6 Some conclusions

The work described in this chapter comes nowhere near to exhausting the topic, yet certain conclusions can clearly now be drawn. The first of these is that linear mathematical programming models of location are inherently unrealistic. The implied number of nonzero elements in a solution of a linear programming model is one aspect of this problem. The fact that the least-cost zone will get all possible locators, even if the zone of next-to-least cost is only marginally more expensive, is another aspect of the problem. The above-discussed sensitivity problem implies a third problem, in that an arbitrary difference in units of measurement, say between hundreds of dollars or thousands of dollars for annualized rent, can result in one component of a model solution being dominant over another.

To a considerable degree the constraint equations of a linear programming model can ameliorate some of the above difficulties. This is, however, a mixed blessing. That they can ameliorate some of the difficulties also points up the considerable extent to which they determine the model solution. The constraints, for example, available residential land per zone, must be exogenously determined, yet their determination, in and of itself, implies a difficult forecasting problem. The reasoning is circular here, perhaps best exemplified in some of the early work with the Herbert–Stevens model where it was necessary to use a Lowry-like model to estimate some of the Herbert–Stevens inputs. The availability of data for constraints during the development of a mathematical programming location model can give a false sense of confidence in the productive power of the model if the problems of forecasting the constraints are not taken into account.

Another major issue is that of obtaining the necessary data. In the above discussion of the Herbert–Stevens model (chapter 4) it was quite clear that the estimation of utilities and preference functions was an extremely difficult matter and yet absolutely critical to the success of the model at predicting location patterns. With the TOPAZ model it has often been so difficult to estimate the benefits of location or interaction that only costs were used. Yet, in the Herbert–Stevens examples given above, it was clearly shown that the use of location costs without location benefits gave unacceptable results. The linear programming versions of TOPAZ would certainly have given different results if some locational advantage variable had been used to yield a 'net' location-cost variable for the location-cost term in the objective function. The rather considerable difference in the results from the Herbert–Stevens model for Archerville, as compared with the TOPAZ results, stems in part from use of a locational advantage term in the Herbert–Stevens model and none in TOPAZ.

Finally, there is the computational problem. The GAMS package was used for all the Archerville tests reported here. The linear formulations were run on an IBM personal computer (PC) with 640K of memory and a hard disk drive. These runs took just a few minutes each. The nonlinear formulations were problematic on the PC, with some taking two or more hours to solve, and others not running at all. Last, a version of GAMS for the IBM 3081 GX mainframe was used for the nonlinear formulations. In Archerville there were 5 zones and 2 household types. For the Herbert–Stevens model there were 3 house types. This means that the objective function for the final versions of the Herbert–Stevens model had 150 variables. The objective function for the final nonlinear version of TOPAZ had 110 variables.

These numbers do not seem too bad, but if there were a 30-zone region to be analyzed, then the nonlinear Herbert–Stevens model would have 5400 variables in the objective function, and the nonlinear TOPAZ model would have 3660. These are rather sizeable problems, and yet a 30-zone spatial interaction model is really too highly aggregated for most policy analysis purposes. Further, in both the examples presented in this chapter, only 2 household types were used, with 3 house types in the Herbert–Stevens model and no housing stock consideration at all in TOPAZ. Current transportation and location modelling applications tend to have 200 to 300 zones. With no increase in activity types, 250 zones would yield 375 000 variables in the Herbert–Stevens model and 250 000 in TOPAZ. Problems of this size are really quite impractical for direct solution.

Despite these concerns there are several important points to be learned from these experiments. Perhaps the most important is that the developing of these model formulations and the testing of their behavior gives wonderful insight into various hypotheses about locational behavior. The eminent sensibility of describing locational behavior as an optimizing process is beyond reproach. The effects, and general importance, of constraints in such formulations became clearly evident in these experiments. At the same time the experiments clearly illustrated the need for inclusion of dispersion terms in such models. The inference is that although locational behavior may be said to be, in principle, an optimizing process, in actuality there are obviously other factors which result in a dispersion of locations around a simple 'least-cost' optimum. Yet the optimizing process provides a model-building rationale which can be particularly helpful in understanding the implications of model structure. This point, in particular, will become evident again in later chapters, in the discussions of trip assignment and model integration.

The problems of solving optimizing models of a realistic (from the applications point of view) size can be dealt with by decomposition procedures if one wishes to maintain the mathematical programming formulations. The possibility of transforming programming models into spatial interaction models, as briefly introduced in this chapter, offers another avenue of approach. Both these approaches will be examined in subsequent chapters. Before that, however, the focus will shift from problems of activity location to problems of assigning trips to networks, the subject of the next chapter.

Transportation models in optimizing frameworks: 1

6.1 Introduction

Under this rubric there is a whole cluster of problems and potential solution techniques. In the most general sense the traffic assignment problem is one of assigning to the links of a network some set of trips or flows from a set of origin zones (or nodes) i to a set of destination zones j. The zones, used as a discrete representation of a continuous spatial distribution of activities, are connected by a set of links which are intended to represent the transportation facilities. We accept as given that the zones and the links are adequate representations of reality for our current purposes, but we acknowledge that in other contexts these matters are open to question. It is also helpful to begin by accepting as given a matrix of flows or trips from each origin to each destination, while noting that the essence of integrated modelling is in the fact that these flows are not fixed in the context of the overall model system. With these stipulations the traffic assignment problem, then, is concerned with the routes or paths taken by these trips in traversing the network from their origins to their destinations.

In this chapter the issues of trip assignment are introduced and discussed. Simple numerical examples are given and are followed by numerical experiments with larger data sets. Some potentially controversial conclusions are reached.

6.2 Unconstrained assignment

This is one of several branching points in the problem definition. Here the issue is whether the characteristics of each link, for example, travel time or cost, are functions of link flows? In other words, is this to be capacity-constrained trip assignment? If it is not, then the problem becomes a rather straightforward matter of (a) path enumeration and (b) path selection. The most frequently encountered form of the path-enumeration problem is that of finding the minimum, or shortest, path between each origin–destination pair. Efficient solutions for this problem usually draw upon the Moore algorithm and are discussed in Sheffi (1985) and Van Vliet (1978). In any case, once a path-enumeration procedure is available the problem becomes one of constructing a decision rule for assigning the flows (or trips) to the path (or paths).

Recalling that we are dealing here with link characteristics which are independent of link flows, we see there are only two reasonable assignment procedures. The first, commonly called 'all-or-nothing' assignment, assigns all trips from zone i to zone j to the links on the shortest path between i and j. This may also be seen as the minimum-cost flow problem as described in chapter 3. The second procedure, commonly called 'multipath' assignment (or stochastic assignment, for reasons to be discussed shortly), disperses the trips over the first, second, third, fourth, ..., rth best paths in inverse proportion to the path lengths.

There is an implicit set of assumptions about human behavior which underlie the decision as to which of these procedures to select. If it is assumed that all tripmakers (a) perceive the link characteristics identically, (b) have complete information on all alternative paths, and (c) are behaving 'rationally', then they will all select the 'shortest' path. If one or more of these assumptions is relaxed then the trips will be dispersed over several different paths between i and j. The usual procedure for accomplishing this dispersion is to consider that variations in individuals' perceptions of the link characteristics (as well perhaps as variations in information availability and in their individual rationality) result in behavior which implies that the link characteristics are perceived as random variables. If this is followed by the proper assumptions regarding the shape of the random distribution, a logit model of path choice is led to. This takes the form

$$T_{ij}^p = T_{ij} \exp(-\theta c_{ij}^p) \bigg/ \sum_n \exp(-\theta c_{ij}^n) , \qquad (6.1)$$

where
T_{ij} is the total number of trips from origin i to destination j,
T_{ij}^p is the number of trips taking path p from origin i to destination j,
c_{ij}^p is the 'cost' of path p between origin i and destination j,
θ is an empirically determined parameter.

Because of the fact that this formulation results from a probabilistic interpretation, the procedure is sometimes called stochastic or probabilistic multipath assignment. It should be noted that alternate assumptions regarding the probability distribution of the link c_{ij}'s could lead to other forms for equation (6.1), such as the probit model. An excellent discussion is given in Sheffi (1985)

6.3 Capacity-constrained assignment

Whereas in the unconstrained case the assignment problem is strictly a question of the path choice procedure, the imposition of link-capacity constraints changes the situation considerably. We are specifically concerned here with the situation where one of the prespecified link characteristics is a 'design' capacity which serves, by means of some volume–capacity function, to increase link travel time or cost as a function of link flows. Although previous work on this topic is far from definitive, there are some standard forms for link volume–capacity functions (Branston, 1976). The direct consequence of such functions is that each time a trip is assigned to a link the travel time or cost incurred in traversing that link increases. As a consequence, after a certain number of trips are assigned to the links of a particular path from i to j, that path may no longer be the minimum path. Consider the situation of having two alternate paths of almost equal 'length' when empty.

An 'all-or-nothing' assignment would put all the trips on the shorter of these paths and none on the slightly longer one. Once the congestion effects were calculated the originally shorter path could be much 'longer' than the now unused originally longer path. The solution of this aspect of the assignment problem requires resolution of (a) procedures for trip reassignment, and (b) criteria for determining the adequacy of such reassignments.

This reassignment problem is not a new one. Capacity-constrained assignment procedures have been plagued by what are often called uneven or unstable link flows for more than twenty years. For most of that time attempts to deal with the problem have involved any number of more-or-less ad hoc procedures for implicit reassignment. One of the first of these procedures used the all-or-nothing path-choice procedure in an iterative scheme of link time and/or cost averaging and trip reassignment. This scheme and others like it were notorious for their failure to converge to a stable solution. The procedure consisted of the following steps: (1) all-or-nothing assignment, (2) recalculation of link characteristics as a function of the flows resulting from the first step, (3) link-by-link calculation of a weighted average of the original link characteristic (for example, cost or time) and the value of that characteristic calculated in the second step, (4) setting all previous link flows (volumes) to zero and beginning again at step 1. The criterion for stopping this procedure was that the flows stabilize—a criterion rarely achieved in practice. A secondary criterion was that of the total travel cost for the network:

$$C = \sum_k T_k c_k , \qquad (6.2)$$

where
C is the total travel cost for the network,
T_k is the number of trips assigned to link k (note that this is the sum of the trips from any origin to any destination, made on paths of which k is one link),
c_k is the cost of travelling on link k when a volume T_k is already on the link.

The idea was that the lower the value of C the better the assignment results. It should be noted that in these procedures the criteria were *not* used in a formal mathematical minimization procedure, as it was generally believed that expressing the problem in such a form was not computationally tractable.

Subsequent developments continued along the line of devising heuristics for what it was hoped would be procedures leading to a 'smoother' or more uniform assignment of trips to the network links. One such development was the incremental assignment procedure. This procedure differed from the one described above, as the matrix of trips to be assigned was divided into several increments of, for example, 25% of the total trips. The steps then followed were: all-or-nothing assignment of the first increment of 25% of the trips; recalculation of link characteristics as a function of the flows resulting from the first step; all-or-nothing assignment of the second increment of 25% of the trips to the partly 'congested' links of the network; recalculation of link characteristics as a function of the sum of the flows from the first and third steps; and so forth until all trips were assigned.

Although this procedure did often produce 'smoother' loadings by spreading the trips more evenly over the links in the network, it still failed to converge, and was subsequently shown to be congenitally incapable of reliable convergence (Florian and Nguyen, 1974). A further modification was used for many years as the trip-assignment algorithm in ITLUP (Putman, 1983a). This method, called incremental tree-by-tree assignment, is done as follows: (1) all-or-nothing assignment of the first increment (again, say 25%) of trips, but for only one origin, to all destinations, (2) recalculation of link characteristics along the paths used for the trips from this first origin, as a function of the resulting flows, (3) all-or-nothing assignment of the first increment of trips from the second origin, to all destinations. Steps 1, 2, and 3 are repeated until the first increment of trips, from all origins to all destinations, has been assigned. Then the whole process is repeated with the additional increments. This procedure, although producing a more even spread of trips over the links in the network may also fail to converge, even when the trip increments are made as small as 10% each.

The reason that these incremental methods fail to converge is that they make no provision for explicit reassignment of trips (Ferland et al, 1975). Thus once trips are assigned to a link, no matter what increase is caused in the travel time or cost of the link, those trips remain assigned to the link. This need for an explicit trip-reassignment technique leads directly to the question of criteria for success with network assignment. A more recently popular criterion is that of user equilibrium (UE).

6.4 Equilibrium assignment
Returning to the question of assumptions, if the three assumptions about human behavior given above are accepted, then a stable trip assignment (that is, an equilibrium) will have been achieved when no trip can be accomplished in a shorter time or at a lower cost by switching paths from the current assignment. More specifically, UE is achieved when for each origin–destination pair the travel time (or cost) on all used paths is equal and this travel time is less than or equal to the length of any unused path. If any of the above assumptions about human behavior are relaxed, then the statement of equilibrium changes to, 'no tripmaker *believes*, or *perceives*, that his or her trip can be accomplished in a shorter time or at a lower cost by switching paths'. This latter is the statement of the condition for stochastic user equilibrium (SUE) (Daganzo and Sheffi, 1977; Sheffi and Powell, 1981b). To accomplish UE or SUE assignment it is necessary to have an algorithm which (a) incorporates the criterion, and (b) provides an explicit procedure for trip reassignment in order to achieve the equilibrium condition.

It should be obvious to the reader that UE or SUE are only relevant in the context of capacity-constrained assignment. In the absence of link volume–capacity functions the UE problem collapses to all-or-nothing assignment, and SUE collapses to multipath assignment. It should also be clear that the UE, or SUE, criterion implies a rather complex set of calculations. The number of paths between origins and destinations in any operationally interesting network of, say, 30 or more nodes and 1000 or more links, is astronomical. Even multipath assignment, without consideration of equilibrium conditions, is only practical when using algorithms which obviate the need for full path enumeration. Put aside the SUE problem for now; to solve the UE problem requires that (a) the problem be formulated mathematically, and (b) that a procedure for its solution be developed. How these may be accomplished is now briefly described.

Assume a set of origins and destinations connected by a set of paths through a network. The paths are made up of sets of links. First define x_ℓ as the flow (volume of trips) on link ℓ and let t_ℓ represent the travel time on link ℓ. As mentioned above, the UE problem is interesting only when t_ℓ is a function of x_ℓ, so let this function be represented as $t_\ell = t(x_\ell)$. Further, let x_p^{rs} represent the flow (volume of trips) on path p connecting origin r to destination s, while t_p^{rs} is the travel time for path p. Then we have

$$t_p^{rs} = \sum_\ell t_\ell \delta_{\ell,p}^{rs} , \tag{6.3}$$

which states that the travel time for path p is the sum of the travel times for all the links composing the path. The $\delta_{\ell,p}^{rs}$ is 1 if the link ℓ is on path p, and 0 if it is not. In the same fashion we have

$$x_\ell = \sum_{r,s,p} x_p^{rs} \delta_{\ell,p}^{rs} , \tag{6.4}$$

which states that the flow on link ℓ is the sum of all the flows from all origins r to all destinations s on all paths p which include link ℓ. Note that to simplify the presentation we have left out the specification of the sets of origins, destinations, links, and paths over which the summations in equations (6.3) and (6.4) are to be taken. With this notation the UE problem may be represented, given a matrix q^{rs} of origin–destination flows, by the following mathematical program:

$$\text{minimize } Z(x) = \sum_\ell \int_0^{x_\ell} t_\ell(x_\ell) \, dx , \tag{6.5}$$

subject to

$$\sum_p x_p^{rs} = q^{rs} , \tag{6.6}$$

$$x_p^{rs} \geq 0 , \tag{6.7}$$

$$x_\ell = \sum_{r,s,p} x_p^{rs} \delta_{\ell,p}^{rs} . \tag{6.8}$$

Here the objective function, equation (6.5), says to minimize some value Z as a function of x, where the functional form is the sum of the integrals (from 0 flow to x_ℓ flow) of the link volume–delay, or volume–capacity, functions. The first set of constraint equations, equation (6.6), simply requires that the sum of flows on all paths connecting origin r to destination s must equal the prespecified flow q^{rs} between r and s. The set of constraints, equation (6.7), prohibit negative path flows. The last set of constraints, equation (6.8), are simply the definitional requirements for link flows as given earlier in equation (6.4).

The mathematical program defined by equations (6.5)–(6.8) is a constrained nonlinear optimization problem. As given above there is an enormous number of constraints. There will be a constraint arising from equation (6.6) and another from equation (6.7) for each *path* between all origin–destination pairs. In addition there will be a constraint arising from equation (6.8) for each *link* in the network. This could easily lead to a problem with millions, or even tens of millions, of constraint equations. This problem definition, developed thirty years ago (Beckman et al, 1956), can be transformed into an equivalent problem which is computationally much more tractable (LeBlanc, 1973). The reformulation draws upon specific properties of the equations which, to gloss over some rather clever mathematical analysis, result in one's being able to deal with link flows without having to keep track of path flows. In operation the algorithm which does this, called the convex-combinations method, is in effect an efficient procedure for reassignment of trips from congested links to less congested ones.

The objective function of the UE mathematical program [equation (6.5)] has a precise form which depends on the volume–capacity function that is selected. For the convex-combinations algorithm to work, the volume–capacity functions, also called link performance functions, must increase monotonically. This means that t_ℓ must increase, continuously, with increasing x_ℓ. The standard function, long used by transportation planners, was developed by the US Bureau of Public Roads and is of the form

$$t_\ell = t_\ell^{(0)}\left[1 + \alpha\left(\frac{x_\ell}{k_\ell'}\right)^\beta\right], \tag{6.9}$$

where
t_ℓ is the 'congested' travel time over link ℓ,
$t_\ell^{(0)}$ is the travel time for free flow, that is, the uncongested travel time,
x_ℓ is the flow (volume) over link ℓ,
k_ℓ' is the 'design' or 'practical' capacity of link ℓ.
α, β are parameters whose usual values are 0.15 for α and 4 for β.

Substitution of this function into equation (6.5), yields, by integration and by using the numerical values of α and β,

$$Z(x) = \sum_\ell \left[t_\ell^{(0)} x_\ell + t_\ell^{(0)} \frac{0.15}{5}\left(\frac{x_\ell}{k_\ell'}\right)^4 x_\ell\right]. \tag{6.10}$$

This is the function to be minimized in solving the UE mathematical program. Most authors agree that this function appears to have no economic interpretation (Eash et al, 1979). This is not to be wondered at, as the volume or capacity function [equation (6.9)] is a strictly empirical construct, developed more than twenty-five years ago. The objective function [equation (6.10)] is an algebraic artifact which results from formulating the UE conditions as a mathematical programming problem.

In the earlier discussion of capacity-constrained assignment, mention was made of the notion of the total travel cost (or time for the network [equation (6.2)] as an alternative criterion by which the results of a trip-assignment solution could be evaluated. It is possible to formulate this as an alternative objective function:

$$\text{minimize } W(x) = \sum_{\ell} x_\ell t_\ell(x_\ell) dx \ . \tag{6.11}$$

This new objective function [equation (6.11)] minimizes some value W as a function of x, where the functional form is the sum of the link volumes x_ℓ multiplied by the link travel times (or costs) t_ℓ which are themselves a function of x_ℓ. This problem is different from the UE problem in that it makes no assumptions about human behavior, but, rather, attempts to find an overall trip-assignment pattern which minimizes the travel time or cost of the total system. This is known as the system optimization (SO) problem and is quite different from the UE problem.

This raises the more general, and currently unanswerable, question of what the objective functions mean in the various mathematical programming formulations of the assignment problem. The SO problem is of public policy interest, as it represents a potential societal goal which might result from a desire to reduce energy consumption or air pollution. It suffers from the problem that it will probably not result in an equilibrium solution. After a trip assignment has been made according to the SO criterion it will usually be possible for tripmakers to reduce their individual travel times by changing paths. Thus the SO solution is unlikely to be stable. The UE solution is an equilibrium solution, but it depends critically on the three behavioral assumptions made earlier. A more likely approach, from the standpoint of replication of actual tripmaking behavior, is the SUE approach. The SUE objective function may involve the UE and the SO objective functions along with the addition of a function of 'expected perceived travel time', the inclusion of this third function will depend upon whether the problem is expressed in terms of a constrained or an unconstrained optimization problem. Here, however, the mathematics again become more difficult, though solution procedures for this problem have been developed (Powell and Sheffi, 1982) and will be described and tested in the next chapter. Next, before proceeding to some simple numerical examples, the method used in solving the UE assignment problem will be described in detail.

6.5 Discussion of solution procedures for the UE problem

First, recall that the UE problem as expressed in equations (6.5)–(6.8) is a constrained nonlinear optimization problem. The difficulty with this problem is the millions of paths (through the transportation network) which are involved. The use of the Frank–Wolfe algorithm to solve this problem, as proposed by LeBlanc (1973), results in a spectacular reduction in the calculations necessary by eliminating the need for path enumeration during the problem solution. The difficulty is that the Frank–Wolfe algorithm, one of a group of procedures known as convex-combinations algorithms, can often be very slow to converge to an optimum solution.

The solution of the mathematical programming formulation of the UE problem involves minimization of a nonlinear objective function [equation (6.10)], subject to linear constraints [equations (6.6)–(6.8)]. Solution of this nonlinear programming (NLP) problem involves calculating a direction in which to search and a step size to make, both of which must remain within the problem constraints. Methods for doing this are called 'feasible-direction' methods. The Frank–Wolfe algorithm may perhaps be best understood by describing it in terms of its being a linear approximation method. The steps of the algorithm are as follows.

Step 0 Find a numerical feasible starting point or 'initial feasible solution' to the NLP problem.

Step 1.1 Take (algebraically) the partial derivatives of the original objective function and evaluate them (numerically) at the initial feasible solution.

Step 1.2 The evaluated partial derivatives will be the coefficients in a linear equation. Solve for the minimum of this equation subject to the same constraints as in the original NLP problem. The line connecting the initial feasible solution to this newly calculated minimum gives the direction of search, or the descent direction.

Step 2 Find the minimum of the original NLP objective function along the line determined in step 1.2. The location of this minimum thus gives the length of the step to move in the descent direction. This minimum is the new solution to the NLP problem.

Step 3 Calculate the convergence measure. If this prespecified convergence tolerance is met, the process is completed, if not, the new solution is substituted for the previous solution (or the initial feasible solution) and the procedure repeated from step 1.1.

There are several characteristics of this procedure which should be noted. First the Frank–Wolfe algorithm is guaranteed to find an optimal solution after an infinite number of iterations. Although being a reassuring property of the algorithm, this guarantee is not helpful in practice unless the procedure can come usefully close to the optimum in a tractably finite number of steps. The extent to which it can is a function both of the individual NLP being solved and of the initial feasible solution. Thus it is necessary to have some notion of how well the algorithm is proceeding towards the optimum.

It should be recalled that in step 1.1 above the partial derivatives of the objective function from the NLP are evaluated at a feasible solution. This gives the slope of the objective function (curve) at that point. The line determined in step 1.2 will be tangential to the objective function at that point. As we are dealing with a minimization problem, we may imagine this line as being external to a downward pointing convex surface. Thus the line will be wholly outside the curve of the NLP objective function. At the minimum point calculated in step 1.2 the line will still be below the NLP objective function. Thus this minimum determines what is called a 'lower bound' to the objective function, along the line between the previous feasible solution and the minimum-value solution to the linear objective function in step 1.2. This lower bound should always be smaller (lower) than the value of the NLP objective-function minimum found in step 2. In general the percentage difference between the NLP objective function minimum and the lower bound should *tend* to decrease with the number of iterations of the Frank–Wolfe algorithm.

It is also possible to consider the 'convex-combinations' aspect of the Frank–Wolfe algorithm in evaluating its progress towards the optimum. The convex-combinations name results from what is happening in step 2. Here a minimum value of the NLP objective function is determined, along the line between the previous feasible solution (which may be called prior-minimum A) and the minimum of the LP problem of step 1.2 (which may be called prior-minimum B). This minimum, being found along a straight line projected onto the NLP objective function is, in fact, a weighted average of the two minima (that is, prior-minimum A and prior-minimum B). As the algorithm nears the NLP minimum the relative weighting of the two minima should tend towards ever smaller contributions by the new LP minimum. It should be noted that neither the percentage differences nor the magnitude of the weighting of the LP minimum will be decreasing monotonically, but that the tendency will be that way.

Last, in the particular case of the Frank–Wolfe algorithm as applied to the UE problem, the change in the flow on a specific link should decrease as the number of Frank–Wolfe iterations increases. This may be measured by calculating a mean percentage flow deviation for all links for each iteration. This too should tend to get smaller with increasing numbers of iterations.

6.6 Numerical experiments with assignment algorithms

In order to both illustrate and examine the behavior of some of these algorithms a series of computer experiments was done making use of the Archerville data.[1] Numerous tests were conducted by using different assignment algorithms, in order to compare their performances. A computer program called NETWRK, which is a component of ITLUP, was augmented

---

[1] Most of these Archerville computer runs were done by Vincent Patterson, one of the author's students in the Department of City and Regional Planning.

so that any of several assignment algorithms could be utilized with identical input and output files and formats. The first comparisons were between the following three algorithms: (1) all-or-nothing with tree-by-tree volume–capacity calculation, (2) stochastic multipath, with use of the double-pass method, (3) UE, using the convex-combinations method.

The all-or-nothing (AON) algorithm was a modified version where instead of first assigning all trips from all origins to all destinations and then calculating the congestion effects, congestion was calculated as the assignment procedure was progressing. Thus an origin node would first be selected. The minimum paths from that origin to all destinations would then be traced. The trips from that origin to all destinations would be assigned to the minimum paths and the 'congested' link times calculated by using equation (6.9). Then the next origin node would be selected and the process repeated. Although this procedure still suffered from the fact that trips were never reassigned, the assignments were built up gradually as the tree-by-tree congestion calculations were done.

It should be noted that several configurations using the AON procedure were tested. The first, AN−100, simply assigned all of the trips by use of the tree-by-tree form of the AON algorithm. The final network statistics were obtained by a postassignment tree tracing over the congested (that is, trips loaded or assigned) network. In the second form, AN−40, 40, 20 the incremental tree-by-tree form of the AON algorithm was used. Here, 40% of the trips were assigned with use of the tree-by-tree method. The partially congested network which resulted was then used as input to the subsequent assignment of an additional 40% of the trips, again by using the tree-by-tree method. Finally, the last 20% of the trips were assigned to the network which had link travel times reflecting the prior presence of the 80% of the trips which had already been assigned. The link statistics given as final results were obtained from a final tracing of trees over the fully loaded network. In the third form, AN−34, 33, 33, three increments were again used—of 34%, 33%, and 33%, respectively, in this case. In the fourth form, AN−25, 25, 25, 25, four increments of 25% each were used.

It is worth noting here that one issue concerning this procedure is the question of the possibility of bias being introduced to the network loadings as a function of the order in which the origins are selected (Ferland et al, 1975). To examine the importance of the sequence of origin selection a separate series of experiments was conducted by using a somewhat larger test network, SF30, which will be described later in this chapter (McBride, 1985). In these tests the origin nodes were put in a random sequence prior to the trip-assignment calculations. Fifteen computer runs were done, where the only difference was that there was a different random sequence used in each case. The effect on several network-wide measures was insignificant. For each origin or destination node the average volume (or capacity) ratio of all links entering or leaving the node was calculated. There appeared to be a low (but statistically significant) negative correlation between the position of a node in the node-selection sequence and its mean volume (or capacity) ratio. This was a curious result and should be explored further. The assumption is made here that the node selection order is of no consequence.

The stochastic multipath (SM) algorithm used was a standard double-pass procedure which did not require path enumeration (Dial, 1971). Although there are known problems with Dial's algorithm when there are overlapping paths, it was hoped that it was safe to assume that this would not cause significant difficulties in these tests (Dalton and Harmelink, 1974). As the way in which the procedure was used, described below, involved nonconstant link times, that is, times were a function of volumes, this was probably a safe assumption (Fisk, 1980). In the sense that these multipath procedures are actually logit models of path choice, a subsidiary set of experiments examined the sensitivity of the assignment results to variations in θ as in equation (6.1). There were no observed data against which to test different values of θ or to calibrate the model (Daganzo, 1977a; Fisk, 1977; Robillard, 1974). Yet, in the test runs, when θ was varied from 0.1 to 0.9, rather little variation in network-wide statistics resulted. Neither of the objective-function values calculated [from equation (6.2) and equation (6.10)] changed by more than one or two percent. Note that four different configurations of the SM algorithm were used for each of two values of θ (0.2 and 0.5). The first of these four configurations, SM—100, combined the notions of the SM algorithm with the tree-by-tree assignment procedure. For each origin–destination pair the SM algorithm was used to assign all the trips from the given origin to the given destination, and the links to which trips had been assigned were congested by using the volume–delay function given in equation (6.9). Another origin–destination pair was then selected and the procedure repeated. The second configuration, SM—40, 40, 20, combined SM with the incremental tree-by-tree procedure, again by using three trip increments of 40%, 40%, and 20% as for the AN—40, 40, 20 procedure. In the third and fourth configurations for SM the same combinations of increments were used as for the third and fourth configurations of the AON algorithm.

The user equilibrium (UE) algorithm was the convex-combinations procedure implemented in the form of the Frank–Wolfe algorithm (LeBlanc, 1973; LeBlanc et al, 1975). As mentioned above, special properties of the assignment problem described in terms of equations (6.5) to (6.8) allow a marvelous simplification of the problem. The algorithm begins with an AON assignment of trips to the network. The resulting trip volumes and congested-link times are saved. A second assignment is done, using these congested times, and this second set of volumes is saved as well. A search procedure is then used to find the minimum value of equation (6.10) in the interval between the first and second sets of volumes. With this new set of intermediate volumes as a starting point, the procedure is repeated. Approximate convergence usually takes place after just a few iterations.

An interesting point is that the solution to the UE problem is unique (in terms of link flows, though not in terms of path flows): there is one and only one equilibrium solution. The ease with which this solution is found depends to a limited extent on how different the AON assignment results are from the UE solution. Use of the AON tree-by-tree method as a starting point and for subsequent assignments seems to produce a more rapid convergence than use of the standard AON procedure as reported in the literature. This method is an extended version of what has been called 'extended quantal' assignment by Van Vliet and Dow (1979). It should be noted that there are, however, arguments as to the stability of such equilibria, and these arguments raise the general issue of starting points (Horowitz, 1984). As has been reported earlier (Putman, 1983a; 1985), the starting point used for the trip-assignment calculation had a significant effect on the stability of previous versions of the integrated model system (ITLUP). More will be said about this later, in chapter 8. Two configurations of the UE method were used. The first, UE—20/3, allowed up to twenty iterations of the line search (step 2 in the description of the method given above) and three main iterations of the whole procedure. The second configuration, UE—20/10, again allowed up to twenty iterations of the line search, but increased the number of main iterations to ten.

The trip matrix used in these simple examples with the Archerville data was generated by the simplified residence location model described in chapter 2, and is given again in table 6.1. Before describing the results of the algorithm tests, we need to define the terms used in the tables of results. The row headings have the following meanings. 'OBJ 1' is the user-equilibrium objective function [equation (6.10)] which results from the UE criterion as described above. 'OBJ 2' is the total travel cost for the system [equation (6.11)] which results from the SO criterion, also discussed above. 'Unused links' is the number of network links which remain unused, that is, have zero volume, after completion of the trip assignment. 'Mean V/C' is the mean, over all links (including those with zero volume) of the volume–capacity ratio. 'Std dev. V/C' is the standard deviation of the volume–capacity ratios. 'Mean c_{ij}' is the mean, over all origin–destination $(i-j)$ pairs, of the travel time traced over the final loaded (congested) network. 'Mean trip length' is the weighted (by number of trips) mean travel time between all origins and destinations, that is, the mean of the number of trips from each origin to each destination multiplied by the travel time between that origin–destination pair.

Table 6.1. Archerville: trip matrix for the assignment algorithm tests.

Zone	Zone				
	1	2	3	4	5
1	67	51	140	12	3
2	40	99	113	24	4
3	10	10	115	9	2
4	35	75	260	46	9
5	10	18	77	11	11

The results of these tests are given in table 6.2. There are several surprises in these results. First, and perhaps more striking than surprising, the single-pass runs (that is, with 100% increment) of the AON and SM runs yield objective-function values which are much greater than the corresponding

Table 6.2. Archerville: test results for various assignment algorithms.

Term	Algorithm			
	AN—100	AN—40, 40, 20	AN—34, 33, 33	AN—25, 25, 25, 25
OBJ 1	43613	13914	15966	13783
OBJ 2	191389	39145	49428	36368
Unused links	12	6	6	5
Mean V/C	0.908	1.014	1.012	1.102
Std dev. V/C	1.390	0.960	1.018	0.950
Mean c_{ij}	14.0	19.2	17.2	20.9
Mean trip length	20.2	30.4	26.3	29.5
	SM—100	SM—40, 40, 20	SM—34, 33, 33	SM—25, 25, 25, 25
$\theta = 0.2$				
OBJ 1	75651	14161	14576	14255
OBJ 2	349706	36838	39159	37292
Unused links	0	3	3	3
Mean V/C	1.079	1.140	1.153	1.153
Std dev. V/C	1.403	0.924	0.929	0.917
Mean c_{ij}	13.7	23.7	23.5	23.7
Mean trip length	19.4	34.3	33.2	33.8
	SM—100	SM—40, 40, 20	SM—34, 33, 33	SM—25, 25, 25, 25
$\theta = 0.5$				
OBJ 1	65054	13473	14214	13830
OBJ 2	297275	36421	38276	36692
Unused links	0	4	4	4
Mean V/C	1.042	1.050	1.111	1.111
Std dev. V/C	1.356	0.936	0.933	0.923
Mean c_{ij}	13.1	19.6	20.4	21.4
Mean trip length	18.6	30.7	28.6	30.8
	UE—20/3	UE—20/10		
OBJ 1	14709	13695		
OBJ 2	40035	35747		
Unused links	4	4		
Mean V/C	1.135	1.106		
Std dev. V/C	0.954	0.951		
Mean c_{ij}	22.5	23.0		
Mean trip length	32.5	35.5		

Note: AN, all-or-nothing; SM, stochastic multipath; UE, user equilibrium; SO, system optimization, for definitions of the terms, see text, page 127.

multiple increment runs. Second, the two SM—100 runs give poorer results (that is, higher values of OBJ 1 and OBJ 2) than the AN—100 run. The multiple increment SM runs give results which are only marginally different from the multiple increment AON runs. Last, the UE runs do not give the best results in terms of the OBJ 1 measure, but are in fact worse than eight out of the nine other multi-increment runs.

The summary results of table 6.2 offer no clue as to the source of the problem. The mean volume–capacity ratios of the various runs are not strikingly different, nor are any of the other measures. As an aside, note that in the SM runs the value of θ determines the amount of dispersion of path choices away from the minimum path. The greater the value of θ, the less the dispersion. The runs with θ equal to 0.5, which should yield less dispersion, do give lower OBJ 1 and OBJ 2 values than do the corresponding runs for θ equal to 0.2. Nonetheless, these results are counter to what was expected. Only by looking at the individual link volumes and volume–capacity ratios can this seeming paradox be understood.

Figures 6.1, 6.2, and 6.3 show the volume–capacity ratios which exceed 2.0, for three selected assignment algorithm tests. The rather high ratios on the direct links from zone 4 to zone 3 are the cause of the peculiar results from these assignment algorithm tests. When a volume–capacity ratio exceeds 5, the resulting link cost or time is equal to more that 100 times the original, uncongested link cost or time. For ratios in the order of 3, the congested link cost or time is equal to more than 10 times the uncongested value. The consequence of these facts is that the measures of algorithm performance are completely dominated by these few highly congested links.

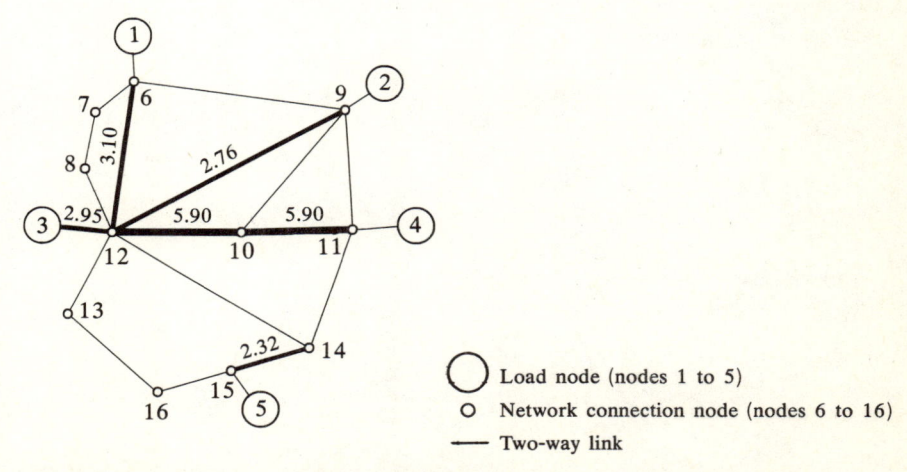

Figure 6.1. Archerville: volume–capacity ratios greater than 2.00, with use of 'original' link capacities for AN—100 algorithm.

The AN—100 run, shown in figure 6.1, produces more links with volume-capacity ratios in excess of 2 than does the run of SM—100 ($\theta = 0.2$) shown in figure 6.2. However, the SM—100 run shows extremely high ratios on two of its three highly congested links owing to the fact that some trips are allocated even to the longest paths (note: a volume-capacity ratio just under 7 yields a congestion multiplier greater than 300). Thus even though the SM—100 runs disperse the trips over more of the network links, their performances, as measured by OBJ 1 and OBJ 2, is worse than that of AN—100. This is a direct consequence of the form of the volume-delay

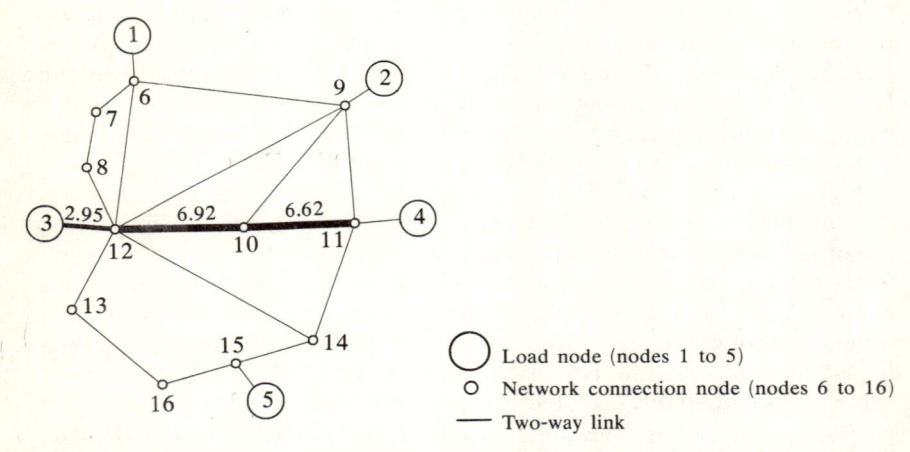

Figure 6.2. Archerville: volume-capacity ratios greater than 2.00, with use of 'original' link capacities for SM—100 algorithm (with $\theta = 0.02$).

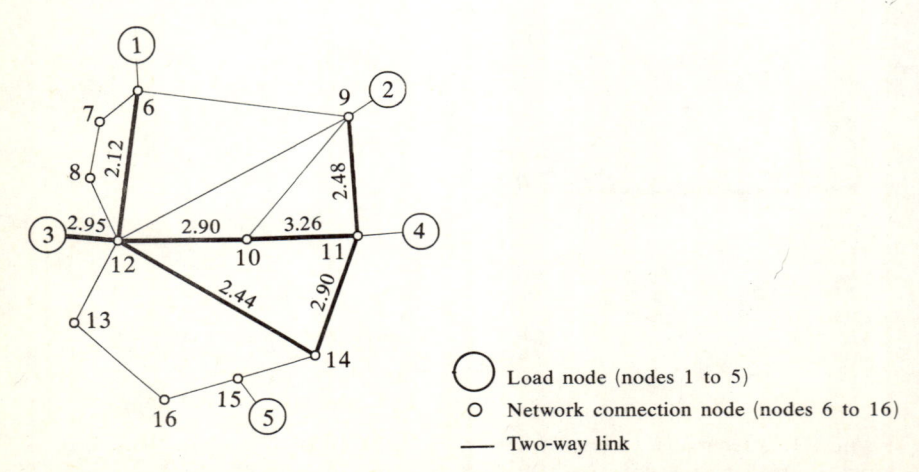

Figure 6.3. Archerville: volume-capacity ratios greater than 2.00, with use of 'original' link capacities for UE—20, 3 increments.

function used, of the network configuration, and of the particular trip matrix being assigned.

The UE—20/3 run shows, in figure 6.3, a substantial improvement over the AN—100 and both SM—100 runs. This is for the simple reason that during its successive iterations the UE algorithm reassigns trips away from the most heavily congested links. In doing this, however, it produces a higher number of moderately congested links. Note, however, that all the incremental assignment runs, both AN and SM, also show substantial improvement over the AN—100 and SM—100 runs. This, too, is because of the reduced numbers of trips on the most heavily congested links. In these algorithms, of course, this end is not accomplished by reassignment, but rather by the fact of the incremental loadings. Thus after the problematic links first begin to be heavily congested they are not used in subsequent incremental assignments, and the additional trips are put on alternate, now minimum, paths.

Though it appeared clear that the Archerville network was not suitable for definitive tests of the assignment algorithms, one further set of tests was done in order to verify the above interpretation. The link capacities of the problem links were increased from the original value of 50 traffic units to a new value of 200 units. The results of this change in link capacity on the performance of the assignment algorithms are shown in table 6.3. Here things are much more as might be expected. AN—100 gives the poorest results, and UE—20/3 gives the best results. As expected, the SM results are in the middle, with the SM result for θ equal to 0.5 (less dispersion off the minimum paths) being better than that for a value of θ equal to 0.2. Clearly these experiments needed to be repeated with larger networks. In the next section of this chapter several sets of experiments with larger networks are described.

Table 6.3. Archerville: test results for various assignment algorithms with increased capacities on problem links.

Term	Algorithm			
	AN—100	UE—20/3	SM—100	SM—100
OBJ 1	11490	8794	9228	8885
OBJ 2	30408	16685	17644	16715
Unused links	11	4	0	0
Mean V/C	0.684	0.795	0.818	0.790
Std dev. V/C	0.841	0.656	0.635	0.623
Mean c_{ij}	12.8	13.0	13.4	12.7
Mean trip length	16.6	17.8	18.2	17.3
θ	–	–	0.2	0.5

Note: see table 6.2.

6.7 Numerical experiments with assignment algorithms: larger networks

The first of the larger networks to be used for these new tests was a network, still rather highly aggregated, representing the highway system in the nine-county metropolitan region of San Francisco. This network will be referred to here as SF30. There were thirty load nodes (that is origins or destinations) and 1592 one-way links. A set of seven different trip-assignment procedures was run, differing slightly from the Archerville tests given above, with the hope of providing a better range of test coverage.

The trip matrix used in these experiments was produced by the residential location model (DRAM), which is used in ITLUP to forecast residence location and trip distribution simultaneously (Putman, 1983a). The matrix itself is multiplied by a scalar to adjust it so that the average impedance on the highway network, after assignment, matches that of exogenously provided zone-to-zone impedance data. This adjustment is a crude form of system calibration and will be discussed further in chapter 7. The first assignment algorithm test results are given in table 6.4.

The results from this first set of trip-assignment experiments for larger networks show that from the worst (highest) value of OBJ 1, produced by AN—1 increment, to the best (lowest) value of OBJ 1, produced by SM—3 increments, there is only a 7.7% difference. For OBJ 2 the AN—1 increment is again the worst, with SM—3 increments again being the best. The difference between best and worst for OBJ 2 is 18.2%. Interestingly enough the UE algorithm is third best on OBJ 1 and second best on OBJ 2. Note that two results are again given for the UE algorithm, with the difference being in the number of iterations permitted in its use of the Frank–Wolfe algorithm. The run UE—20/3 allows up to twenty search iterations within each of three main iterations, whereas in UE—20/10 there are up to twenty search iterations within each of ten main iterations. It is worth mentioning that after the second main iteration the algorithm shows an improvement in OBJ 1 of less than 0.1% per iteration.

Some of the other results in this first series of tests with a larger network should be commented on. In the SM algorithm, as θ approaches 0, all reasonable paths between a given origin–destination pair become equally likely to be chosen despite variation in their lengths. This shows up in the results for SM—θ = 0.1 which has the fewest unused links of any of the test runs. The lowest mean volume–capacity ratio is for the SM—3 increments run, with AN—1 increment and both the UE runs all being tied for 'last place' according to this measure. According to the mean c_{ij} measure the UE—20/10 run produces the best result, with SM—3 increments being second best. Overall these first tests with a larger network produce one major conclusion, which is a negative one. The UE run results are *not* clearly superior to the other algorithms tested.

The second set of tests with a larger network was done with the volumes of trips in the original trip matrix increased by 50% to see whether with increasing network congestion there will be more significant differences in performance for the different algorithms. According to the OBJ 1 criterion (user equilibrium) SM—3 increments was the best procedure, with UE—20/10 being 3.5% worse. According to the OBJ 2 criterion (system optimization), UE—20/10 was best, with SM—3 increments being second best and 1.2% worse. The other measures of performance also gave mixed results. These tests are given in table 6.5.

In the third set of tests with a larger network, again a scaled up version of the original trip matrix was used. In this case the trip volumes from the original matrix were increased by 100%. Again, the procedure termed SM—3 increments was the best procedure as measured by OBJ 1. In these tests, however, UE—20/10 was third best, being 7.2% worse, with AN—3 increments being between, and 5.3% worse than, SM—3 increments. As measured by OBJ 2, the SM—3 increments results were best, followed by

Table 6.4. San Francisco: test results for various assignment algorithms with use of SF30 base-level trip matrix.

Term	Algorithm		
	AN—1 increment	AN—3 increments (40%, 40%, 20%)	
OBJ 1	64 708 032	61 405 056	
OBJ 2	84 975 560	72 201 440	
Unused links	526	561	
Mean V/C	0.317	0.310	
Std dev. V/C	0.445	0.422	
Mean c_{ij}	688.3	686.4	
Mean trip length	405.5	405.4	
	SM—θ = 0.1	SM—θ = 0.5	SM—θ = 0.5, 3 increments
OBJ 1	62 220 304	62 337 888	60 562 976
OBJ 2	78 534 200	79 546 680	69 527 960
Unused links	463	520	543
Mean V/C	0.320	0.311	0.305
Std dev. V/C	0.438	0.436	0.416
Mean c_{ij}	689.5	687.3	684.9
Mean trip length	404.9	404.7	404.2
	UE—20/3 iterations	UE—20/10 iterations	
OBJ 1	62 081 744	61 876 256	
OBJ 2	70 591 200	70 089 040	
Unused links	476	492	
Mean V/C	0.317	0.317	
Std dev. V/C	0.408	0.408	
Mean c_{ij}	683.7	682.8	
Mean trip length	402.2	401.8	

Note: see table 6.2.

UE—20/10 which was 5.6% worse and by AN—3 increments which was 8.2% worse. In this case SM—3 increments was best as measured by all criteria. The results of this test series are given in table 6.6.

Thus, the conclusion from this set of three series of tests with a larger network is rather surprising. With increasing overall levels of network congestion the stochastic multipath algorithm, run in three increments of 40%, 40%, and 20% of the total trips, emerges as the best method. The user-equilibrium method is second best as measured by many criteria and is only marginally better than a simple all-or-nothing (tree-by-tree) algorithm, run in three increments of 40%, 40%, and 20%, when the congestion levels are at their highest for the three tests. Because these results were not what the literature would lead one to expect, a second series of tests was conducted for a different data set on a large network. These are described below.

Table 6.5. San Francisco: test results for various assignment algorithms with use of SF30 1.5 × base-level trip matrix.

Term	Algorithm	
	AN—1 increment	AN—3 increments
OBJ 1	128 322 608	102 176 384
OBJ 2	274 805 720	145 397 200
Unused links	494	413
Mean V/C	0.501	0.494
Std dev. V/C	0.662	0.582
Mean c_{ij}	803.5	794.7
Mean trip length	504.2	509.6
	SM—$\theta = 0.5$	SM—$\theta = 0.5$, 3 increments
OBJ 1	120 541 520	99 191 056
OBJ 2	245 434 400	140 448 040
Unused links	474	415
Mean V/C	0.490	0.481
Std dev. V/C	0.646	0.572
Mean c_{ij}	795.7	788.3
Mean trip length	500.5	507.4
	UE—20/3 iterations	UE—20/10 iterations
OBJ 1	104 197 600	102 681 232
OBJ 2	141 488 000	138 776 000
Unused links	358	364
Mean V/C	0.521	0.510
Std dev. V/C	0.563	0.557
Mean c_{ij}	784.8	782.9
Mean trip length	504.3	500.6

Note: see table 6.2.

The second data set used in these examinations of a larger network was for the Minneapolis–St Paul metropolitan area. This network, too, had thirty load nodes (MN30), though with fewer one-way links (1100). Again, the trip matrix was produced by the residential location model (DRAM) from ITLUP. As before, this matrix was multiplied by a scalar to adjust it so that the average impedance, after trip assignment, matched that of an exogenously provided zone-to-zone travel time matrix. A series of simulation runs was done for these Minneapolis–St Paul data to see whether the results would match those obtained for the San Francisco data. The number of separate runs made was somewhat reduced, as only a comparison of 'final' runs was necessary here.

The results from assigning the original trip matrix to the network according to the three different algorithms are given in table 6.7. Regardless of which of the two objective functions [equation (6.10) or equation (6.11)] is used as the criterion, the SM—3 increments procedure produced the best results.

Table 6.6. San Francisco: test results for various assignment algorithms with use of SF30 2.0 × base-level trip matrix.

Term	Algorithm	
	AN—1 increment	AN—3 increments
OBJ 1	200 748 960	161 696 960
OBJ 2	496 683 200	304 398 400
Unused links	425	345
Mean V/C	0.710	0.703
Std dev. V/C	0.835	0.745
Mean c_{ij}	990.9	10005.4
Mean trip length	707.4	732.5
	SM—$\theta = 0.5$	SM—$\theta = 0.5$, 3 increments
OBJ 1	179 505 808	153 457 072
OBJ 2	404 721 040	281 390 246
Unused links	396	351
Mean V/C	0.692	0.678
Std dev. V/C	0.796	0.729
Mean c_{ij}	975.6	977.2
Mean trip length	701.9	709.9
	UE—20/3 iterations	UE—20/10 iterations
OBJ 1	168 540 592	164 641 008
OBJ 2	312 344 680	297 251 400
Unused links	290	278
Mean V/C	0.752	0.736
Std dev. V/C	0.729	0.716
Mean c_{ij}	1012.5	998.8
Mean trip length	750.1	738.1

Note: see table 6.2.

Table 6.7. Minneapolis: test results for various assignment algorithms with use of MN30 base-level trip matrix.

Term	Algorithm		
	AN—3 increments	SM—$\theta = 0.5$, 3 increments	UE—20/10 iterations
OBJ 1	27 955 712	25 292 224	35 338 400
OBJ 2	35 271 750	31 717 050	44 874 750
Unused links	484	520	309
Mean V/C	0.265	0.239	0.332
Std dev. V/C	0.411	0.393	0.434
Mean c_{ij}	608.5	602.9	629.1
Mean trip length	612.8	600.9	657.2

Note: see table 6.2.

Table 6.8. Minneapolis: test results for various assignment algorithms with use of MN30 1.5 × base-level trip matrix.

Term	Algorithm	
	AN—1 increment	AN—3 increments
OBJ 1	78 554 304	57 649 456
OBJ 2	185 888 100	106 624 050
Unused links	242	286
Mean V/C	0.524	0.454
Std dev. V/C	0.669	0.585
Mean c_{ij}	796.7	782.6
Mean trip length	985.2	1005.9
	SM—$\theta = 0.5$	SM—$\theta = 0.5$, 3 increments
OBJ 1	67 497 536	49 066 048
OBJ 2	141 243 750	80 868 900
Unused links	234	319
Mean V/C	0.495	0.413
Std dev. V/C	0.628	0.550
Mean c_{ij}	755.5	724.1
Mean trip length	882.9	853.1
	UE—20/10 iterations	
OBJ 1	66 633 248	
OBJ 2	118 913 100	
Unused links	175	
Mean V/C	0.533	
Std dev. V/C	0.596	
Mean c_{ij}	841.7	
Mean trip length	1098.1	

Note: see table 6.2.

The AN—3 increments procedure produced the second-best results, and the UE procedure produced the worst results.

In table 6.8 the results from additional test runs, in which trip volumes 1.5 times the base trip matrix were used, are shown. It is quite clear that although the UE procedure performs better than either AN—1 increment or SM—1 increment, it is worse than either AN—3 increments or SM—3 increments. Thus these Minneapolis–St Paul results show nearly the same pattern as the San Francisco results. As such, a new series of questions arises. First, are these results, like those for Archerville, an artifact of the test networks used? The networks are rather highly aggregated, and show considerable variation in congestion levels on individual links. If the results are artifacts of the specific networks, then a set of questions, about the circumstances under which such results can arise, must be answered. If these

Table 6.9. Houston: test results for various assignment algorithms with use of HU98 base-level trip matrix.

Term	Algorithm	
	AN—1 increment	AN—3 increments
OBJ 1	2 774 984 450	1 578 661 890
OBJ 2	2 564 119 990	1 786 316 290
Unused links	442	445
Mean V/C	1.121	1.081
Std dev. V/C	1.387	1.241
Mean c_{ij}	1734.0	1957.2
Mean trip length	2430.9	3020.6
	SM—θ = 0.5	SM—θ = 0.5, 3 increments
OBJ 1	2 030 591 740	949 417 472
OBJ 2	1 764 921 090	1 166 892 030
Unused links	460	491
Mean V/C	0.977	0.877
Std dev. V/C	1.235	1.046
Mean c_{ij}	1433.7	1381.2
Mean trip length	1833.2	1841.2
	UE—20/10 iterations	
OBJ 1	1 032 751 100	
OBJ 2	4 319 887 360	
Unused links	247	
Mean V/C	1.340	
Std dev. V/C	1.262	
Mean c_{ij}	3065.7	
Mean trip length	7742.1	

Note: see table 6.2.

results are not artifacts of the test networks then a set of questions about the performance of these assignment algorithms remains to be addressed. To try to determine whether these results do stem from the particular networks, a further set of test runs was made with more detailed networks.

A data set was available for the eight-county Houston–Galveston area. The network had 98 load nodes and 2514 one-way links (HU98). The trip matrix was not, in this case, generated by DRAM, but was an actual work-trip matrix obtained from the Houston Galveston Area Council, who developed it from Bureau of the Census UTTP data. It was again necessary to scale the trip matrix in order to compensate for the differences between actual on-the-ground highway capacity and the capacity of the abstract network representation used in the modelling experiments. Here, however, the network was allowed to develop rather high levels of congestion in order to give the UE algorithm the possibility of doing a greater number of link flow reassignments. The results of the tests with the Houston network are given in table 6.9. Once again, the same general pattern of performance results appears. According to the UE criterion (OBJ 1), SM—3 increments is best, with UE being second, and AN—3 increments being third. By the SO criterion (OBJ 2), SM—3 increments is still the best method, but now AN—3 increments is second best, and UE gives the worst results of all methods used.

Although it is planned to extend these experiments to additional data sets in the next chapter, it appears unlikely that the main conclusion will change. On a simple comparison, with use of three different larger networks, of UE assignment to SM—3 increments, the best (lowest) value of the UE objective function is always achieved by the SM—3 increments algorithm. In some, but not all, of the tests made, even the AN—3 increments procedure achieved better results than the UE algorithm. In theory, these results should not have been obtained. As they were, several questions are raised. First, are the constraints of equations (6.6), (6.7), and (6.8) being met in all the assignment procedures? The answer to this is yes. Second, is the computer program correct? The answer to this, too, seems to be yes. Third, is there something wrong with the theoretical statement of user equilibrium? The answer to this is, probably not. One conclusion which could then be drawn is that there is some gap between the theory and the practical application. These results strongly suggest that there is further empirical research to be done on traffic assignment algorithms. In the next chapter further exploration of assignment algorithms will be described, with particular emphasis on a combination, in principle, of the logic of stochastic assignment with that of equilibrium assignment. In addition there will be brief discussions of several related matters.

7

Transportation models in optimizing frameworks: 2

7.1 Introduction
In the previous chapter the traffic-assignment problem was introduced. Three general types of algorithm for solving the problem were introduced: (1) all-or-nothing assignment (AON), (2) stochastic multipath assignment (SM), and (3) user equilibrium assignment (UE). Numerical examples were given, along with the results of numerical tests in which larger networks were used. The conclusion was reached that even though the UE approach was, from a conceptual point of view, more likely to give the best assignments of trips to networks, in practice the incremental versions of both AON and SM assignment could often give better results. This was true regardless of whether the UE objective function was used as a criterion of the best, or whether the SO (minimum total transport cost) objective function was used.

In this chapter, after a brief look back at the workings of the UE solution procedure, the concept of stochastic user equilibrium (SUE) is introduced. Again, numerical experiments are reported. This is followed by brief discussions of several important, but lesser, matters regarding trip assignment.

7.2 Properties of the Frank–Wolfe solution to the UE problem
To demonstrate the properties of the Frank–Wolfe algorithm as applied to the UE problem, the SF30 network was run for the same conditions as used to produce the results given in table 6.6. Several different numerical values, as well as measures of the performance of the algorithm for twenty iterations of the Frank–Wolfe algorithm are given in table 7.1.

There are a few points worth noting with regard to the values in table 7.1. First, the value of the NLP objective function in column five is steadily decreasing for all twenty iterations of the Frank–Wolfe algorithm. Note that from iteration 9 to iteration 10 the change in the value of the NLP objective function is only 0.47%. From iteration 10 to iteration 20 the change is only 1.65%. The value of the NLP objective function becomes generally closer to the lower bound with increasing numbers of iterations. The mean percentage flow deviation and λ, the convex-combinations weight, also show a general tendency to decrease over the twenty iterations. An unusual result is that the value of the NLP objective function falls below the lower bound after the sixteenth iteration. Strictly speaking this result is inconsistent with the proper functioning of the Frank–Wolfe algorithm. Careful checking of the computer program showed that this result, where the algorithm is apparently quite near the global NLP optimum, is a result of the use of integer arithmetic and the consequent rounding error in the calculations.

With this closer examination of the Frank–Wolfe algorithm, and with the demonstration given in chapter 6 that the results of the first series of network experiments were not just artifacts of the network configurations, there remains the challenge of developing some resolution of what could be considered their surprising results.

Table 7.1. San Francisco: Frank–Wolfe algorithm for UE assignment, with use of the SF30 network with 2.0 times the base-level trip volumes.

It.	Main	Aux.	LB	New	$\frac{\text{New}-\text{LB}}{\text{New}} \times 100$	λ	MPFD
1	200 748 960	241 951 344	10 801 968	169 627 136	93.6	0.4444	2.47
2	169 627 136	306 177 280	151 759 520	168 708 336	10.0	0.0935	2.48
3	168 708 336	250 848 016	162 334 160	168 540 592	3.7	0.0530	1.63
4	168 540 592	250 688 176	163 926 480	168 357 456	2.6	0.0519	1.92
5	168 357 456	188 722 352	155 732 656	167 218 272	6.9	0.2329	1.79
6	167 218 272	269 180 416	152 840 848	166 159 360	8.0	0.1342	1.56
7	166 159 360	739 030 272	158 138 576	165 820 384	4.6	0.0658	1.74
8	165 820 384	203 755 088	162 772 432	165 645 808	1.7	0.0417	1.50
9	154 645 808	189 906 528	160 861 840	165 419 584	2.8	0.1111	1.58
10	165 419 584	197 090 672	156 488 320	164 641 008	5.0	0.1724	1.82
11	164 641 008	240 501 152	157 256 638	164 234 704	4.2	0.0943	2.40
12	164 234 704	191 185 248	160 798 320	164 037 152	2.0	0.0909	1.62
13	164 037 152	178 767 952	156 348 880	163 298 544	4.2	0.2083	1.43
14	163 298 544	318 652 160	159 646 272	163 216 304	2.2	0.0528	2.04
15	163 216 304	196 477 264	157 365 088	162 977 952	3.4	0.1029	2.14
16	162 977 952	263 115 616	162 718 576	162 900 080	0.1	0.0270	2.94
17	162 900 080	211 487 280	163 507 840	162 836 752	−0.4	0.0137	1.69
18	162 836 752	233 894 112	163 562 000	162 729 392	−0.5	0.0227	2.15
19	162 729 392	223 984 048	164 943 920	162 686 416	−1.4	0.0149	1.61
20	162 686 416	204 191 712	162 469 184	162 659 088	0.1	0.0364	1.87

Note: It., iteration number; Main, starting value of the nonlinear programming (NLP) objective function for that iteration; Aux., value of the auxiliary linearized approximation, linear programming (LP) objective function for that iteration; LB, value of the lower bound; New, new minimum of the NLP objective function along the line between the starting value of the NLP objective function and the LP objective function; λ, relative weighting of the LP minimum in the calculation of the convex combinations; MPFD, mean percentage flow deviation for all links that had flows.

7.3 The stochastic user equilibrium (SUE) problem

In terms of underlying theory, perhaps the most interesting approach to the traffic-assignment problem is that of stochastic user equilibrium (SUE). The techniques discussed in chapter 6 make the basic assumption that all tripmakers, in travelling from origin i to destination j, are attempting to maximize some personal utility measure. It is virtually always assumed that this is identically equivalent to minimizing some generalized measure of travel cost. In lieu of more complete data, out-of-pocket cost is often used, but this is not of direct concern here. What is important is that in early work on traffic assignment it was generally assumed that minimization of total system travel cost was equivalent to minimization of individual tripmaker's travel costs, and that such a minimum would be an equilibrium solution to the problem. The later work on the UE problem showed that neither of these last two assumptions were correct. An alternative formulation was developed [for example, equations (6.5)–(6.8)] which, when solved, gave both a minimization of travel costs for individual tripmakers and an equilibrium solution.

A difficulty with the UE solution to the problem is the stringency of the three underlying assumptions: all tripmakers (a) perceive the link characteristics identically, (b) have complete information on all alternative paths, and (c) are behaving rationally, that is, choosing the shortest path. It is virtually certain that tripmakers are *not* perceiving the link characteristics identically, and completely certain that they do not have complete information on all alternative paths. Although it may reasonably be assumed that all tripmakers are choosing the shortest paths, it is obvious that these are the paths *perceived* by the tripmakers to be the shortest, and, further, that some tripmakers will be diverted from their perceived shortest paths by noncost considerations such as, say, pleasant scenery.

The SUE formulation is specifically intended to deal with minimizing the tripmakers' perceived shortest paths. In general the formulation of the SUE problem is accomplished by considering the consequences of the UE conditions, as discussed earlier, when the link costs are some probabilistic function of link volumes. Obvious questions arise as to the specific distribution to be used for the perceived link times. One possibility is to assume that the perceived link times are themselves normally distributed which leads, after some manipulation, to a probit model of path choice which may then be incorporated in the SUE formulation (Sheffi and Powell, 1981a). There are, however, rather substantial computational problems associated with probit-based stochastic user equilibrium. In particular, it has proved difficult to formulate an objective function which could be efficiently solved for the SUE conditions. Even approximating procedures involve the use of Monte Carlo methods (for generating randomly distributed link time or cost values) in the course of solving the SUE nonlinear optimization problem.

An alternative approach, which appears to be much more tractable, is to assume a somewhat different shape for the underlying perceived travel-time distribution, which will yield a logit-based path-choice model. Fisk (1980) defines the following objective function:

$$\text{minimize } Z(x) = \frac{1}{\theta} \sum_{r,s,p} x_p^{rs} \ln x_p^{rs} + \sum_{\ell} \int_0^{x_\ell} t_\ell(x_\ell) \, dx \,, \tag{7.1}$$

subject to

$$\sum_p x_p^{rs} = q^{rs} \,, \tag{7.2}$$

$$x_p^{rs} \geq 0 \,, \tag{7.3}$$

$$x_\ell = \sum_{r,s,p} x_p^{rs} \delta_{\ell,p}^{rs} \,. \tag{7.4}$$

Note that these constraints are identical to equations (6.6)–(6.8), thus the only difference between the UE problem and this formulation of the SUE problem is the addition of the first term in the SUE objective function to account for the logit model of path choice. This, however, is an important difference in that it puts path flows, x_p^{rs}, in the objective function whereas previously there were only link flows x_t. Note that it is necessary to define

$$x_p^{rs} \ln x_p^{rs} = 0, \quad \text{when } x_p^{rs} = 0,$$

in order to avoid a discontinuity in the objective function owing to the fact that ln 0 is indeterminate. Also note that θ is the dispersion parameter from the logit-based path-choice model. When θ becomes extremely large, the SUE objective function becomes equivalent to the UE objective function, as the first term in equation (7.1) approaches zero. As the value of θ approaches 0 the first term in equation (7.1) becomes very large, dominating the objective function and implying that path choice is independent of path cost.

These notions can be better understood with a glance back at the logit model of path choice, given in equation (6.1) and revised and repeated here for convenience:

$$p_{rs}^p = \exp(-\theta c_{rs}^p) \bigg/ \sum_n \exp(-\theta c_{rs}^p), \qquad (7.5)$$

where
p_{rs}^p is the probability that a trip from origin r to destination s will use path p,
c_{rs}^p is the cost of travelling path p between origin r and destination s,
θ is a parameter.

The parameter θ determines the dispersion of the trips away from the shortest path. It can be interpreted as a parameter which scales the importance (to the tripmaker) of the path travel costs. When θ is zero then all paths are equally likely to be chosen, regardless of their costs. When θ is small, but not zero, then travellers have widely varying perceptions of travel costs and many will choose paths which have greater than the minimum cost. When θ is large most travellers will take the shortest paths, as the perceived travel costs will not vary much around the minimum travel cost. Last, as should by now be clear, the θ in equation (7.5) is the same θ which appears in equation (7.1) and thus all these qualitative interpretations of its effect on path choice are consistent with its effects on the SUE objective function.

Before we move on to the problems of solving the SUE problem, there are some comments to be made on the logit model of path choice. The major matter has to do with what is known as the independence-of-irrelevant-alternatives (IIA) property of the logit choice model in general. The practical consequence of this property may best be seen in an example frequently used in the literature (Daganzo and Sheffi, 1977; Sheffi, 1979; Van Vliet, 1976). The example is one of 'parallel' paths as shown in the sample network in figure 7.1. Note that the cost of travelling path AB is 1, the cost of path CB is x, and the costs of both paths from A to C are $1-x$. In this example, we consider trips from A to B. When $x = 0$, there are three 'parallel', equal-cost paths, and equation (7.5) would put a third of the total trips from A to B on each of the three paths. When $x = 1$ there are, in effect, two 'parallel', equal-cost paths and equation (7.5) should split the trips evenly between them. But, if three paths are enumerated, even though the AC portions of the two lower paths are both of zero length, equation (7.5) will put a third of the trips on the upper (AB) path and two thirds of the trips on the two lower (ACB) paths. Stated in other words, equation (7.5) is quite sensitive, in an unreasonable way, to the specific definition or configuration of the network to which the trips are being assigned, that is, to the network topology.

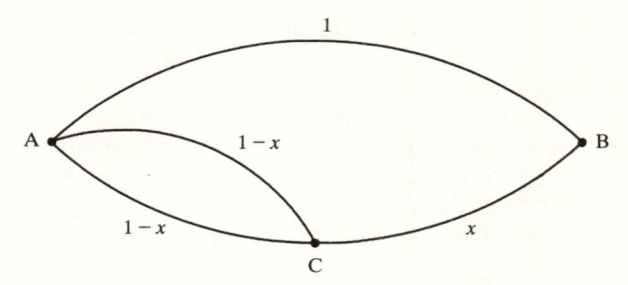

Figure 7.1. An example of a network with 'parallel' paths. Travel costs are shown beside each route.

Other stochastic assignment algorithms have been proposed and are discussed by Dial (1971), Sheffi (1979), and Van Vliet (1976). All of them tacitly assume that link costs are independent of link flows even though it is perfectly obvious that this is not the case. Van Vliet (1976, page 155), citing Murchland, suggests that "one might argue that an assignment method which relates route choice to link costs and link costs in turn to capacity but does not explicitly include capacity as a factor in route choice is inherently unstable". This leads back to the SUE problem and the work of Fisk (1980).

Again, the network configuration of figure 7.1 is used, but now the costs on each link are made a function of the flows. It now becomes possible to use a number of methods of trip assignment to the network of figure 7.1. Sheffi has done two versions, logit path choice and probit path choice, both with *link costs independent of link flows* (1985), figure 7.2 gives a plot of the probability that a trip from origin A to destination B will take the direct AB path. The probability is shown (expressed as a proportion) on the vertical axis. The horizontal axis gives a measure of the overlap of the parallel links AC along the path ACB. In particular, the horizontal axis shows the value of x in the network of figure 7.1. Obviously as $x \to 0$ we approach a complete overlap of the 'parallel' paths. The long-dash line in figure 7.2 shows the constant value of $p_{AB} = 0.33$ which results from the logit model. The solid line shows the more-or-less exponentially increasing probability resulting from the probit model. Note how the probit model begins with a value of $p_{AB} = 0.33$ when the lower paths are fully overlapping ($x = 0$) and moves to 0.5 when there is, effectively, only one lower path ($x = 1$).

To this graph we now add the results of Fisk's analysis (1980). The short-dash line shows the result from a standard UE assignment, with an approximately linearly increasing probability from $p_{AB} = 0.33$ to $p_{AB} = 0.5$ as x increases from 0 to 1. The dotted line results from the SUE assignment for which the logit model formulation has been used (here *link cost is dependent upon link flows*) and again there is a nearly linear increase from $p_{AB} = 0.33$ to $p_{AB} = 0.44$ as x goes from 0 to 1. Last, note the vertical line, Δ, from $p_{AB} = 0.33$ to $p_{AB} = 0.49$ at $x = 0.9$, which is the range of p_{AB} for the SUE method as θ varies from 0 to 1. When $\theta = 0$, as discussed above, the logit model assigns trips to paths independent of their costs and thus SUE degenerates to a simple logit-based assignment. When $\theta = 1$ there is no dispersion and the SUE model becomes equivalent to the UE model.

Overall, figure 7.2 is a most informative diagram. Albeit, some of the lines are approximated as not all the points are given in the work by Sheffi or Fisk. Further, the points for Fisk's UE and SUE curves would be somewhat different as a result of different relationships between cost and flow, but overall the results would not change. Fisk's SUE example, except at the

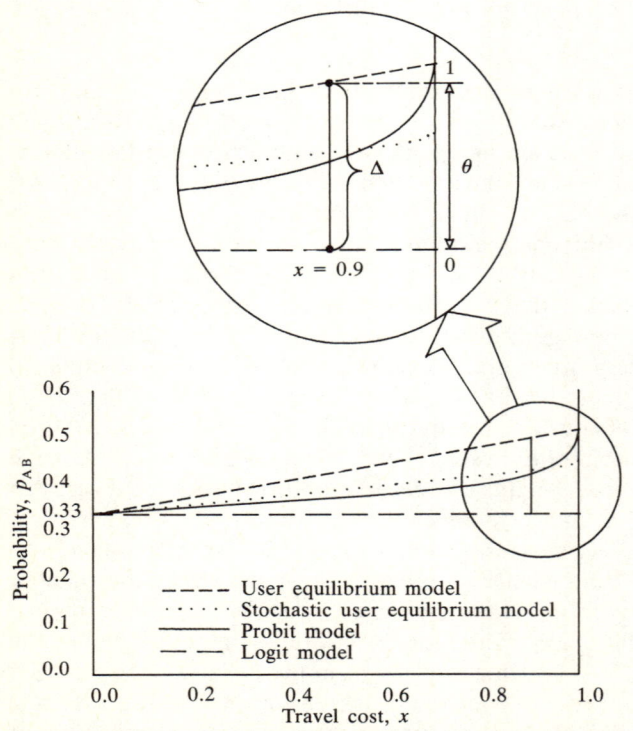

Figure 7.2. The probability, p_{AB}, that a trip from A to B will be along the direct path AB (see figure 7.1).

extreme when $x \to 1$, is a reasonable approximation to the probit assignment procedure. Although further work remains to be done, the implication is that one might do reasonably well using a logit-based SUE assignment procedure when link costs are a function of link flows. A further question as to the frequency with which, in practical application, such fully overlapping paths with equal costs arise also remains to be settled. Fisk contends that, in general, if link costs are a function of link flows then the SUE model, with logit-based path choice, no longer suffers from the IIA problem.

Some final comments are in order here regarding SUE versus UE generally. The major point is that it is stated in the literature that SUE or SM are to be preferred to UE when congestion levels are low (Daganzo and Sheffi, 1977; Van Vliet, 1976). Such a situation will tend to be like the situation where link costs are independent of link flows and thus UE assignment, in effect, produces the same results as AON (all-or-nothing) assignment. Few practical situations are so simple. In the example shown in figure 7.1, AON assignment would put all trips on the shortest path. If all paths were of equal length some special rule would be needed to determine the procedure to follow. The rule could be to divide the trips equally among equal length paths, but one could easily imagine a path AB with a cost of 1.00 and paths ACB with costs of 1.01, which would put all trips on path AB if the AON algorithm were to be used. A more realistic comparison should be made, but it must take the value of θ into account as well. Clearly as congestion decreases, and thus as link flows have less effect on link costs, UE approaches a simple AON assignment. This corresponds to SM when θ is at or near 0. On the other hand the SM results will be quite different from those of the AON algorithm, when θ is at or near 1. Properly stated, we find that for a fixed value of θ the SUE solution will tend to approach the UE solution as congestion increases.

To cover this point in more detail, several items should be noted. First, as stated, when a network becomes more congested the equilibrium flow patterns for UE and probit-based SUE become increasingly similar (Sheffi and Powell, 1982). In Sheffi and Powell's analysis this phenomenon appears rather rapidly. When travel times were only 20% (on average) greater than free-flow times, the flows for SUE and UE were within 10% of each other. Although it must be borne in mind that these results were for a small sample network of two load nodes and thirty-four links, a numerical substitution of a 20% increase in travel time into the standard volume–capacity function of equation (6.9) yields a corresponding average volume–capacity ratio of only 1.074. It has also been noted for the same test network (Sheffi and Powell, 1981a) that as $\theta \to 0$ then the SUE flows become increasingly similar to the UE flows. For a given value of θ, at high levels of congestion there will be large differences between the optimum and nonoptimum paths; thus the trip-makers will, presumably, be better able to perceive the differences. This should lead to less dispersion off the optimum path and a solution more closely approximating the UE flows, and is consistent with the analyses above.

7.4 Solution procedures for the SUE problem

In general the solution of the SUE problem is complicated by the presence of path flows in the objective function. The use of a logit-based path-choice function as was discussed above, at least allows the writing of a deterministic objective function as in equation (7.1). The use of other probability distributions would result in the first term of equation (7.1) being replaced by an expected-value function and would force the use of a Monte Carlo simulation for the evaluation of the objective function. It appears more than likely that the potential improvements in the equilibrium flows would not justify the costs of solving what appear to be intractable computational difficulties. Further, the principle objection to the logit-based SUE, the IIA property, is greatly diminished if not eliminated by using a system in which link costs are a function of link flows (Fisk, 1980). Thus the remaining discussion here will be directed to logit-based SUE with variable link costs.

Even the solution of logit-based SUE presents difficulties. In particular, we are again confronted with the problem of path flows. The great beauty of the Dial (1971) algorithm used in SM assignment is that it obviates path enumeration. If path enumeration were required to solve a particular formulation of the SUE problem, it would be likely that the problem would fall into what is known as the class NP (nondeterministic polynomial time) (Karp, 1982; Lewis and Papadimitriou, 1978). This is of concern for practical purposes as it implies that all algorithms for the SUE problem will require solution times that increase exponentially with the problem complexity, that is, the numbers of nodes and links. Solution procedures not requiring path enumeration require solution times that increase as a polynominal function of problem complexity. Although the computational times for such problems are far from trivial, they are possible where the other class of problems may not be. Given this, and given that there are algorithms for the logit path-choice problem that obviate path enumeration, and given that Fisk has presented a reasonable argument against the existence of the IIA problem when logit choice models are used with flow dependent link costs, it is worth discussing some heuristic approaches to the solution of the logit-based SUE problem.

The heuristic approach to SUE (HSUE) has seen some prior investigation. Sheffi and Powell (1981a) examined three procedures, all of which were modifications of the convex-combinations approach discussed above. There were two differences from that already described. One difference was common to all three methods, and one was different for each. What was common to each of their analyses was that the actual assignment of flows to the network was according to a probit-based path-choice function. The difference was the way in which the convex combinations were calculated. In particular, the revised flows were previously calculated as follows:

$$X_\ell^{n+1} = X_\ell^n + \alpha(Y_\ell^n - X_\ell^n), \tag{7.6.1}$$

$$= (1.0 - \alpha)X_\ell^n + \alpha Y_\ell^n, \tag{7.6.2}$$

where
X_ℓ^{n+1} is 'new' or revised 'main' flow at iteration $n+1$ on link ℓ,
X_ℓ^n is the 'prior' main flow at iteration n on link ℓ,
Y_ℓ^n is the 'auxiliary' flow at iteration n on link ℓ.

In the usual use of this method the initial set of link flows X_ℓ^n is obtained by performing a simple AON assignment. The link costs are then adjusted to reflect these link volumes. This is followed by another AON assignment which yields the auxiliary flow Y_ℓ^n. These two sets of flows, X_ℓ^n and Y_ℓ^n are then combined, with a weighting α, to yield the next iteration flows X_ℓ^{n+1} as in equation (7.6.2). At issue here, of course, is how the value of α is to be determined.

In the usual solution method for the UE problem, which is to use the Frank–Wolfe algorithm, the value of α must be calculated. The method starts with an initial feasible solution, the set of link flows X_ℓ^n which result from the initial AON assignment. Then a feasible descent direction must be chosen. This choice is based on both the steepness of descent (the rapidity with which the objective function improves) and the distance which can be traversed in this direction without violating the problem constraints. To accomplish this task the algorithm looks for another (auxiliary) feasible solution (in addition to the initial one). In this case the auxiliary solution is simply an AON assignment to the links whose costs have been updated to reflect the flows X_ℓ^n of the initial feasible solution. The auxiliary flows are the Y_ℓ^n mentioned above.

These two sets of flows define the descent direction. The next step is to calculate the step size. This is done by finding the minimum value of the original objective function along the descent direction line. This minimization is done as a function of α and thus the value of α is determined at the minimum of the function along this line. Given α, the convex combination can be made according to equation (7.6.2), to yield a new vector of main flows.

These steps may also be shown in equation form. First, the UE problem is repeated for convenience:

$$\text{minimize } Z(x) = \sum_\ell \int_0^{x_\ell} t(x) \, dx , \tag{7.7}$$

subject to

$$\sum_p x_p^{rs} = q^{rs} \tag{7.8}$$

$$x_p^{rs} \geq 0 \tag{7.9}$$

$$x_\ell = \sum_{r,s,p} x_p^{rs} \delta_p^{rs} . \tag{7.10}$$

If it is assumed that there exists some initial solution set of flows, X^n, then finding the line of maximum descent in the direction of some other feasible solution Y, is a matter of maximizing the gradient of the objective function in that direction. This gives

$$\text{maximize } -\nabla Z(X^n)(Y-X^n)^T. \tag{7.11}$$

Multiplication by -1 converts this to a minimization problem, yielding

$$\text{minimize } \nabla Z(X^n)(Y-X^n)^T. \tag{7.12}$$

Note that at the point $X = X^n$, the gradient $\nabla Z(X^n)$ is constant, so the term $\nabla Z(X^n)(X^n)^T$ may be dropped. Equation (7.12) may be simplified, and then written in terms of partial derivatives:

$$\text{minimize } Z^n(Y) = \nabla Z(X^n) Y^T = \sum_{\ell} \left[\frac{\partial Z(X^n)}{\partial x_\ell}\right] y_\ell, \tag{7.13}$$

where we note that X and Y are vectors, and x_i and y_i are scalars. Now the gradient of the objective function $Z(X^n)$ with respect to the link flows is simply the link-cost (or travel-time) vector, as per equation (7.7):

$$\frac{\partial Z(X^n)}{\partial x_\ell} = t_\ell^n = t_\ell(x_\ell^n). \tag{7.14}$$

So this minimization problem becomes,

$$\text{minimize } Z^n(Y) = \sum_{\ell} t_\ell^n y_\ell, \tag{7.15}$$

subject to

$$\sum_{p} g_p^{rs} = q^{rs}, \tag{7.16}$$

$$g_p^{rs} \geq 0, \tag{7.17}$$

$$y_\ell = \sum_{r,s,p} g_p^{rs} \delta_{\ell,p}^{rs}. \tag{7.18}$$

The three constraint equations are identical to the main problem constraint equations [equations (7.8)–(7.10)] expressed in terms of link flows y_ℓ instead of x_ℓ and path flows g_p^{rs} instead of x_p^{rs}. This problem, equations (7.15)–(7.18), is solved by assigning all trips to the minimum paths, yielding the desired auxiliary flows solution Y^n. As this problem is also linear, and its solution is thus a solution to a linear programming problem, it will always fall at a boundary of the solution space. With the descent direction determined as

$$D^n = Y^n - X^n \tag{7.19}$$

and given that Y^n will be at a boundary, then the optimum flow vector will be between Y^n and X^n. This optimum may be found by solving

$$\text{minimize } Z[X^n + \alpha(Y^n - X^n)], \tag{7.20}$$

subject to

$$0 \leq \alpha \leq 1.$$

This is a line-search problem for which there are many efficient algorithms. Basically, what is involved is, for different values of α, substitution of the bracketed part of equation (7.20) into equation (7.7), until a minimum value of Z as a function of α is found.

In using this method for solving the UE problem, the descent direction can be uniquely determined, because of the pleasant way in which equation (7.12) simplifies to equation (7.15). This simplification *does not* apply to the SUE problem, so several alternatives were tried for the calculation of α as discussed above with respect to equations (7.6.1) and (7.6.2). One way is to use the UE calculation of α even though the main and auxiliary flow vectors are obtained by SM assignment rather than by AON assignment. A second way is to use a predetermined series for α, such as $\alpha = 1/n$ for each of n iterations. A third way would be to set α to a constant. Sheffi and Powell (1981a) tested all three of these approaches.

This third method is equivalent to iterative capacity-constrained assignment as described above, with neither incremental adjustments nor quantal loading. This procedure is notorious for not converging and it lived up to this dubious distinction in the Sheffi and Powell experiments, where it did not converge.

For the second method, use of a predetermined series of values for α (the particular series suggested above) $\alpha = 1/n$, was selected. In this case the vector of main link flows X^n will be, at every iteration, the average of the initial feasible solution and all succeeding auxiliary flow solutions. This algorithm is consequently called the method of successive averages (MSA). For probit-based assignment the MSA algorithm seemed to converge to a solution, and it has been proven that the MSA algorithm does converge for the UE and logit-based SUE problems (Powell and Sheffi, 1982).

The remaining method of calculating α uses the same procedure as is used for the UE problem, minimizing equation (7.15) subject to equations (7.16)-(7.18) and then minimizing equation (7.20). According to Sheffi and Powell this procedure worked just about as well as the MSA. The difficulty is that the descent direction calculated for the UE objective function is not the best direction for the SUE objective function. Thus although one might hope that there will be a general improvement from one iteration to the next, there is no guarantee that this will be so. In fact, their results indicate that the MSA algorithm does produce somewhat better results for the SUE problem than does use of the UE algorithm.

As the results given by Sheffi and Powell were based on a very small sample network it seemed sensible to reexamine these procedures with a somewhat more realistic problem. The SF30 network was used (as described above in section 6.7). The base-level trip matrix was used also, so the following results bear comparison with table 6.4 in the previous chapter. The first of these tests involved use of the UE algorithm to calculate descent direction, but the SM algorithm was used to calculate the initial feasible solution and the auxiliary flows. The SM algorithm, as described in the earlier discussion of the SF30 tests uses the standard link delay function of equation (6.9) to congest the links after trip volumes are assigned. This is done in a tree-by-tree or quantal fashion, the links for each origin-destination pair being congested after the trips are assigned and before the next origin-destination pair is considered. The results from ten iterations of this procedure are given in table 7.2. Note first that for iteration 1, the main flow value of the objective function [the UE objective function as specified in equation (6.10) and referred to in all the prior tables] is identical to the result of SM—1 increment ($\theta = 0.5$) in table 6.4. The value of the new objective function at the tenth iteration is only slightly less (0.25%) than that from SM—3 increments in table 6.4. The entire improvement for the ten iterations is only 3.1%, most of which is achieved in the first few iterations. These results hardly seem worth the substantial increase in computational effort which is required to achieve them.

The second test involved the use of the MSA algorithm for the same problem. Again ten iterations were run, with the results being given in table 7.3. Here we have a slightly better performance. The final value of the objective function is the lowest achieved by all procedures tested on this data set. Yet, here too, the improvement is rather limited, being only 1.06% better than the SM—3 increments result. It is interesting to compare the results of the MSA algorithm with those of the Frank-Wolfe algorithm for the logit-based SUE problem. MSA clearly gives the better result, though the difference between them is less than 1% according to the UE objective-function criterion. The MSA procedure achieves after

Table 7.2. San Francisco: results of SUE assignment with use of the Frank-Wolfe descent direction and the SF30 network with base-level trip volumes.

It.	Main	Aux.	LB	New	$\dfrac{\text{New} - \text{LB}}{\text{New}} \times 100$	λ	MPFD
1	62 337 888	62 463 808	53 251 232	61 021 360	12.7	0.3512	1.90
2	61 021 360	61 012 528	60 076 160	60 849 744	1.3	0.3855	1.14
3	60 849 744	61 344 336	60 136 736	60 674 304	0.9	0.3857	1.88
4	60 674 304	61 100 800	60 236 432	60 633 408	0.7	0.1538	0.75
5	60 633 408	60 497 984	60 387 776	60 493 856	0.2	1.0000	1.36
6	60 493 856	61 542 048	60 006 800	60 427 600	0.7	0.2577	1.21
7	60 427 600	61 068 480	60 653 408	60 430 624	−0.4	0.0177	0.87
8	60 430 624	61 053 200	60 732 128	60 425 664	−0.5	0.0089	1.16
9	60 425 664	61 052 960	60 736 768	60 421 952	−0.5	0.0091	0.78
10	60 421 952	61 056 432	60 748 864	60 413 728	−0.6	0.0092	0.45

OBJ 1	60 413 728
OBJ 2	68 389 328
Unused links	448
Mean V/C	0.313
Std dev. V/C	0.402
Mean c_{ij}	681.5
Mean trip length	400.7

Note: see table 7.1; for definitions of terms, see page 127.

five iterations a criterion which is lower than that of the Frank–Wolfe procedure after ten iterations. It takes the MSA procedure four iterations to achieve a criterion lower than that of SM—3 increments. Evaluated overall, although the MSA procedure for the logit-based SUE does achieve a better final objective function value that SM—3 increments, it requires at least 40% more computational effort. Whether or not it is worth the additional cost is not at all clear. From a practical point of view, probably not. Without a formal conclusion, we turn, before any further network tests, to discussion of several related matters.

Table 7.3. San Francisco: results of SUE assignment with use of the method of successive averages (MSA) descent direction and the SF30 network with base-level trip volumes.

It.	Main	Aux.	LB	New	$\frac{\text{New}-\text{LB}}{\text{New}} \times 100$	λ	MPFD
1	62337888	62463808	53251232	62463808	14.7	1.0000	170.78
2	62463808	63063712	57644464	61047664	5.6	0.5000	101.93
3	61047664	60607600	60022272	60672448	1.1	0.3278	7.59
4	60672448	60202832	59997600	60459408	0.8	0.2447	5.49
5	60459408	60148480	60060912	60307616	0.4	0.1739	2.75
6	60307616	60189712	60136512	60200784	0.1	0.1176	2.23
7	60200784	60137344	60092032	60102656	0.0	0.0714	2.03
8	60102656	60162016	60143472	60042256	−0.2	0.1176	1.32
9	60042256	60150464	60133424	59976816	−0.3	0.0714	1.45
10	59976816	60181136	60164000	59920752	−0.4	0.0625	1.64

OBJ 1   59920752
OBJ 2   67532496
Unused links   449
Mean V/C   0.311
Std dev. V/C   0.400
Mean c_{ij} 679.3
Mean trip length   399.3

Note: see table 7.1; for definitions of terms, see page 127.

7.5 Congestion functions

Throughout the above experiments the standard BPR (Bureau of Public Roads) volume–capacity function shown in equation (6.9) was used whenever link costs (or times) were dependent upon link flows (or volumes). As most of the interesting uses and questions regarding traffic assignment involve flow-dependent link characteristics, the function relating these characteristics to the flows is clearly a critical determinant of the results of the analyses. Oddly enough there has been rather little recent research on these functions. The work by Branston (1976) cited earlier is a good overall survey.

The BPR function is an almost wholly empirical construct, at least two aspects of which should give the user pause. First, there is the question of the meaning of the 'free-flow' travel time $t_\ell^{(0)}$ and the 'design' or 'practical capacity' k_ℓ' for each link. Second, there is the question of both the meaning and the derivation of the function parameters α and β. As has already been pointed out (Boyce et al, 1981) the definition of capacity is ambiguous. This ambiguity results principally from a practical inability to determine the vehicle capacity of a length of roadway. Although there are some guidelines for such determinations (HRB, 1965), they are far from definitive. The difficulties with the function parameters result both from the virtual impossibility of obtaining, by any statistically reliable method, link-specific values, and from the general inability to represent different classes of roadway facility. Thus rather general approximations of the values are, of necessity, customarily accepted. The value of β is almost universally taken to be 4, yet only a few simple numerical experiments were necessary to show that variation in the value of β will produce noticeable changes in trip-assignment results (Ellerman and Gibbons, 1984). In an AN—3 increment assignment procedure the reduction of β can have the curious result of increasing the final value of the SO criterion of total network travel cost. This results from it taking, in effect, a higher link volume on congested links before trips are diverted to less congested links. This higher allowable volume results in more trips being on congested links and thus the product $t_\ell x_\ell$ becomes larger. In addition, there is an increase in the number of unused links in the trip assignment. The effects on UE assignment are somewhat less definitive, but similar.

An alternative volume–capacity function was proposed by Davidson (1978). This function has the following form:

$$t_\ell = t_\ell^{(0)}\left(1 + J_\ell \frac{x_\ell}{x_\ell^{\text{sat}} - x_\ell}\right), \qquad (7.21)$$

where, as before,

t_ℓ is the 'congested' travel time over link ℓ,
$t_\ell^{(0)}$ is the travel time for free-flow, that is, the uncongested travel time,
x_ℓ is the flow (volume) over link ℓ,
and, in the case of this function,
J_ℓ is a 'delay' parameter,
x_ℓ^{sat} is the 'saturation' or maximum link flow for link ℓ.

The value of J_ℓ is to be estimated empirically, and is capable of accounting for different types of roadway (for example, freeway or arterial) and for different locations in the urban area (for example, urban core or rural). A difficulty with this function is that for values of $x_\ell \geq x_\ell^{\text{sat}}$ it is undefined. A suggestion as to modification of the function in order to avoid this discontinuity has been made by Akcelik (1978).

Some tests have been done which evaluate the effect of using alternate volume–capacity functions (Boyce et al, 1981; Taylor, 1984). The experiments by Boyce et al conclude that the Davidson function should only be used with 'extreme caution in any planning application'. More particularly, even though this function resulted in a greater number of divergent paths being used in the assignment procedure, these paths often had unrealistically low trip volumes assigned to them. Further, Boyce et al concluded that, given the inherently unreliable estimates of link capacity and saturation-flow volume, the link travel times and volumes produced from the assignments were not likely to be very reliable either.

Taylor's tests were made with use of Akcelik's modification of Davidson's function, which eliminates the discontinuity when $x_\ell \geqslant x_\ell^{\text{sat}}$, whereas Boyce et al simply truncated the function, limiting the maximum value of the term within the large brackets in equation (7.21) to 10, regardless of any value of x_ℓ being greater than $x_\ell \geqslant x_\ell^{\text{sat}}$. In addition, Taylor also tested the BPR function and a stepwise linear function, as well as briefly discussing procedures for estimating the parameters of these functions. One major conclusion of these investigations was that the modified version of the Davidson function performed much better than the original version of the function as tested by Boyce et al. Perhaps the most important conclusion was that *all three* congestion functions yielded reasonably good assignment results (measured in several ways, including comparison with observed link flows). This is a reassuring result which suggests that use of any of three reasonable approaches to the definition of volume-capacity functions will produce acceptable final results from the UE traffic assignment algorithm.

Note the consequence of arbitrarily limiting the volume-capacity function as was done by Boyce et al. Defining an arbitrary limit on this function has the effect of making the link flow function nonconvex at its limit. As a result the Frank-Wolfe algorithm for solving the UE problem may no longer produce a proper solution. This is not surprising, as in the formulation of the UE problem strict convexity of these functions is assumed. A modified formulation of the UE problem which incorporates link-flow constraints has been proposed and examined by Daganzo (1977b; 1977c). Nonetheless, the Boyce et al result should serve to illustrate that a simple 'fix' to a practical problem can produce unwanted results if it violates the assumptions on which the procedure is based. It further points up the fact that although several different congestion functions did yield good results, there is certainly a need for further research into the formulation and parameter estimation of such functions, as well as on their effects on traffic-assignment procedures.

7.6 Levels of network congestion

Throughout all the above discussions there have been mentions of network congestion. Various authors have suggested that the proper choice of assignment algorithm for a particular application may depend to a considerable degree on the level of congestion on the network. Yet in the new experiments described here as well as in experiments described by other authors, it is clear that 'network congestion' is a rather nebulous concept. This is because any realistic network, even the rather highly aggregated SF30 and MN30 networks used here (and described above), will show a rather spread out frequency distribution of link volume-capacity (V/C) ratios which will not be at all well described simply by stating the mean

V/C ratio of the network. In the SF30 tests described above, a frequency distribution of the link V/C ratios shows more than half of them are below 0.6, with the remainder spread out over a substantial range above 0.6, to the maximum, which was above 2.0.

In the above-mentioned work by Taylor (1984), a network for Melbourne with 50 load nodes and 796 one-way links was used. For the observed link flows the mean V/C ratio was 0.32, whereas in the various assignment algorithm tests the mean V/C ratio was about 0.44. The standard deviations were 0.22 and 0.32, respectively. Only about 4% of the links in the network were assigned sufficient volume for their V/C ratios to exceed 1.0. In the SF30 experiments, in which twice the base-level trip volumes were used, depending upon which assignment algorithm was used, the mean V/C ratio ranged from 0.68 to 0.75, with standard deviations of roughly the same magnitude. In the MN30 experiments, with twice the base-level trip volumes, the mean V/C ratio ranged from 0.88 to 1.34 with standard deviations slightly greater. In the HU98 experiments the mean V/C ratios ranged from 0.88 to 1.34 again with standard deviations of approximately the same values. In all cases the overwhelming majority of the links had assigned volumes which yielded V/C ratios less than 1.0.

In the experiments whose results are given here and in chapter 6, the SM−3 increments procedure, though not performing quite as well as the two SUE procedures, gave better results for the UE objective function than did any of the other assignment procedures tested. This result is contrary to what much of the literature would lead one to expect. It probably is a consequence of the fact that even in a highly aggregated network, which nonetheless attempts to replicate reality, there will be only a few links with high V/C ratios. This may be a consequence of the extreme nonlinearity of the link-congestion function. When the SM assignment procedure disperses trips away from a very congested link on the shortest path to less congested links (even though they are on longer paths) there will be a consequent net decrease in the UE objective function. This is despite the fact that in all the tests done here, excepting those of the two SUE procedures, the UE algorithm assignments result in the lowest numbers of unused links. What is happening here is that although the UE procedure is using more links, it is reassigning less trip volume off highly congested links than is accomplished by dispersion in the SM assignments. As an aside, it is interesting to note that for a given SF30 trip matrix the simple correlation between link volumes resulting from any one of the assignment algorithms tested, as compared with link volumes resulting from any other of the algorithms, is always above 0.975. There is a clear need for further examination here in order to develop a better understanding of the relationship between the frequency distribution of the link V/C ratios and the performance of the assignment algorithms.

It is appropriate to put in a word here about the magnitude of the multiplicative effect of the volume−capacity function. Take the usual parameters of $\alpha = 0.15$ and $\beta = 4$ and consider the term inside the large brackets of equation (6.9). This term acts as a multiplier for $t_\ell^{(0)}$. Tabulated below are the values of this term for increasing values of the x_ℓ-to-k_ℓ' ratio.

x_ℓ/k_ℓ'	Multiplier	x_ℓ/k_ℓ'	Multiplier
1.0	1.15	3.5	23.51
1.5	1.76	4.0	39.40
2.0	3.40	4.5	62.51
2.5	6.86	5.0	94.75
3.0	13.15		

One has the sense from the literature that, for the most part, these multipliers should not get too large in a run of a network assignment algorithm. In fact, if one considers a network link where $t_\ell^{(0)}$ is 5 minutes, the resulting link time when the congestion ratio reaches 3 would be 66 minutes, and for a congestion ratio of 4 the time would be 197 minutes. The appearance of such values in a model run almost certainly indicate problems either in the network representation or in the estimated trip volume being assigned. Yet these high values of x_ℓ/k_ℓ' do occasionally show up, especially in UE assignment where each iteration is in effect an AON assignment. There are ways to modify the UE algorithm to limit these flows, the easiest of which is to constrain the value of α in equation (7.20) of the convex-combinations procedure (Daganzo, 1977a). The difficulty in imposing such constraints is that if the flows are constrained the step size and direction in the Frank–Wolfe algorithm are no longer optimal and thus convergence is, at best, retarded and, at worst, may not occur. More practical experience with such constraints is necessary before clear recommendations for practice can be made, but it is known that the computer code for some traffic-assignment algorithms will experience difficulty with multiplier values greater than 10. Thus the occurrence of such high congestion on links being modelled should be cause for careful scrutiny of the input data to any exercise in which they appear.

7.7 The effects of levels of detail
Implicit in the discussion of link-capacity functions is the further question of the level of detail of network representation. In general the problem of level of detail results from the necessity for aggregating the actual 'on-the-ground' network into an abstract representation suitable for use in models. Even for an actual section of roadway the problem of determining capacity is formidable. When several or many roadway segments are aggregated, the problem becomes even more difficult, as one has then to define a capacity for a representation of a bundle of roadway segments.

If the capacity estimates are in error, then the flow-dependent link times or costs will also be in error. Virtually no information exists on the specific response or sensitivity of trip-assignment results to such errors. There has, however, been a series of analyses of the error properties of the components of the trip-assignment process which suggest a direction for continued research (Bovy, 1984; 1985; Bovy and Jansen, 1983a). One important finding of this work was that "rather small unbiased variations in travel times lead to considerable variations in the [link] load estimates" (Bovy and Jansen, 1983a, page 16). In subsequent analyses it was found that variations and/or errors in estimates of link travel times act directly through the mechanism of minimum path tracing on the trip-assignment process. Thus from this point of view one might expect AON assignment, done without incremental or quantal loading techniques, to be most susceptible to the propagation of such errors. With the addition of an incremental process and/or quantal loading (either of which, in essence, allow subsequent trips to shift to paths which were near-optimal prior to congestion, and optimal after other links became congested) the overall procedure should become less sensitive to errors in link travel time. The various stochastic assignment techniques should be even more robust in this regard, as the distribution of trips to multiple paths in proportion to the path lengths should result in a cancelling out of small random errors in estimates of link travel time.

There have been several studies of the effects of aggregation of network descriptions on trip-assignment results (Bovy and Janson, 1983b; 1983c; Eash et al, 1983; Janson and Bovy, 1982; Wildermuth et al, 1972). In all the studies it was found that there were negative effects on the quality of traffic-assignment model outputs as a result of network aggregation. More important perhaps was the finding that although greater detail in network representation did produce more accurate results generally, the relationship between level of detail and accuracy of output was not at all linear. A shift of the analysis from very coarse levels of spatial detail and network representation to a 'middle' level of detail was usually accompanied by substantial improvement in the accuracy of output. A further move from the 'middle' level to a 'fine' level of detail often produced only minor improvements in the accuracy of output.

Eash et al (1983) used data for the Chicago area, and aggregated from 1797 zones, 12040 nodes, and 37065 one-way links, to 317 zones, 820 nodes, and 2422 one-way links. The experiments were made by using the UE assignment algorithm. Even with this quite considerable reduction in level of detail the authors concluded that "overall results from the sketch planning assignment compared reasonably well with the regional assignment, and zone level assignment quantities were well correlated with regional assignment counterparts" (page 27). The level of detail used here corresponds to what is referred to as a 'middle' level of detail. It should be noted that in doing the network aggregation rather careful attention was paid to the capacity determination of the aggregated links. This was done as a function of the characteristics of the detailed-level links which were being collapsed to one link in the aggregated network.

Bovy and Jansen made use of data for Eindhoven, The Netherlands, aggregated from 1286 zones, 4312 nodes, and 12871 one-way links to two different levels. The medium level had 183 zones, 826 nodes, and 2490 one-way links, and the coarse level had 47 zones, 204 nodes, and 544 one-way links. There is quite a lot of uncertainty associated with the decision as to exactly how such aggregations should be done, and there is only limited development of theoretical justification for the possible procedures (Hearn, 1984). In the Bovy and Jansen studies the network aggregation was accomplished by retaining some links of the detailed network and deleting others. Capacities of links retained in the more aggregated network were identical to their capacities in the more detailed network(s). The most general conclusion of this study was that the coarse-level outputs were much less accurate than the medium-level outputs, but the fine-level outputs were only slightly better than the medium-level outputs. In addition it appears, though the authors do not emphasize the point, that if it is necessary to work at the coarse level of detail, the link capacities can be a problem in capacity-constrained assignment (in this case UE was used). This is probably a result of greater trip volumes being assigned to far fewer network links, with no change in link capacities. That these authors had no greater problems than they did is undoubtedly because of the low levels of traffic congestion in the study region overall. Most important, however, is the authors' conclusion that, "it does not make sense to use very fine networks. A medium-level network, consisting of all arterials and collectors, appears to give results which can hardly be improved" (Jansen and Bovy, 1982, page 317).

7.8 System calibration and empirical verification

In the above discussion of the effects of levels of detail, mention was made of the problem of determining link capacity, for individual network links as well as for bundles of links in aggregated networks. In either case this can be a very difficult process. The direct consequence of being systematically wrong about capacities is that regional distributions of link V/C ratios will be too high or too low. One will then be faced with several equally distasteful alternatives: (1) do a link-by-link recalculation of capacities, (2) do a more general scaling of link capacities at a regional or subregional level, or (3) arbitrarily scale the regional trip volumes to compensate for the errors in capacities. The first of these alternatives is to be preferred in theory, but is often too costly in practice. Some combination of the second and third alternatives will usually be taken. Once these adjustments are made some network paths should be traced and the resulting zone-to-zone travel times compared with the best available information on actual travel times.

There has so far been no discussion of the ability of the various assignment algorithms to estimate observed link volumes. The evaluation and discussion given above has been solely in terms of a comparison of the algorithms with themselves. In point of fact, there has been very little work done on the comparison of link volumes, as estimated by the various traffic-assignment algorithms, to actual 'ground counts' of traffic volumes on roadways. In addition to the Wildermuth and the Bovy and Jansen work cited above, three studies report such comparisons (Florian and Nguyen, 1976; Janson et al, 1986; Tobin, 1979). In general all the reports indicate that for those cases where there are large samples of observed link volumes to compare with assigned volumes the comparisons are quite good. Where the samples are smaller, the results are less good, but this may well be owing to these being less-comprehensive and/or more coarse-level analyses. In only one set of studies, those by Bovy and Jansen (Bovy, 1985) is there a comparison of observed link volumes with the results of different assignment algorithms. Here, although UE assignment is somewhat better, it is compared with AON assignment without capacity constraint, a method unlikely to be much used in practice in any case. Thus overall, there is a clear need for a thorough examination of the ability of alternative assignment algorithms to produce reliable estimates of observed link flows.

7.9 A note on the effects of starting points

In the discussions of the mathematical programming formulations of the assignment problem it is always assumed that the algorithm begins with a 'feasible solution'. It should be clear that the closer the initial feasible solution is to the optimum, the more rapidly will that optimum be found (or approximated). This is an issue of considerable importance not only for the assignment problem considered alone, but for combined location and transportation models as well. In all the tests described here the traditional AON assignment procedure in its simplest form was *never* used. Instead a quantal loading procedure was used whereby as each shortest path from a given origin to a given destination was traced, the trips were assigned to the path links and the travel times of those links were recalculated with a volume-capacity function. Although this will not have affected the final results of the algorithm-to-algorithm comparisons, it may well have affected the rates of convergence for the iterative procedures. Although some work has been reported regarding the efficacy of quantal loading within the Frank-Wolfe algorithm for UE traffic assignment, some additional work should be done here to evaluate the efficiency of what is in effect a procedure within an algorithm (Arezki and van Vliet, 1985). For the present it is tacitly assumed that following this method does accelerate convergence of the various programming algorithms.

It should also be noted that researchers have recently identified the presence of cyclic flows in networks to which traffic was assigned by using the Frank-Wolfe procedure for finding the UE solution. Cyclic flows are flows (trips) which originate from (or terminate at) a particular location and pass each other in opposite directions between some pair of network nodes. They are called cyclic because a complete loop (or cycle) can be traced along the network links being used by them. These cyclic flows, if they appear in significant quantities in the initial feasible solution, can cause the

Frank–Wolfe algorithm to proceed very slowly in its approach to the network equilibrium solution (Janson and Zozaya-Gorostiza, 1985). Further work needs to be done to evaluate the effects of such cyclical flows on assignment results and the extent to which the several algorithms are or are not prone to the generation of these flows.

7.10 Final tests of large-scale networks

As a final set of empirical examinations of the various assignment algorithms, two larger networks were analyzed. The first of these, for Houston, TX, had 199 load nodes and 5100 one-way links. The second large network, for Washington, DC, had 182 load nodes and 5272 one-way links. For each of

Table 7.4. Houston: results from the assignment algorithm test with use of the H199 base-level trip matrix.

Term	Algorithm	
	AN—1 increment	AN—3 increments (40%, 40%, 20%)
OBJ 1	211 877 280	192 987 456
OBJ 2	342 578 688	306 960 640
Unused links	249	302
Mean V/C	0.807	0.757
Std dev. V/C	0.759	0.720
Mean c_{ij}	994.9	983.0
Mean trip length	671.5	674.5
CPU time (seconds)[a]	11.32	18.75
	SM—θ = 0.5, 3 increments (40%, 40%, 20%)	
OBJ 1	172 456 384	
OBJ 2	260 469 728	
Unused links	309	
Mean V/C	0.699	
Std dev. V/C	0.684	
Mean c_{ij}	904.8	
Mean trip length	614.3	
CPU time (seconds)[a]	54.65	
	UE—10/5 iterations	SUE—10/5 iterations
OBJ 1	206 615 312	177 656 672
OBJ 2	332 519 680	262 693 824
Unused links	212	252
Mean V/C	0.815	0.727
Std dev. V/C	0.711	0.662
Mean c_{ij}	1013.7	901.7
Mean trip length	691.3	604.6
CPU time (seconds)[a]	34.69	105.17

[a] Computer usage times on an IBM 3090/180E computer.
Note: for definitions of terms see page 127.

these a trip matrix was estimated, and all the various assignment algorithms were tested. In table 7.4 the results are given for Houston, and in table 7.5 they are given for Washington. For both networks, the UE results are *not* the best, in terms of the UE objective function. For the Houston network, the lowest value of the UE objective function is achieved by the SM—3 increment algorithm. For the Washington network, the lowest value of the UE objective function is achieved by the SUE−MSA algorithm. For both networks, the AON—3 increment results are better than the UE results. Again, the question is, 'how can this happen?' It appears that the Frank−Wolfe algorithm is at fault. In the UE assignment runs, the algorithm converges quickly to a certain level of the objective function. Then, many iterations are computed with very little change in objective function. Eventually, additional iterations produce no change in the objective function, and even show slight oscillations, as the effects of round-off error (owing to the use of integer numbers of trips on links) become larger compared to the very small changes in link flows.

Table 7.5. Washington: results of the assignment algorithm test with use of the W182 base-level trip matrix.

Term	Algorithm	
	AN—1 increment	AN—3 increments (40%, 40%, 20%)
OBJ 1	315 978 496	272 372 480
OBJ 2	477 953 024	427 395 584
Unused links	173	169
Mean V/C	0.995	0.942
Std dev. V/C	0.876	0.804
Mean c_{ij}	863.5	878.6
Mean trip length	1029.1	1059.3
	SM—$\theta = 0.5$, 3 increments (40%, 40%, 20%)	
OBJ 1	242 272 864	
OBJ 2	376 476 416	
Unused links	171	
Mean V/C	0.877	
Std dev. V/C	0.769	
Mean c_{ij}	813.9	
Mean trip length	987.7	
	UE—10/5 iterations	SUE—10/5 iterations
OBJ 1	277 907 712	235 866 480
OBJ 2	523 131 904	399 267 584
Unused links	89	108
Mean V/C	1.037	0.942
Std dev. V/C	0.792	0.739
Mean c_{ij}	978.5	820.2
Mean trip length	1185.0	960.2

Note: for definitions of terms see page 127.

Recent research shows that the UE algorithm gives, even though it fails to find the 'true' minimum of its objective function, marginally better estimates of actual link flows than do the other algorithms (Almaani, 1988). Clearly there is a need here for some improvement in the algorithm for calculating the UE assignment. As it is, UE cannot be said to be, unequivocally, the best approach. In practice few users will worry about whether the algorithm has converged. In such cases they might have been better off to have used SM−3 increments or AN−3 increments. An investigation into different ways of starting the UE algorithm might prove fruitful, as well as some procedures for 'breaking-out' of the slow convergence path taken by the Frank–Wolfe algorithm.

7.11 Conclusion

The equilibrium assignment approach seems clearly superior to a traditional all-or-nothing approach. However, as a result of subjecting several new approaches to the traffic-assignment problem to tests with actual data sets (as distinct from contrived small examples), some conclusions arise which are not quite in accord with what current literature would lead one to expect. In particular it appears that stochastic assignment techniques when combined with link volume–capacity functions come closer to a user equilibrium solution faster than a straight forward user equilibrium technique. In addition, heuristic solutions to stochastic user equilibrium techniques appear to be the best of all practical approaches to traffic assignment, though only marginally better than incremental stochastic assignment with capacity constraint. All of these results are based on algorithm-to-algorithm comparison and await future research to confirm the findings when compared with observed network link volumes.

It is clear that in practice the equilibrium assignment methods will yield good estimates of traffic assignment. It is not clear just how much better these will be than those which can be obtained from quantal or incremental tree-by-tree or stochastic multipath assignment. The answer to this question will only come from further research and may, in the final analysis, turn out to be determinable only on a case-by-case basis.

Even so, the development of the equilibrium assignment approach brings an undeniable benefit to the field by virtue of its structuring and formulating the assignment problem. With the problem developed according to this approach its mathematical representation is easily written and analyzed. What is more, in this form it leads directly to a clear depiction of the relationship between trip distribution and traffic assignment and, taking one more step, to the relationship between location, distribution, and traffic assignment. Thus even if the practical consequences are modest, and they may indeed be much more, the consequences in terms of theory development are substantial and more than worth the effort. In the next chapter, use is made of the theory-development aspect of this work in order to begin an examination of combined distribution and assignment models.

8

Simultaneous location–transportation models: 1

8.1 Introduction

Activity location models formulated as mathematical programs were presented in chapters 4 and 5. Models for assigning trips to networks were discussed, in their simplest form, in chapter 3. More sophisticated network assignment models were presented in considerable detail in chapters 6 and 7. In this chapter the location models and assignment models are combined.

The chapter begins with the development of a simple combined location and trip-assignment model. It is then shown that what appears to be a simple joining of two models leads to significant difficulties in both formulation and solution of the combined model. The Archerville data are used to illustrate the difficulties. The resolution of this problem involves dealing with intractable numbers of variables and equations, and thus requires the partitioning and decomposition of the model formulation. Several alternative problem formulations are then explored in an attempt to cope with the issue of problem size. The Archerville data are then used to illustrate the performance of these alternative model formulations and to explore possible modifications and extensions.

8.2 A simple combined location and trip-assignment model

The final form of the location model discussed in chapter 5 was as follows [note these are equations (5.49)–(5.54), and are reproduced here for the reader's convenience]:

$$\text{minimize } S = \frac{1}{\beta} \sum_{i,j,l} S_{ijl} \ln S_{ijl} + \sum_{i,j,l} S_{ijl} c_{ijl} + \sum_{i,j} b_{ij} X_{ij} , \tag{8.1}$$

subject to

$$\sum_{l} S_{ijl} - s_i X_{ij} = 0 , \tag{8.2}$$

$$\sum_{j} S_{ijl} - r_i X_l = 0 , \tag{8.3}$$

$$\sum_{j} X_{ij} = A_i , \tag{8.4}$$

$$\sum_{i} X_{ij} \leq L_j , \tag{8.5}$$

$$S_{ijl} \geq 0 , \quad X_{ij} \geq 0 , \tag{8.6}$$

where
- A_i is the regional total of activity i,
- b_{ij} is the unit cost less the benefit of locating activity i in zone j,
- c_{ijl} is the unit (travel) cost less the benefit of interaction between zone j and zone l for activity i,

s_i is the level of interaction (trips generated) per unit of activity i,
r_i is the number of trips attracted by employment, per unit of activity i,
S_{ijl} is the level of interaction (trips) of activity i between zone j and zone l,
X_{ij} is the amount of activity i to be allocated to zone j,
L_j is the capacity (area available) of zone j.

Recall that the first term in the objective function is the entropy or dispersion term which results from the assumption that the work trips are exponentially distributed according to the doubly-constrained spatial interaction model of equations (5.40)–(5.42). If the value of β, which can be considered to be a travel-deterrence or trip-dispersion parameter, becomes very large, then this term of the objective function vanishes. Conversely, when β approaches 0 the dispersion term becomes very large and dominates the objective function.

The second term in the objective function is simply the transport-cost term, the numbers of trips multiplied by their individual costs. It is important to note that the cost is assumed to be the minimum cost (path) from zone j to zone l, along the path taken through the network which connects them. The third and final term in the objective function is simply the location-cost term, being the product of the amount of locating activity multiplied by its zone-specific location costs.

The first two constraints, equations (8.2) and (8.3), are the trip-end constraints. The first constraint requires that the sum over all destinations l of all trips originating at (that is produced at) j, must equal the total number of trips produced (or generated) at j. The second constraint requires that the sum over all origins j of all trips terminating at (that is, attracted to) l, must equal the total number of trips attracted to l.

Now recall the minimum-cost flow model as described in chapter 3. The structure is as follows [note that these are equations (3.8)–(3.10), and are reproduced here for the reader's convenience]:

$$\text{minimize } Z = \sum_{i,j} c_{ij} X_{ij}, \tag{8.7}$$

subject to

$$\sum_j X_{ij} - \sum_k X_{ki} = \begin{cases} O_i, & \text{if } i \text{ is the origin,} \\ -D_i, & \text{if } i \text{ is the destination,} \\ 0, & \text{otherwise,} \end{cases} \tag{8.8}$$

and subject to the nonnegativity constraints

$$X_{ij} \geq 0, \quad \forall i,j, \tag{8.9}$$

where
O_i is the net number of trips leaving node i,
D_i is the net number of trips arriving at node i.
Note: in this chapter, X_{ij} has been substituted for x_{ij}, used in chapter 3 to represent the flow between zones i and j.

It is worth noting that the cost variable in the minimum-cost flow model is link cost, not minimum path cost. Of course, the whole model is cast in terms of link costs and flows. Now, in order to make the minimum-cost flow model consistent with the location model, let the link costs equal d, and the link flows equal F. The node indices are defined as m and n, to distinguish them from the zone indices j and l in the location model. If it is assumed that there are i types of trip (or tripmaker), but that the link costs are independent of trip type, then the objective function of the minimum-cost flow problem becomes

$$\text{minimize } Z = \sum_{i,m,n} d_{mn} F_{imn} \, . \tag{8.10}$$

The principal constraint of the minimum-cost flow problem is the set of flow-balance, or node-balance, equations. These serve to ensure that no trips enter or leave the network except where permitted. The network nodes are considered to be of two types, load nodes and ordinary, nonload (or transshipment) nodes. Trips can get on or off the network only at load nodes. No trips can originate or terminate at the nonload (transshipment) nodes. Thus the constraints require that for nonload nodes the sum of all trips arriving must equal the sum of all trips departing. For load nodes the difference between all trips arriving and all trips departing must equal the trips originating or terminating at the load node. It is customary to assume that if there are M nodes in the network and J zones in the region, that the first J nodes in the network are the load nodes.

In the location model the total trips leaving each zone (load node) equal $s_i X_{ij}$, and the total trips arriving at each zone (load node) equal $r_i E_l$. If it is assumed that each load node is both an origin (trips are generated) and a destination (trips are attracted) then the net flow at each load node is the difference between the trips arriving and the trips leaving. Thus the node-balance constraints may be written:
subject to

$$\sum_{i,n} F_{imn} - \sum_{i,n} F_{inm} = \begin{cases} \sum_i (s_i X_{im} - r_i E_m), & \text{for } m \leq J \, , \\ 0, & \text{for } m > J \, . \end{cases} \tag{8.11}$$

It should, of course, be clear that there is an implicit assumption that no interzonal trip from zone i can have a lower cost than an intrazonal trip in zone i. The consequence of this assumption is that the employment 'needs' of each zone will, in effect, have first claim on the residents in the zone, as they will be the least-transportation-cost workers for the jobs in that zone. If it were to be desired to remove this assumption, the load nodes would have to be redefined so that there was a separate 'origin' node and 'destination' node for each zone. There would also need to be a set of links to represent intrazonal travel costs. Last, the node-balance constraints would have to be of the form given in equation (8.8), rather than that of equation (8.11).

Even so, it is this assumption which, as will soon be shown, is not valid, and which obfuscates the intractable aspects of the computation of solutions to this model formulation.

Note now that the c_{ijl} in the objective function of equation (8.1) is the minimum (path) cost for activity (trips) type i between zones (load nodes) j and l. Thus the second term in the objective function is the minimum cost (path) between nodes j and l, multiplied by the total trips from j to l. In the objective function of equation (8.10) d_{mn} is the travel cost for the links connecting nodes m and n. Taken together with the node-balance constraints, only the links on the shortest paths will be used. Thus the objective function of equation (8.10) also gives the minimum-cost (path) between nodes j and l (the load node subset of all nodes) multiplied by the total trips between j and l. The difference is that, in this case, these costs are expressed in terms of the individual links on the minimum paths and the flows on those links. Numerically, the results will be identical.

It appears, then, that the system optimal (SO) combined location, distribution, and assignment model may be represented by the following set of equations:

$$\text{minimize } S = \sum_{i,m,n} d_{mn} F_{imn} + \frac{1}{\beta} \sum_{i,j,l} S_{ijl} \ln S_{ijl} + \sum_{i,j} b_{ij} X_{ij} , \qquad (8.12)$$

subject to

$$\sum_{i,n} F_{imn} - \sum_{i,n} F_{inm} = \begin{cases} \sum_i (s_i X_{im} - r_i E_m), & \text{for } m \leq J , \\ 0, & \text{for } m > J . \end{cases} \qquad (8.13)$$

$$\sum_l S_{ijl} - s_i X_{ij} = 0 , \qquad (8.14)$$

$$\sum_j S_{ijl} - r_i E_l = 0 , \qquad (8.15)$$

$$\sum_j X_{ij} = A_i , \qquad (8.16)$$

$$\sum_i X_{ij} \leq L_j , \qquad (8.17)$$

$$S_{ijl} \geq 0 , \qquad (8.18)$$

$$X_{ij} \geq 0 , \qquad (8.19)$$

$$d_{mn} > 0 , \qquad (8.20)$$

where the variable names d and F (link cost and link flow) are substituted for the c and X used in the minimum-cost flow problem equations above. J is the number of zones (load nodes). In addition the employment in zone l is represented by E_l in equation (8.15) to avoid confusion.

8.3 A simple combined model: numerical examples of intractable solutions

The Archerville data given in chapter 2 can be used, as it has been all along, for a simple numerical example of the operation of the combined model. Here, as was the case in the examples of chapter 5, employment location is taken as given. The value of β is taken to be 0.05. In order that all three terms of the objective function have an effect on the model solution, the location-cost term is multiplied by 0.7 (as in the examples in the latter sections of chapter 5). The transportation-cost term is multiplied by 10.0, again as in the prior examples. Both the need for and the effects of these scalings will be discussed later in this chapter. The results of the model solution are given in table 8.1 and figure 8.1.

Once the combined model is formulated, it is not too difficult a matter to add in the effects of congestion—thus giving link costs which are a function of link flows. The congestion function used in these examples was given in equation (3.42) and is repeated here with notation consistent with the combined model formulation. The form is:

$$d_{mn} = d_{mn}^{(0)}(1.0 + \chi F_{*mn}^2) , \qquad (8.21)$$

where

d_{mn} is the 'congested', or flow-related cost of traversing link mn,
$d_{mn}^{(0)}$ is the 'uncongested', or 'design' cost of traversing link mn,

Table 8.1. Archerville: location of households, based on the combined system optimal (SO) trip-assignment and activity-location model (without congestion), with $\lambda_2 = 0.70$ and $\beta = 0.05$ ($\lambda_3 = 20$).

Zone	Number of households		
	LI	HI	total
1	76	232	308
2	282	235	517
3	35	126	161
4	407	57	464
5	0	0	0

	Number of LI households[a]					Number of HI households[a]				
	1	2	3	4	5	1	2	3	4	5
1	16	18	27	11	5	48	56	80	32	16
2	58	68	97	39	19	49	57	81	32	16
3	7	8	12	5	2	26	31	44	17	9
4	84	98	140	56	28	12	14	20	8	4
5	0	0	0	0	0	0	0	0	0	0

[a] A breakdown of households by zone of work. There are slight differences because of rounding off.
Note: λ_2, location-cost multiplier in the objective function; λ_3, entropy term multiplier in the objective function see equation 5.55; LI, low-income; HI, high-income.

$F_{*mn} = \sum_i F_{imn}$ is the sum of all trip types travelling on link mn,

χ is a parameter, set to 0.0002 in this example.

If equation (8.21) is substituted into the objective function of the combined model, it becomes

$$\text{minimize } S = \sum_{i,m,n} (d_{mn}^{(0)} F_{imn} + \chi d_{mn}^{(0)} F_{imn}^3) + \frac{1}{\beta} \sum_{i,j,l} S_{ijl} \ln S_{ijl} + \sum_{i,j} b_{ij} X_{ij} \,. (8.22)$$

With this revised objective function, the minimum-cost flow term becomes a cubic function of link flow. Of course, the location patterns and trip flows in the numerical example will be altered in the solution of this form of the combined model. These results are shown in table 8.2 and figure 8.2 (see over).

It is immediately apparent from a comparison of these tables and figures that the inclusion of flow-dependent (or variable) link costs in the model results in increased dispersion of household location and of trips on links. This is entirely owing to the effect of increased travel cost in the, now, non-linear minimum-cost flow term of the objective function. In both model solutions exactly the same quantity of land was used in each zone, even though the types and numbers of households locating in the zones were obviously different. The available-land constraint was binding (operational) in zones 2, 3, and 4. The link flows, as can readily be seen in the figures, were quite different for the links in the top half (in terms of the figures) of the network.

Before showing what is wrong with all this, it may be recalled from chapter 6 that there is a difference between the system optimal (SO) trip-assignment approach and that of user equilibrium (UE). In particular, the

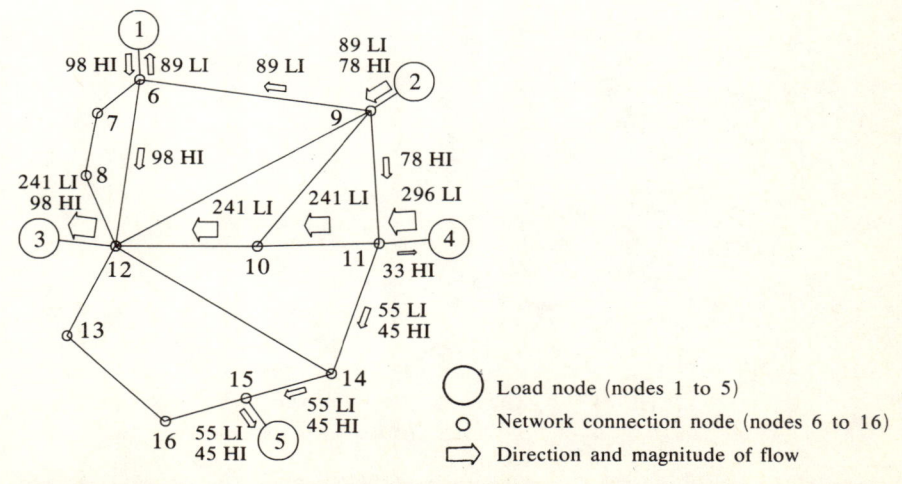

Figure 8.1. Link flows from the system optimal (SO) combined model with constant link costs, by low-income (LI) and high-income (HI) households.

SO solution is what might be termed the least-'societal-cost' solution whereas the UE solution is the least-'individual-cost' solution. SO assignment with variable costs would, given the flow-cost (congestion) function of

Table 8.2. Archerville: location of households, based on the combined system optimal (SO) trip-assignment and activity-location model (with congestion), with $\lambda_2 = 0.70$ and $\beta = 0.05$ ($\lambda_3 = 20$).

Zone	Number of households		
	LI	HI	total
1	160	164	324
2	211	293	504
3	190	0	190
4	239	193	432
5	0	0	0

	Number of LI households[a]					Number of HI households[a]				
	1	2	3	4	5	1	2	3	4	5
1	33	39	55	22	11	34	40	57	23	11
2	44	51	73	29	15	61	71	101	40	20
3	39	46	67	26	13	0	0	0	0	0
4	49	58	82	33	18	40	47	66	27	13
5	0	0	0	0	0	0	0	0	0	0

[a] A breakdown of households by zone of work.
Note: see table 8.1.

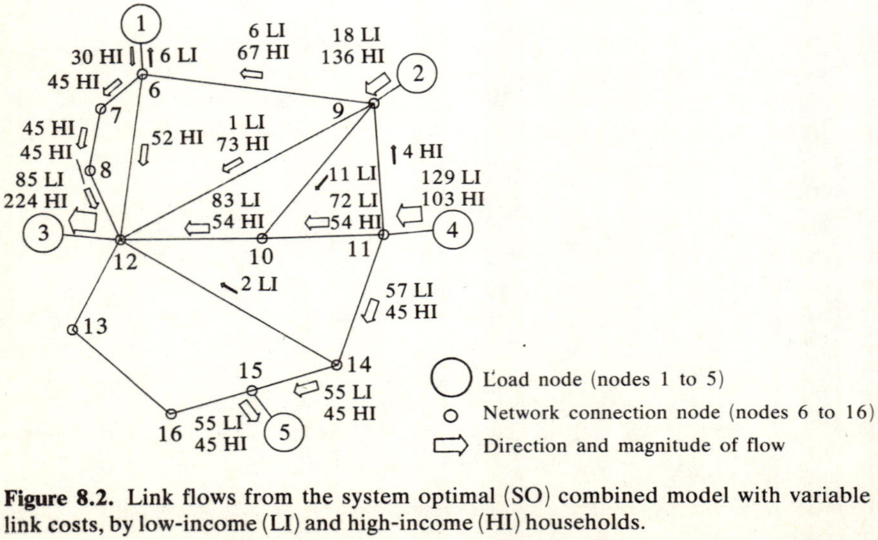

Figure 8.2. Link flows from the system optimal (SO) combined model with variable link costs, by low-income (LI) and high-income (HI) households.

equation (8.21), use as its objective function the first term of equation (8.22),

$$\text{minimize } S = \sum_{i, m, n} (d_{mn}^{(0)} F_{imn} + \chi d_{mn}^{(0)} F_{imn}^3) \,. \tag{8.23}$$

UE assignment has as its objective function, as discussed in chapter 6, the following:

$$\text{minimize } S = \sum_{\ell} \int_0^{x_\ell} t_\ell(x_\ell) \, dx \,, \tag{8.24}$$

where
x_ℓ is the flow (trips) on link ℓ,
t_ℓ is the time or cost of traversing link ℓ,
and where the summation is over all the network links for which the flow is greater than 0. In the current case, the function $t_\ell(x_\ell)$ is given by equation (8.21). After substitution of this congestion function into equation (8.24), (and also after converting the variables into the same terms being used in this example), integration yields

$$\text{minimize } S = \sum_{i, m, n} \left(d_{mn}^{(0)} F_{imn} + \frac{\chi}{3} d_{mn}^{(0)} F_{imn}^3 \right) \,. \tag{8.25}$$

An interesting point here is that the only difference between the SO objective function and the UE objective function is the scalar 1/3 which is multiplied by the second term of equation (8.25). It is worth further note that if one were using the standard BPR flow–cost function of equation (6.9), then the only difference between the SO and UE objective functions would again be a scalar resulting from the integration of the flow–cost function, which would be multiplied by the second term in the objective function. This may be seen in equation (6.10), where the scalar is 1/5.

The user equilibrium (UE) combined location, distribution, and assignment model is thus identical to the SO combined model, with the single exception of the division by 3 of the χ in the objective function. The results of the numerical example of the UE combined model are given in table 8.3 and figure 8.3.

The results in table 8.3 and the link flows shown in figure 8.3 are, of course, different from those of table 8.2 and figure 8.2, the SO (variable-cost) solution. But, how can one tell whether these new results do indeed represent a UE (variable-cost) solution? Recall from chapter 6 that at the optimum for the UE model there are certain conditions which must be met. In particular, for a given origin–destination pair between which there is trip flow, all used paths must be of equal cost. This also means that all other paths between that origin–destination pair must be both larger and unused. By inspection of figure 8.3 one finds that from zone (origin) 2 to zone (destination) 3 there are a total of four different paths used by high-income (HI) households. In terms of the nodes passed, these paths are: (1) 2-9-12-3; (2) 2-9-10-12-3; (3) 2-9-6-12-3; and (4) 2-9-6-7-8-12-3.

The effects of the flows on the paths, by virtue of the congestion function, are to give a total path cost of 33.9 in each case. Thus, indeed, for the trips shown on the network there is a UE assignment.

Table 8.3. Archerville: location of households, based on the combined user-equilibrium (UE) trip-assignment and activity-location model (with congestion), with $\lambda_2 = 0.70$ and $\beta = 0.05$ ($\lambda_3 = 20$).

Zone	Number of households			Number of LI households[a]					Number of HI households[a]				
	LI	HI	total	1	2	3	4	5	1	2	3	4	5
1	102	211	313	21	25	35	14	7	44	51	73	29	15
2	256	256	512	53	62	88	35	18	53	62	88	35	18
3	131	49	180	27	32	45	18	9	10	12	17	7	3
4	311	134	445	64	75	107	43	21	28	32	46	19	9
5	0	0	0	0	0	0	0	0	0	0	0	0	0

[a] A breakdown of households by zone of work.
Note: see table 8.1.

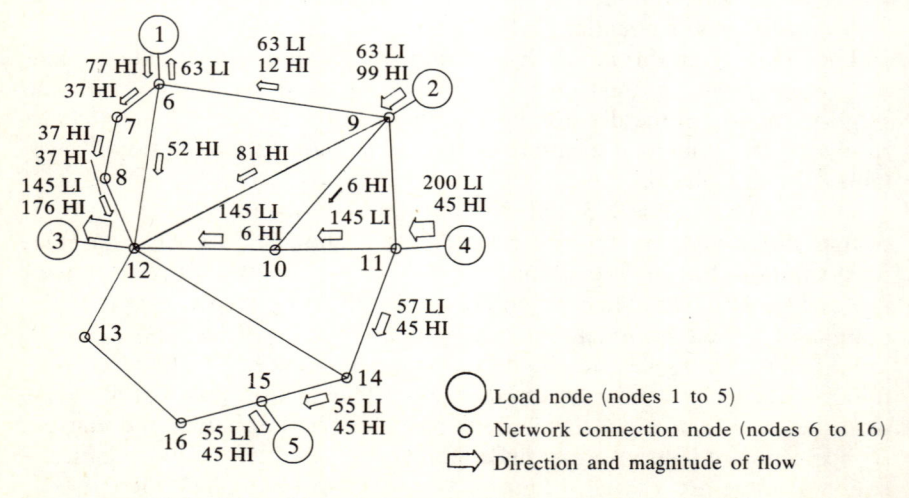

Figure 8.3. Link flows from the user-equilibrium (UE) combined model with variable link costs, by low-income (LI) and high-income (HI) households.

Having been forewarned, the reader should now be asking what is wrong with all this. The answer can readily be seen by noting that in figure 8.3, from zone (origin) 1 there are no trips to zones 2, 4, or 5, and from zone 2, there are no trips to zones 4 or 5. Somehow, the trip-distribution portion of the combined model is not showing up in the link flows. Inspection of table 8.3 shows that such zone-to-zone trips were generated. For example, of the 102 low-income (LI) household heads living in zone 1, 21 work in zone 1, 25 work in zone 2, 35 work in zone 3, 14 work in zone 4, and 7 work in zone 5. Yet the map of link flows (figure 8.3) shows only 63 LI trips arriving. What is happening is that only the net flows are showing up on the network.

The total number of heads of LI households who work in zone 1 is 165 (0.55 LI households per employee, multiplied by 300 employees), as calculated by the household location and trip distribution portion of the model. The total number of LI households living in zone 1 is 102. The trip-assignment portion of the model then simply deals with the net requirement of 63 LI trips into zone 1. In effect, once the trips get on the network the assignment component of the model ignores the paired origin–destination attributes of the trips, simply considers them as LI travellers, and balances out the total zonal originating or terminating trip requirements.

This problem may be traced to the node-balance constraints. As they are presently stated, there is one such constraint, equation (8.8) or (8.11), for each network node. To eliminate the current difficulty would require a node-balance constraint for each origin–destination pair for each node or, the number of origins multiplied by the number of destinations and the number of network nodes. By these means the trips between different origin–destination pairs would be kept separate. An unfortunate consequence of this strategy is that it would also be necessary to distinguish these flows throughout the model. Thus the flow variable F_{imn} would become F_{imnjl} and would give a very substantial increase in the number of variables in the objective function too. For the Archerville data the existing objective function has 942 variables. With the new formulation this would increase to 22110 variables. The model now has 36 constraint equations. The new formulation would increase this to 420 constraint equations. This problem is, like the path-variable problem discussed in the previous chapter in connection with the equilibrium assignment algorithms, not solvable by a direct approach. In the next section of this chapter, approaches to solving this problem will be discussed and will be followed by numerical examples.

8.4 Reformulating and solving the combined model: the Florian approach
Several different approaches to solving the combined distribution and assignment problem have been published. All of them are, in effect, variations on the same general scheme. Notable amongst the proposed approaches and discussions are those of Erlander (1977), Evans (1976), and Florian et al (1975).

Florian et al define the combined distribution and assignment problem in terms of a trip-distribution component and a UE trip-assignment component. The objective function of their model has the following form:

$$\text{minimize } S = \gamma \sum_{j,l} T_{jl} \ln T_{jl} + \sum_{\ell} \int_0^{x_\ell} t_\ell(x_\ell) dx . \qquad (8.26)$$

Clearly the first term is the trip-distribution model and the second term is the UE trip-assignment model. The constraints are the combined constraints from both models:

subject to

$$\sum_l T_{jl} = O_j , \qquad (8.27)$$

$$\sum_j T_{jl} = D_l , \qquad (8.28)$$

$$-T_{jl} + \sum_{\wp} F_{\wp,jl} = 0 , \qquad (8.29)$$

$$F_\ell = \sum_{k,j,l} \delta_{\ell\wp,jl} F_{\wp,jl} , \qquad (8.30)$$

$$T_{jl} \geq 0, \quad F_{\wp,jl} \geq 0 ,$$

where
- T_{jl} is the number of trips from origin j to destination l,
- O_j is the total number of trips originating at origin j,
- D_l is the total number of trips terminating at destination l,
- γ is $1/\beta$ as per equation (8.1),
- $F_{\wp,jl}$ is the number of trips on path $\wp$ between origin j and destination l,
- F_ℓ is the number of trips (volume of flow) on link ℓ,
- t_ℓ is the travel time (costs) on link ℓ,
- $\delta_{\ell\wp,jl}$ is equal to 1 if link ℓ is on path $\wp$ between zones j and l, and is equal to 0 otherwise.

Thus the first two constraints, equations (8.27) and (8.28), are the usual trip-end constraints. The third constraint, equation (8.29), is a conservation-of-flows-on-paths constraint, and the fourth constraint is a link-flow accounting, which simply defines the total flow on each link as being equal to the sum of the flows on all paths which include the link. The usual non-negativity constraints apply as well.

It is the appearance of the paths in these constraints which differentiates this formulation from the invalid ones discussed in the preceding section of this chapter. By posing the problem formulation in terms of paths, the error of assigning net flows only is thus precluded. At the same time, as discussed in chapter 6, inclusion of the path-defined variables vastly increases the size of the problem.

Simultaneous location–transportation models: 1

Consider the very small example used by Florian et al as shown in figure 8.4. There are 100 trips originating at zone 1 and zone 2, and 150 trips terminating at zone 3 and 50 trips terminating at zone 4. The link volume–delay functions are linear and of the form

$$t_\ell = \eta + \mu F_\ell. \tag{8.31}$$

The values of η and μ are tabulated for each link in figure 8.4. As this is a *very* small problem and as each path consists of only one link, the problem can be solved directly, as a nonlinear mathematical programming problem.

The objective function becomes:

$$\begin{aligned}\text{minimize } S = {}& \gamma(T_{13}\ln T_{13} + T_{14}\ln T_{14} + T_{23}\ln T_{23} + T_{24}\ln T_{24}) \\ & + \left(\eta_1 + \frac{\mu_1}{2}F_1\right)F_1 + \left(\eta_2 + \frac{\mu_2}{2}F_2\right)F_2 + \left(\eta_3 + \frac{\mu_3}{3}F_3\right)F_3 \\ & + \left(\eta_4 + \frac{\mu_4}{2}F_4\right)F_4 + \left(\eta_5 + \frac{\mu_5}{5}F_5\right)F_5 + \left(\eta_6 + \frac{\mu_6}{2}F_6\right)F_6 \\ & + \left(\eta_7 + \frac{\mu_7}{2}F_7\right)F_7 + \left(\eta_8 + \frac{\mu_8}{2}F_8\right)F_8. \end{aligned} \tag{8.32}$$

The first set of terms, all multiplied by γ, is the trip-distribution portion of the objective function. The last eight terms are, in each case, the definite integral of the volume–delay function for each of the eight paths (each only one link long). The constraints may be expanded and written out in a similar fashion. The problem may then be solved directly by any of a number of nonlinear mathematical programming approaches. In this case MINOS (Murtaugh and Saunders, 1983) was used. The equilibrium link flows are shown in table 8.4, along with the path (one link long) costs. Note that in every case, all paths used between a given origin–destination pair are, after

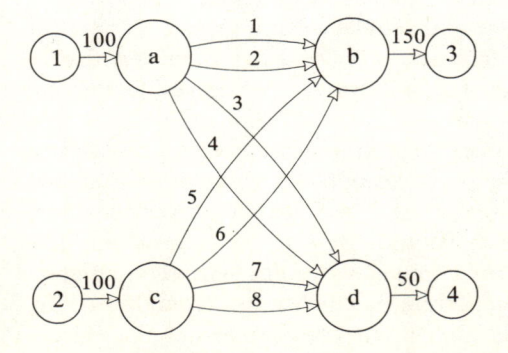

Link	Link parameter	
	η	μ
1	1	10
2	2	11
3	2	13
4	1	8
5	1	9
6	2	12
7	1	10
8	2	9

Figure 8.4. An example of the combined distribution–assignment problem used by Florian et al (1975), with the associated link parameters.

calculating the volume-delay function results, of equal length. In addition, all constraints are satisfied. Thus a combined distribution and assignment equilibrium has been achieved.

This problem would be far more complex if there were more links, origins, and destinations in the problem. In the case of Archerville, for example, there are fifteen paths possible between origin 1 and destination 5. There are seven paths from origin 4 to destination 3. Given that there are twenty-five origin-destination pairs, there would be several hundred paths to enumerate in the objective function.

Table 8.4. An example of the Florian approach: the MINOS solution, giving link flows and origin-destination flows (by zone).

Link	Flow	Cost	From (zone)	To (zone)	Flow
1	39.28	393.8	1	3	75
2	35.62	393.8	1	4	25
3	9.51	125.5	2	3	75
4	15.58	125.6	2	4	25
5	42.95	387.6			
6	32.13	387.6			
7	11.84	119.4			
8	13.05	119.5			

Because of the near impossibility of a direct solution of the problem for any meaningful problem size, Florian et al propose a decomposition approach. The method was first developed by Benders (1962), and then extended to nonlinear problems by Geoffrion (1972). Florian et al modify this approach somewhat in order to achieve guaranteed convergence by use of an adaptation of the Frank-Wolfe (1956) algorithm. The steps in this algorithmic approach are as follows.

Step 1 Calculate (to begin the process), or assume that there is known a feasible, but not optimal, solution to the problem (from a prior iteration if this is not the first). This solution consists of an OD trip matrix, with elements t_{jl}, and the corresponding set of path flows, $f_{\wp,jl}$, which result from the assignment of those trips to the network.

Step 2 The purpose of this step to determine the descent direction, that is, the direction in which to change the variables in order to further reduce the objective function. The descent direction will be from the current problem solution towards a new approximate solution now to be determined. The new approximate solution is found by minimizing the gradient of the original objective function (evaluated at the current problem solution) multiplied by the new approximate solution. This becomes, in equation form,

$$\text{minimize } Z = \sum_{j,l} \left[\gamma(1 + \ln t_{jl}) Y_{jl} + \sum_{\wp} u_{\wp,jl} X_{\wp,jl} \right], \qquad (8.33)$$

where

t_{jl} is the previously determined number of trips from origin j to destination l,

Y_{jl} is the new number of trips (to be determined) from origin j to destination l,

$u_{\wp,jl}$ is the length of the previously determined path $\wp$ from origin j to destination l,

$X_{\wp,jl}$ is the new number of trips (to be determined) on path $\wp$ from j to l.

This problem is a linear programming problem, which is somewhat easier to deal with but still difficult inasmuch as path variables are still present. A considerable simplification can be had by recognizing that at the optimal solution to this problem the objective function [equation (8.33)] will use only the shortest path between each origin–destination pair. All paths whose cost is greater than that of the shortest path will have zero flow. By virtue of the requirement of the flow constraint [equation (8.29)] the new OD flows will all be along the minimum paths. Thus the objective function of this descent direction problem reduces to

$$\text{minimize } Z = \sum_{j,l} \left[\gamma(1 + \ln t_{jl}) + u_{*,jl} \right] Y_{jl} , \qquad (8.34)$$

where

$u_{*,jl}$ is the length of the previously determined *shortest* path from origin j to destination l.

This is a classical linear programming problem of the type known as the Hitchcock problem, and may readily be solved by any of a number of algorithms for dealing with such forms (Glover et al, 1974). Thus the first computational tasks are to calculate the link costs which result from the flows produced in step 1, and then to use a path-tracing algorithm to find the shortest paths through the network. Once this problem [equation (8.34)] for the OD flows has been solved, the actual link or path flows $X_{\wp,jl}$ can be found by assigning the OD flows Y_{jl} to the previously determined minimum paths.

Step 3 The purpose of the third step is to check on the progress and convergence of the algorithm. This is done by calculating

$$E = \sum_{j,l} [\gamma(1 + \ln t_{jl}) + u_{*,jl}](t_{jl} - Y_{jl}) . \qquad (8.35)$$

If the absolute value of E is less than some prespecified criterion then the algorithm is terminated and the solution is considered to be at, or acceptably near, the optimum.

Step 4 The purpose of this step is, if the optimum has not yet been reached, to calculate an updated set of t_{jl} and $f_{\wp,jl}$ according to the descent direction. This is done by forming a convex combination of the prior flows and present flows in a manner similar to that described in chapter 6 for the UE trip-assignment problem. This involves a simple minimization problem of the following form:

$$\text{minimize } Q = \sum_{j,l} \{\gamma(1-\rho)t_{jl} \ln[(1-\rho)t_{jl}] + \gamma\rho Y_{jl} \ln\rho Y_{jl}\}$$
$$+ \sum_{\ell} \int_0^{\Omega_\ell} t(\Omega_\ell) d\Omega, \qquad (8.36)$$

where

$$\Omega_\ell = (1-\rho)f_\ell + \rho X_\ell, \qquad (8.37)$$

$$f_\ell = \sum_{p,j,l} \delta_{\ell p, jl} f_{p, jl}, \qquad (8.38)$$

$$X_\ell = \sum_{j,l} \delta_{\ell*, jl} Y_{jl}. \qquad (8.39)$$

Note that the f_ℓ are the flows on the links from the prior solution, and the X_ℓ are the present flows along the links of the previously determined minimum paths as defined by $\delta_{\ell*, jl}$. Note further the difference between the set of all used paths for the prior flows, as defined by $\delta_{\ell p, jl}$ and the minimum paths, defined in terms of the link characteristics which result from the effects of the flows on the links from the prior solution. Thus the minimization of equation (8.36) involves finding a value of ρ which yields the optimal combination of the previous and present flows. This is a one-dimensional optimization problem which can easily be solved by various line-search methods.

Step 5 In this final step, the previous and present flows are combined to give a new set of flows:

$$t'_{jl} = (1-\rho)t_{jl} + \rho Y_{jl}, \qquad (8.40)$$

$$f'_\ell = (1-\rho)f_\ell + \rho X_\ell. \qquad (8.41)$$

The algorithm then returns to step 1 for further iterations.

Florian et al applied this algorithm to their own test problem, and got the results given in table 8.5. It is worth noting that although the OD flows differ only slightly from those of the MINOS solution given in table 8.4, the link flows differ somewhat more substantially. Perhaps most important is the fact that all used paths between a given origin–destination pair *are not* of equal length. Thus the solution given by Florian et al is *not* a solution to equation (8.26). This is a definite consequence of one matter and a possible additional consequence of another. First, in order to avoid certain numerical problems in their solution they felt obliged to change the second term in the objective function, equation (8.26), to yield the following

$$\text{minimize } S = \gamma \sum_{j,l} T_{jl} \ln(T_{jl} + 1) + \sum_{\ell} t_\ell F_\ell. \qquad (8.42)$$

This objective function no longer defines a UE trip assignment to the network links. As a consequence, the solution is not a UE assignment.

The second matter is that the Frank-Wolfe algorithm is known to be slow to converge. To this is added the known property of linear programming solutions to have only a certain number of nonzero elements in the solution. In the case of the Hitchcock problem for n origins and n destinations, only $2n-1$ nonzero entries are required. As there are n^2-n values in the interzonal trip table (with intrazonal trips not included), it may take a number of iterations of this algorithm, of which the Hitchcock problem is a part, before all elements of the trip matrix have a calculated value. Last, to tie this together, note that the second term in equation (8.42) is the SO assignment criterion, being arithmetically equivalent to the total system transport cost. This is known to yield an unstable equilibrium. Taken all together, it is no wonder that the Florian solution of table 8.5 is not a good one.

Table 8.5. An example of the Florian approach: the Florian solution, giving link flows and origin-destination flows (by zone).

Link	Flow	Cost	From (zone)	To (zone)	Flow
1	38.00	381.0	1	3	74
2	36.00	398.0	1	4	26
3	11.50	151.5	2	3	76
4	14.50	117.0	2	4	24
5	41.00	370.0			
6	35.00	422.0			
7	11.50	116.0			
8	12.50	114.5			

8.5 Reformulating and solving the combined model: the Evans approach

In an attempt to solve the same problem, Evans (1976) proposed what turns out to be a somewhat better solution. Before discussing her approach, there is a minor point which must be attended to regarding the trip-distribution component of the combined model.

As has already been mentioned, a portion of the objective function of the combined model can be shown to be a standard doubly constrained spatial interaction or, in this case, trip-distribution model. Showing this involves reference to the entropy-maximizing approach to model development, where an objective function which is a measure of system entropy is maximized subject to constraints on the trip ends and the total system transport cost. In most cases the entropy of the system, S, is defined as

$$S = -\sum_{i,j} \ln T_{ij}! \, . \tag{8.43}$$

Stirling's approximation for the factorial is used, being

$$\ln T_{ij}! = T_{ij} \ln T_{ij} - T_{ij} \, . \tag{8.44}$$

This gives a revised definition of entropy to be maximized, of the form

$$S = -\sum_{i,j}(T_{ij} \ln T_{ij} - T_{ij}) \,. \tag{8.45}$$

There is, however, another definition of entropy, known as Shannon's definition, which is

$$S = -\sum_{i,j} T_{ij} \ln T_{ij} \,. \tag{8.46}$$

For all practical purposes the difference between the two definitions is unimportant, as the last term in equation (8.45), the $-T_{ij}$, is, when summed over all i and all j, a constant. Thus the maximization of equation (8.45) will give the same set of T_{ij} as the maximization of equation (8.46), but the value of the objective functions will differ by the value of T, the total number of trips in the system.

Evans's approach to solving the combined distribution assignment problem makes use of the convex-combinations approach too. What is different is her recognition that a portion of the objective function is the mathematical programming equivalent of the doubly constrained spatial interaction model. This was followed by her proving that the model could then be decomposed and solved, to yield the optimal solution to the original problem as stated in mathematical program form. The steps in the procedure are as follows.

Step 1 In the Evans approach, the first step is identical to the Florian approach, involving getting an initial OD trip matrix and set of path flows, that is, the trip flows on all the links in the network.

Step 2 As in the Florian approach, it is first necessary to calculate the link costs which result from the flows obtained from step 1, and then to find the minimum paths. Evans then makes use of the fact that the first term in the combined-model objective function, equation (8.26) along with the trip-end constraints of equations (8.27) and (8.28) are the mathematical programming statement of the doubly constrained spatial interaction model, as discussed in chapter 5. Thus it is possible to solve the following model to get the OD trip matrix:

$$Y_{jl} = A_j B_l O_j D_l \exp \frac{-1}{\gamma C_{jl}} \,, \tag{8.47}$$

where

$$A_j = \left(\sum_l B_l D_l \exp \frac{-1}{\gamma C_{jl}}\right)^{-1} \,, \tag{8.48}$$

$$B_l = \left(\sum_j A_j O_j \exp \frac{-1}{\gamma C_{jl}}\right)^{-1} \,. \tag{8.49}$$

The nice thing about this, is that these equations can be solved iteratively with no difficulty. Last, the new link or path flows, X_{pjl}, can be found, as in the Florian approach, by assigning the newly calculated OD flows, Y_{jl}, to the previously determined minimum paths.

Steps 3, 4, and 5 are exactly the same as the Florian approach, involving a convergence check, a one-dimensional search for the optimal combination of previous and present trips and link flows, and the combination of the previous and present trips and flows preparatory to the next iteration of the process.

Despite the fact that one iteration of the Evans algorithm requires more computation than one iteration of the algorithm of Florian et al, the Evans algorithm has been shown to converge with far fewer iterations (Frank, 1978; Harker, 1986).

8.6 The Evans approach: a numerical example

To illustrate the operation of the Evans approach, the Archerville data are used as a numerical example. The model to be used is a slightly modified version of the one given in equations (8.12)-(8.20). The modification is that the UE criterion is used for the trip-assignment portion of the model, with equation (8.25) being substituted for the link-cost term in equation (8.12), as was done for the UE combined model discussed earlier in this chapter. This gives, for the objective function

$$\text{minimize } S = \sum_{i,m,n} \left(d_{mn}^{(0)} F_{imn} + \frac{\chi}{3} d_{mn}^{(0)} F_{imn}^3 \right)$$

$$+ \frac{1}{\beta} \sum_{i,j,l} S_{ijl} \ln S_{ijl} + \sum_{i,j} b_{ij} X_{ij} \; . \tag{8.50}$$

For convenience, in the following discussion the first term on the right-hand side of equation (8.50) will be referred to as the UE assignment portion of the model, and the second two terms (taken together) will be referred to as the location-distribution portion of the model. The following steps were followed in solving this UE combined model with the Archerville data. Note that these steps represent a variation on the Evans approach rather than an exact, step-by-step application.

Step 1 For the Archerville highway network, with 0 trip volumes, the minimum-cost path was calculated from each zone to each other zone.

Step 2 With these zero-volume zone-to-zone costs the location-distribution portion of the model was run, producing estimates of residence location and a work-to-home trip matrix.

Step 3 These trips were assigned to the network by using the UE assignment portion of the model. This resulted in a first set of link flows, and revised, congested, link costs for the network. Then a revised set of minimum-cost zone-to-zone paths was calculated.

Step 4 With these revised, congested, zone-to-zone costs the location-distribution portion of the model was rerun, producing a revised set of residence locations and a revised work-to-home trip matrix.

Step 5 The revised trip matrix was assigned to the network by using the UE assignment portion of the model. A second set of link flows, link costs, and zone-to-zone costs was thus obtained.

Step 6 A convex-combinations calculation was done, by using the trip matrices from steps 2 and 4 and the link flows from steps 3 and 5. This results in a set of combined link flows.

Step 7 The link costs which result from these combined link flows are calculated and then, given the congested link costs, the minimum cost zone-to-zone paths are calculated.

Step 8 A test for convergence is made. If the procedure has converged, then it is terminated. If the procedure has not converged the zone-to-zone costs from step 7 are used as input to step 4 and the procedure is repeated.

In the case of this Archerville example the results obtained on the fourth execution of step 4 were identical to those obtained on the third. Thus an equilibrium solution was achieved. The residential location results of this process are given in table 8.6, with a record of the iteration by iteration results for household location being given in table 8.7, and for highway link flows in table 8.8. As can be seen, convergence is achieved rather rapidly. Figure 8.5 shows the equilibrium trip-flow pattern on the network. Figure 8.6 gives the cost of traversing each network link, calculated as a function of the equilibrium trip flows. To confirm that the flow pattern is a UE solution one may examine, for example, trips from zone 2 to zone 5. Two paths are used: (1) 2−9−12−13−16−15−5; and (2) 2−9−11−14−15−5. Both paths have a total cost of 32. Trips from

Table 8.6. Archerville: results from the Evans algorithm, based on the combined UE trip-assignment and activity-location model (with congestion), with $\lambda_2 = 0.70$ and $\beta = 0.05$ ($\gamma = \lambda_3 = 20$).

Zone	Number of households											
	LI	HI	total									
1	78	231	309									
2	293	227	520									
3	137	44	181									
4	293	149	442									
5	0	0	0									
	Number of LI households[a]					Number of HI households[a]						
	1	2	3	4	5	1	2	3	4	5		
1	74	0	4	0	0	121	0	103	0	7		
2	83	189	20	0	0	14	156	53	4	0		
3	0	0	137	0	0	0	0	44	0	0		
4	9	4	115	110	55	0	0	25	90	34		
5	0	0	0	0	0	0	0	0	0	0		

[a] A breakdown of households by zone of work.
Note: see table 8.1.

zone 2 to zone 3 are on two paths also: (1) 2-9-12-3; and (2) 2-9-10-12-3. The length of path 1 is 42 units and that of path 2 is 43 units. The difference is owing only to round-off error in the computations (owing to the fact that the computer program that was used dealt only in integer link costs).

It should be noted that the value of γ plays a surprisingly important role in the solution process. In the actual finding of the optimal combinations of prior and current trips and link flows, the assumption is made that the function being minimized is indeed convex and unimodal in the region of interest. The function was shown in equation (8.36), and it should be recalled that γ is equal to $1/\beta$, and is thus related to the travel 'propensity' of the residents of the region. For the Archerville example, if γ is less than 1.0, the function is not convex, and the step 3 calculation does not provide a step-size calculation which will allow the algorithm to converge.

The appropriate value for γ for these experiments, given the value of β used in the residence location–distribution model component of this combined model, is 1.0/0.05, or 20. Some brief numerical experiments were performed to assess the effects of varying the value of γ. When γ is increased to 100, that is, $\beta = 0.01$, the convex combination calculations are virtually unaffected. When, however, γ is decreased to 10, that is, $\beta = 0.10$, it is clear that there is a significant effect. The rate of convergence of the algorithm is significantly reduced, as shown in table 8.9. Convergence in this

Table 8.7. Archerville: results (number of households) from each iteration of the Evans procedure (for $\gamma = 20$).

Zone	Iteration				
	0	1	2	3	4
Low-income households					
1	45	123	91	78	78
2	319	217	282	293	293
3	20	159	138	137	137
4	416	301	290	293	293
5	0	0	0	0	0
High-income households					
1	258	194	220	231	231
2	205	288	235	227	227
3	138	25	43	44	44
4	49	142	152	149	149
5	0	0	0	0	0
All households					
1	303	317	311	309	309
2	524	505	517	520	520
3	158	184	181	181	181
4	465	443	442	442	442
5	0	0	0	0	0

Table 8.8. Archerville: results (link flows) from the Evans procedure ($\gamma = 20$).

It. 0	It. 1	Convex 1	It. 2	Convex 2	It. 3	Convex 3
132	67	108.6	101	108.0	114	107.9
208	355	260.9	168	253.4	174	253.4
21	0	13.4	0	12.3	0	12.3
327	242	296.4	240	291.8	243	291.7
0	0	0.0	0	0.0	0	0.0
0	0	0.0	0	0.0	0	0.0
27	16	23.0	8	21.7	9	21.6
0	0	0.0	0	0.0	0	0.0
38	53	43.0	48	43.7	51	43.7
38	53	43.0	48	43.7	51	43.7
0	0	0.0	0	0.0	0	0.0
365	315	347.0	319	344.7	320	344.6
27	16	23.0	8	21.7	9	21.6
0	0	0.0	0	0.0	0	0.0
12	0	7.6	6	7.4	5	7.4
24	0	15.3	0	14.0	0	13.9
0	0	0.0	0	0.0	0	0.0
0	0	0.0	0	0.0	0	0.0
0	0	0.0	0	0.0	0	0.0
27	16	23.0	8	21.7	9	21.6
130	49	100.8	90	99.9	106	99.9
38	53	43.4	48	43.7	51	43.7
73	51	65.0	53	64.0	63	64.0
21	23	21.7	0	19.9	0	19.8
100	101	100.3	100	100.2	100	100.2
0	0	0.0	0	0.0	0	0.0
0	0	0.0	0	0.0	0	0.0
167	124	151.5	142	150.7	140	150.6
36	0	23.0	0	21.1	0	21.0
0	75	27.0	0	24.8	0	24.7
52	4	34.7	9	32.6	10	32.5
73	85	77.3	92	78.4	91	78.4
0	0	0.0	0	0.0	0	0.0
60	0	38.4	0	35.3	0	35.2
167	114	147.9	136	146.9	136	146.8
125	89	112.0	101	111.1	102	111.0
47	101	66.4	5	61.4	9	61.3
33	199	92.7	0	85.2	5	85.1
77	100	85.2	81	84.8	71	84.7
118	109	114.7	84	112.2	101	112.1
12	84	37.9	5	35.2	4	35.1
36	63	45.7	3	42.2	2	42.1

Note: It., iteration; Convex, convex-combinations calculation. Flows on all 42 one-way links in the Archerville highway network are given. See table 2.3 and figure 2.2.

case is achieved with the sixth iteration giving identical results to those of the fifth. In the first test of the algorithm, with the 'correct' value of γ, the step-size calculations were more efficient, with convergence being achieved with the fourth iteration being identical to the third. In addition, and to be discussed later in this chapter, note that the final solution values in this test *do not* match those of the previous example.

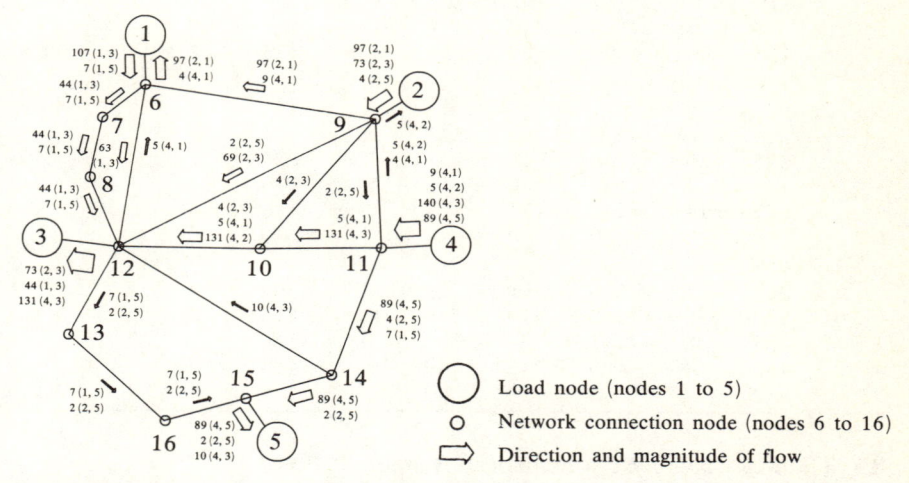

Figure 8.5. Link flows from the corrected user-equilibrium combined model with variable link costs.

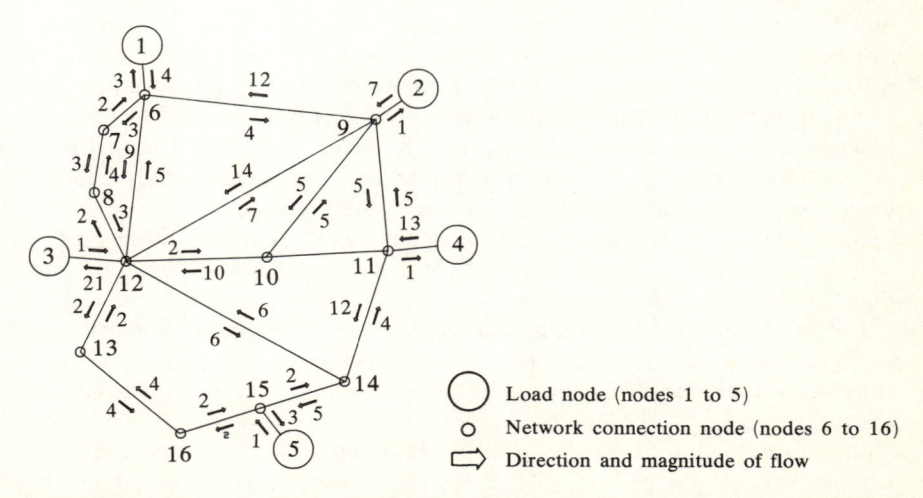

Figure 8.6. Congested link costs from the corrected user-equilibrium combined model with variable link costs.

Table 8.9. Archerville: results (number of households) from each iteration of the Evans procedure (for $\gamma = 10$).

Zone	Iteration						
	0	1	2	3	4	5	6
Low-income households							
1	45	123	76	66	88	41	41
2	319	217	281	295	263	290	290
3	20	159	144	116	119	136	136
4	416	301	299	323	330	333	333
5	0	0	0	0	0	0	0
High-income households							
1	258	194	233	240	223	261	261
2	205	288	236	224	251	229	229
3	138	25	38	60	58	44	44
4	49	142	144	125	119	116	116
5	0	0	0	0	0	0	0
All households							
1	303	317	309	306	311	302	302
2	524	505	517	519	514	519	519
3	158	184	182	176	177	180	180
4	465	443	443	448	449	449	449
5	0	0	0	0	0	0	0

8.7 The MSA approach: a numerical example

The reason for the degraded performance of the Evans algorithm with certain values of γ is that the step size in the convex-combinations algorithm is not of the most efficient length. For example, between the third and fourth iterations of the example of the Evans algorithm using γ equal to 10 (the results given in table 8.9), the value of the objective function increases substantially. Thus the fourth iteration solution is a 'worse' solution to the problem than the third iteration solution. In the fifth iteration the objective function again begins to decline. Although such errors should not (if they are not very large) prevent the procedure from converging, they can, as shown, result in an appreciable reduction in the rate of convergence. More importantly, the performance of the algorithm, as a result of variation in the scaling (the relative magnitudes) of the terms in the objective function and the step-size calculation, can be unpredictable.

Before an alternative solution procedure is discussed, it is worth, for reference purposes, showing the consequences of attempting to achieve convergence by simple repeated iterations of the location–distribution–assignment models. This procedure is similar to that which has been followed in numerous applications of integrated models, including this author's own (Putman, 1983a). The results of this procedure are shown in table 8.10, where it is quite clear that the system is oscillating with a relatively constant amplitude beginning with iterations 3 and 4. It is not at all likely that this procedure would ever converge.

As part of the discussion in chapter 6 of the stochastic user-equilibrium (SUE) approach to trip assignment, the method of successive averages (MSA) algorithm was described. This is a somewhat less sophisticated but quite robust procedure for dealing with trip assignment and, as will be shown, for combined distribution-assignment problems. Basically, the MSA algorithm replaces the step-size calculation of equation (8.36) with a predetermined sequence of step sizes. It has been shown that for appropriate model structures, this algorithm will converge 'almost surely' to the minimum (Sheffi, 1985). The only change necessary in the Evans algorithm to use the MSA algorithm is to change the procedure for determining the ρ in equation (8.36). In particular, for the nth iteration of the algorithm, ρ equals $1/n$. Of course there is a cost to pay for the simplicity and robustness of this algorithm. It converges slowly. With current computer technology this is not so great a problem. Not terribly long ago, the use of a variation on the MSA approach was abandoned by this author owing to the high cost of the many iterations necessary to achieve convergence (Putman, 1973).

Once again the Archerville data were used to test and illustrate the performance of the MSA algorithm. It should be noted that the approach used here is actually a mixture of the MSA algorithm within the overall structure of the Evans approach to solving the combined distribution-assignment problem. The results for Archerville are given in table 8.11.

Table 8.10. Archerville: iteration results (number of households) without convex combinations.

Zone	Iteration						
	0	1	2	3	4	5	6
Low-income households							
1	45	123	32	166	59	166	64
2	319	217	280	200	222	214	213
3	20	159	138	115	174	104	174
4	416	301	350	320	345	317	349
5	0	0	0	0	0	0	0
High-income households							
1	258	194	268	160	290	160	310
2	205	288	237	302	240	291	223
3	138	25	42	62	14	70	13
4	49	142	103	127	107	129	104
5	0	0	0	0	0	0	0
All households							
1	303	317	300	326	349	326	374
2	524	505	517	502	462	505	436
3	158	184	180	177	188	174	187
4	465	443	453	447	452	446	453
5	0	0	0	0	0	0	0

Convergence is reached at the seventh iteration. The eighth iteration (not shown in the table) gave identical results to the sixth. The differences between the sixth and seventh (and eighth) iterations were entirely a result of round-off errors in the model calculations.

Here again the question of the 'correct' solution to the problem arises. If the results of the MSA algorithm in table 8.11 are compared with the results of the original Evans approach given in table 8.7, it can readily be seen that although both procedures converged, they gave slightly different results. Both procedures should have given identical results. It has been reported that the Evans and Florian approaches applied to a common data set gave different solutions (Frank, 1978), but this discrepancy has not been followed up in the literature. In the case of these Archerville results the differences are small enough to be due to computational differences in the two different computer programs and to the 'lumpy' quality of such small numerical examples. This property of small problems has been encountered in other examples discussed in this text. The full-scale data-set experiments to be analyzed in the next chapter should allow a more definitive answer to this question.

Table 8.11. Archerville: results (number of households) from the Evans procedure in which the MSA algorithm was used.

Zone	Iteration							
	0	1	2	3	4	5	6	7
Low-income households								
1	45	123	73	82	65	105	67	73
2	319	217	291	285	283	251	279	281
3	20	159	132	124	133	123	125	122
4	416	301	304	309	319	320	329	324
5	0	0	0	0	0	0	0	0
High-income households								
1	258	194	235	228	241	208	240	235
2	205	288	228	233	234	260	238	236
3	138	25	47	54	47	55	53	56
4	49	142	140	136	128	127	119	124
5	0	0	0	0	0	0	0	0
All households								
1	303	317	308	310	306	313	307	308
2	524	505	519	518	517	511	517	517
3	158	184	179	178	180	178	178	178
4	465	443	444	445	447	447	448	448
5	0	0	0	0	0	0	0	0

8.8 The MSA approach: a numerical example using a spatial interaction model
Before concluding this chapter the robustness of the MSA algorithm will again be demonstrated. In chapter 2 a small sample spatial interaction model for residence location was described. A numerical example of that model was

then developed by using the Archerville data. Here that same model [equation (2.10)] will be linked to the UE trip-assignment model. The coupled models will then be solved for an equilibrium solution with use of the MSA algorithm.

The overall process is virtually the same as that followed in using the MSA algorithm for the mathematical programming model of location-distribution combined with the UE assignment model. The sequence of steps in solving that combined model formulation was given above. The sequence of steps necessary to solve the spatial interaction model in combination with the UE assignment model has only one difference. In each place where the mathematical programming model of location-distribution was used, the spatial interaction model is substituted.

It should be recalled that the MSA algorithm can make, indeed is virtually certain to make, occasional bad steps. The expectation is that the steps will, on average, be in the correct direction. The results of the numerical example are given in table 8.12, where it can be seen that the system has, for all intents and purposes, converged by the seventh iteration. Note, however, the very substantial variations in iterations 1, 2, and 3, after which the system converges slowly but directly through iteration 7. Additional iterations were done, but the results were just small variations around the apparent equilibrium solution. These variations again appear to be owing to the 'lumpy' nature of the Archerville data set, and to the fact that some of the computer

Table 8.12. Archerville: results (number of households) from a spatial interaction model linked to a UE assignment model, in which the MSA algorithm was used.

Zone	Iteration							
	0	1	2	3	4	5	6	7
Low-income households								
1	217	165	278	149	167	177	173	172
2	268	205	234	215	239	220	235	233
3	138	334	147	341	293	297	295	293
4	166	87	100	88	82	91	83	89
5	61	59	91	58	69	65	64	64
High-income households								
1	152	150	187	147	159	168	167	166
2	127	152	129	143	146	136	143	143
3	35	264	43	112	71	72	72	72
4	314	82	236	212	227	234	230	232
5	97	78	130	111	122	114	113	113
All households								
1	369	315	465	296	326	345	340	338
2	395	357	363	358	385	356	378	376
3	173	598	190	453	364	369	367	365
4	480	169	336	300	309	325	313	321
5	158	137	221	169	191	179	177	177

programs deal only in integer link flow volumes, thus inducing a certain amount of round-off error. It is to be expected that both these issues will be resolved in larger-sized problems. The matter of greatest importance here is the clear demonstration of the ability of the MSA algorithm to bring the system of models to a stable equilibrium solution.

8.9 Conclusions

A considerable amount of material has been covered in this chapter. The first point was that, given a mathematical programming formulation for a residence-location and trip-distribution model, and given a mathematical programming formulation for a trip-assignment model, both could be put together in a single combined location–distribution and assignment model. Now it is perfectly obvious that the idea of combining these models is not new (Putman, 1983a). What is different here is that by combining them into a single, consistent, mathematical framework, they can be much more clearly understood.

The second major point in this chapter was that even though a combined model could be formulated in a rather straightforward way, it was quite another matter to produce numerical solutions. Indeed it is only by decomposing the problem that a solution is possible. The approaches proposed by Florian et al (1975) and by Evans (1976) provide the key to numerical solutions to the problem. The Evans approach is the more efficient because she bypasses any attempt to solve the location–distribution problem as a mathematical program. Rather, as was done for the UE trip-assignment problem (and discussed in chapter 6), an equivalent problem is solved. In the small example used here the mathematical programming formulation of the location–distribution model is retained, but this will clearly be impossible for the full-scale examples that are to be discussed later on.

With the formulation of the combined model in view, and considering it in terms of the convex-combinations solution approach, the possibilities of variations on that theme appear. In particular the method of successive averages (MSA) algorithm emerges as a possibly less efficient, but very robust approach to solving such problems. To demonstrate its performance, both the original combined problem and a new one which uses a standard spatial interaction model, are solved. This result is of particular practical importance as it points to the solution of a problem which has plagued users of integrated transportation and land-use models for more than a decade. The MSA algorithm appears to be sufficiently robust to be able to find equilibrium solutions to many different integrated transportation and land-use model system configurations.

This is not to say that no problems remain. First, from results gained with the Archerville data, it appears that different solution methods may give different solutions to the same problem. It cannot now be stated how serious a problem this is. These different solutions may be the result of computational round-off errors. If they are not, then the question is 'what causes them?'

Do these model systems have unique solutions which these procedures are unable to find? Or, do these model systems have multiple local optima? In either case, it is clear that further work is needed. In addition, and in the same vein, mention was made of the effect of scalar multiples of the several terms in the objective function of the convex-combinations calculations. These may not be of concern to the MSA approach, but still bear further consideration.

Finally, no attempt was made here to examine the effects of starting points. In theory the solutions to these model systems are unique and should be independent of the starting points. Yet, these experiments raise questions about the uniqueness of the solutions and therefore open up all the related issues to discussion. These issues will be dealt with in subsequent chapters.

9
Simultaneous location–transportation models: 2

9.1 Introduction
In the last chapter examples were given of methods for solving combined location–distribution–assignment models. The first example of a correct solution to the problem used a nonlinear mathematical programming formulation to solve for the location–distribution portion of the problem (solved by using GAMS) and another nonlinear mathematical programming formulation to solve for the UE (user-equilibrium) trip-assignment portion of the problem (by using the Frank–Wolfe algorithm). The two portions of the problem were linked together by an approach suggested by Evans (1976), in which the Frank–Wolfe algorithm was again used to solve the combined problem.

Following this first example a slightly simpler algorithm, using the MSA (method of successive averages) approach, was applied to the same example. Then, finally, the MSA algorithm was used to solve an example where the NLP (nonlinear programming) formulation of the location–distribution problem was replaced by the simple spatial interaction model of residential location that was introduced in chapter 2.

In this chapter the combined location–distribution–assignment model is again the principal focus. Here, however, it is treated as both an extension of and a variation on the theme of integrated transportation and location models. A good deal of previous discussion of these integration issues and of the author's ITLUP package has already been published (Putman, 1983a), as has a survey of a number of such model systems (Webster et al, 1988). To illustrate here the behavior of different arrangements of model system components, an extensive series of tests was run with each of two full-scale urban-area data sets. The first of these data sets was for the region of Washington, DC, divided into 182 zones. The second data set was for the region of Houston, TX, divided into 199 zones.

In the next section of this chapter the alternate model system configurations used for these numerical experiments are described. This is followed by a section which briefly describes the two data sets. The remainder of the chapter is devoted to descriptions and comparisons of the model runs.

9.2 Alternate model system configurations
A prototype for some of the numerical experiments to be described here has already been described (Putman, 1986). The general procedure followed here involves repeated model runs with use of the two data sets in increasingly complex model system configurations. The purpose of introducing the increased complexity to the model systems is to model interactions between transportation and land use in increased detail with increased reliability. Thus the first model system structure used here describes activity location with fixed transport costs. At another extreme are models with variable

Simultaneous location-transportation models: 2

transport costs (as a function of flows), with fixed activity location. The most complex structure examined here has both variable activity location and variable transport costs.

9.2.1 Variable activity location with fixed transport cost

This is the simplest model configuration used, and is presented solely to provide a basis for comparison with the other model system configurations.

Even so, this first model structure is typical of the way in which 'land-use' models have most often been used in planning. Two linked models are involved. The first is EMPAL, which forecasts the location of four types of employment, at place-of-work, by small area. The second model is DRAM, which forecasts the location of households, usually disaggregated to four income levels, at place of residence, also by small area. Both these models have already been described in some detail (Putman, 1983a). The sequence of operation of the two models for this first system configuration (test 1) was simply EMPAL, DRAM. This sequence is shown along with the data inputs in figure 9.1. For the purposes of these tests the sequence was run through four recursions.

Each recursion begins with the execution of EMPAL. To forecast the location of employment of type k in zone j at time $t+1$, EMPAL uses the following input variables: employment of type k in all zones at time t, population of all types in all zones at time t, total area per zone for all zones, zone-to-zone travel cost (or time) between zone j and all other zones at time $t+1$, and, as the specific lagged variable, employment of type k in zone j at time t. The parameters used are derived from time t and time $t-1$ data. The model requires exogenous regional employment forecasts for time $t+1$.

After the employment location forecasts are produced by EMPAL, a set of residence location forecasts is produced by DRAM. To forecast the location of residents of type h in zone i at time $t+1$, DRAM uses the following

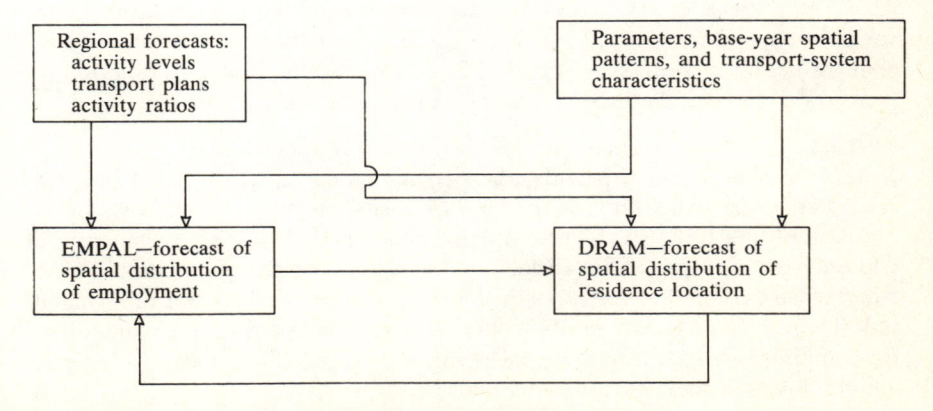

Figure 9.1. Model system configuration for fixed transport cost.

input variables: residents of all h types in zone i at time t, land used for residential purposes in zone i at time t, the percentage of the developable land in zone i which has already been developed at time t, the vacant developable land in zone i at time t, zone-to-zone travel cost (or time) between zone i and all other zones at time $t+1$, and employment of all k types in all zones at time $t+1$. The parameters used are derived from time t data. The model requires exogenous regional population forecasts for time $t+1$.

Even though this first model sequence is quite simple there are still questions left unresolved. One major question, that of sequential versus simultaneous connection of the two models, will be taken up in the next chapter. The possibilities range from simple sequential linking of the models, as was done for these experiments, to fully simultaneous linking of the employment and household models. The question of location-to-transportation feedback, left untouched in this model structure, leads to the discussion of the second structure. In particular, for the first model system structure the zone-to-zone transport costs are constant. This implies that the links in the transport network which connects the zones have costs which are constant (that is, are invariant with respect to link flows). Thus it will suffice to find the minimum zone-to-zone paths in advance, and simply to use the lengths of these paths to describe travel times or costs for all subsequent model system recursions.

9.2.2 Fixed activity location with variable transport cost

In this chapter no separate model system runs are described for this type of system configuration. There are really two cases here. First, there is fixed activity location and fixed trip interchanges with variable transport cost. This is simply the trip-assignment problem as discussed in chapters 6 and 7, where a fixed trip matrix is assigned to a network, and where link costs are a function of link flows. The interesting question is that of which criteria and algorithms are to be used, as was discussed in those chapters.

Second, there is the case of fixed trip origins and destinations, but variable trip interchanges (that is, a variable trip matrix). The trip interchanges are a function of trip costs. These costs are themselves a function of link flows. No specific tests were done of this subproblem either.

9.2.3 Variable activity location with variable transport cost: one mode

In this case, both the trip ends, the origins and destinations, and the trip interchanges (matrix) vary, as do the link costs. Here, too, link costs are a function of link flows. As a consequence, when activity location changes the trip ends are changed. This, in turn, results in a changed trip matrix. These trips, when assigned to the links of the transport network, result in changed link flows and, thus, changed link costs. These changed link costs mean that the minimum-cost zone-to-zone paths will change as well. As a consequence, activity location will change, and so on.

Simultaneous location–transportation models: 2

For this second set of test runs (test 2) an additional model was added to the initial structure, and unexercised options from DRAM were switched on. The additional model was NETWRK, the network assignment model which was used for the tests described in chapters 6 and 7. The overall configuration of this second structure is shown in figure 9.2. In this configuration EMPAL does exactly the same task it did in the first structure—employment location forecasting. DRAM, however, now has an augmented role. A matrix of work-to-home trips is generated simultaneously with the calculation of residence location, and, subsequently, matrices of home-to-shop and work-to-shop trips are calculated. These three trip matrices are then combined to produce a single matrix of trips from each zone to each other zone. These trips are then taken as input by NETWRK. NETWRK then performs a capacity-constrained assignment of these trips to a representation of the highway network in the region (for time $t+1$). The resulting congestion of the network yields, after NETWRK does a final tracing of the minimum paths through the congested network, a revised matrix of zone-to-zone travel costs (or times). These become input to the next recursion of the system.

In this modified sequence, too, there are unresolved questions of sequence and simultaneity of the solutions of the various models in the system. In addition, there appears here the question of convergence of this linked set of models as well as the general issue of equilibrium solutions. There is also the question of modal choice, which leads to the discussion of the third set of test runs.

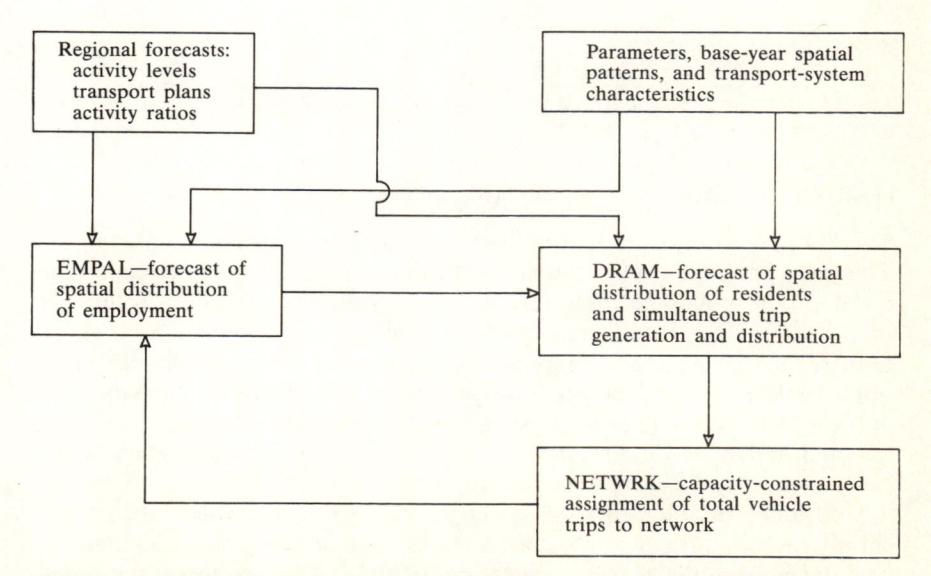

Figure 9.2. Model system configuration for variable transport cost, one mode.

9.2.4 Variable activity location with variable transport cost: two modes

For this third set of test runs (test 3) only one major model was added, MSPLIT. This addition, however, occasioned the need for several subsidiary procedures. The purpose of this addition was to add a representation of the reality of mode switching by travellers. In addition to the several new models, this third structure also required the augmenting of DRAM to provide for the output of matrices of trips by type. The overall structure of this model system is shown in figure 9.3. Not shown, but discussed later in this chapter are the supplementary procedures for calculating composite costs and matrices of total trips (after mode split).

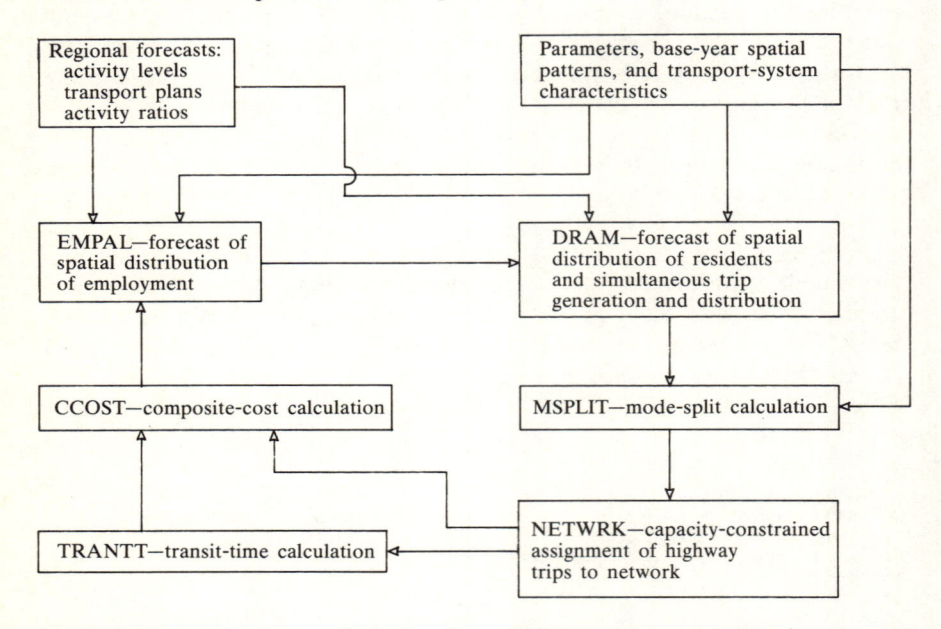

Figure 9.3. Model system configuration for variable transport cost: two modes.

9.2.5 Variable activity location with variable transport cost: equilibrium

The three previous configurations of the model system have in common that no attention is paid to the question of whether an equilibrium solution is found for the combined activity-location and trip-assignment problem. In each recursion the general inputs are the activity levels, employment, and population, at time t. The general outputs are taken to be the activity levels at time $t+1$, by virtue of their being scaled to match exogenously estimated regional activity levels for time $t+1$. Within each recursion each of the component models of the system is run once only.

Consider the elements of a single recursion in which the 'test 2' configuration of figure 9.2 is used. With use of the congested network impedance from the previous recursion, EMPAL is run to forecast employment location. The outputs from EMPAL become the inputs to the forecast of residence location from DRAM. The trip matrices from DRAM are then assigned to the networks to provide congested impedances for inputs to the next recursion. These impedances could, instead, be used to rerun EMPAL and DRAM for the same forecast period. In principle, the procedure could be repeated many times within each recursion in an attempt to reach some form of stable, or equilibrium, solution within each recursion. In practice it appears that without some special procedure to ensure it, convergence is usually not achieved. The MSA algorithm as used to solve the Archerville example at the end of the previous chapter may also be used with the model system configurations of test 2 and test 3 to produce equilibrium solutions. Thus in the last set of tests to be discussed in this chapter, the MSA algorithm is used first to demonstrate that an equilibrium solution can be found for each recursion, and second to solve for these equilibria for a sequence of recursions. The model system configuration for these test 4 runs is shown in figure 9.4.

In the next section of this chapter, prior to a discussion of the various test runs, a brief description of the two data sets which were used are provided.

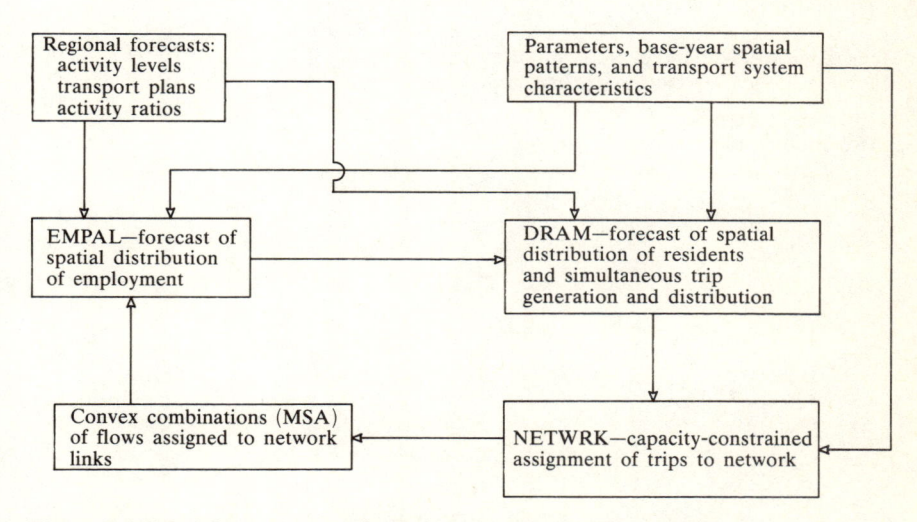

Figure 9.4. Model system configuration for equilibrium solutions.

9.3 The Washington and Houston data sets

As a result of other research activities two different but comparable data sets were available, both for 1980. The first of these was for the Washington, DC, region, and was obtained from the Washington Metropolitan Council of Governments (WASHCOG). The second data set was for the Houston, TX, region and was obtained from the Houston Galveston Area Council (HGAC). The Washington data set had 182 zones, and the Houston data set had 199 zones. Some regional comparisons are given in table 9.1. Note that although the populations of the two regions are roughly the same, as are the total employments in the regions, the physical sizes of the two regions are quite different. This is evident both in their total areas and in their zone-to-zone travel times. The mean zone-to-zone travel time in the Houston area is 50% greater than that in the Washington area. The same is true for the maximum value of the zone-to-zone travel time in each region. Note that these times are based on the estimates of the reporting agencies of peak-hour travel time over the highway networks in the regions. Figures 9.5 and 9.6 show the overall configurations of each region. Figures 9.7 and 9.8 show the 'sketch-planning' highway networks used for each of the regions.

Table 9.1. Regional statistics (1980) for the Houston and Washington metropolitan regions.

Statistic	Houston, TX[a]	Washington, DC[b]
Total population	3 119 825	2 957 899
Total employment	1 102 601	1 084 740
Total land area (acres)	4 891 544	1 326 758
Total unusable land	330 401	284 478
Mean zone-to-zone travel time (minutes)	56.4	39.7
Maximum zone-to-zone travel time (minutes)	191.0	127.0

[a] 199-zone system.
[b] 182-zone system.

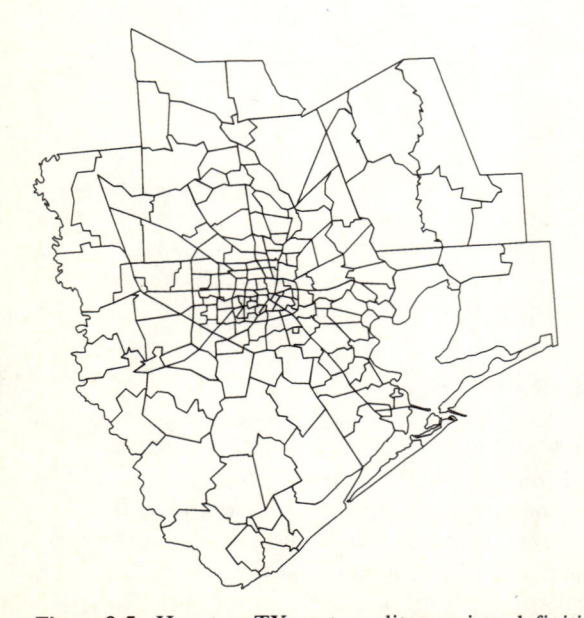

Figure 9.5. Houston, TX, metropolitan region: definition of the 199-zone system.

Simultaneous location-transportation models: 2   197

Figures 9.9 and 9.10 show the 1980 spatial distributions of population and employment for the two regions. Note, of course, that the maps of the two regions are at different scales—the area of the Houston region is nearly four times that of the Washington region.

From an historical perspective, readers may be familiar with the two regions. Washington is an eastern city whose development, beginning in the early nineteenth century, was largely a result of the location there of the US government 'seat', in the District of Columbia at the center of the region. To a very considerable degree the economic development of the region has been stimulated by the growth of government employment and the activities which serve it. With the increased availability of private automobiles in the second half of the twentieth century, coupled with an extensive program of road building, the predominantly rural, agricultural, environs of the District of Columbia were rapidly transformed into suburbs. This has been followed, mostly in the last quarter of the twentieth century, by decentralization of employment and subsequent subnucleation of both employment and population into what in many cases were originally nineteenth-century small towns, arranged in what would then have been described as an agriculturally based hierarchy of central places.

Houston, located in the southwestern United States, was an agricultural central place of rather modest size and importance well into the twentieth century. It experienced a period of explosive growth, largely the result of increased employment in the petroleum and chemical industries in the third

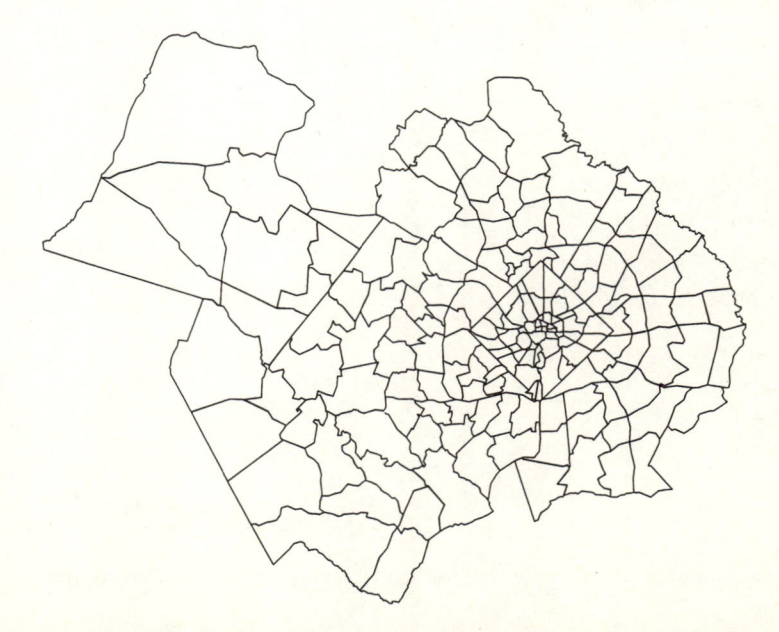

Figure 9.6. Washington, DC, metropolitan region: definition of the 182-zone system.

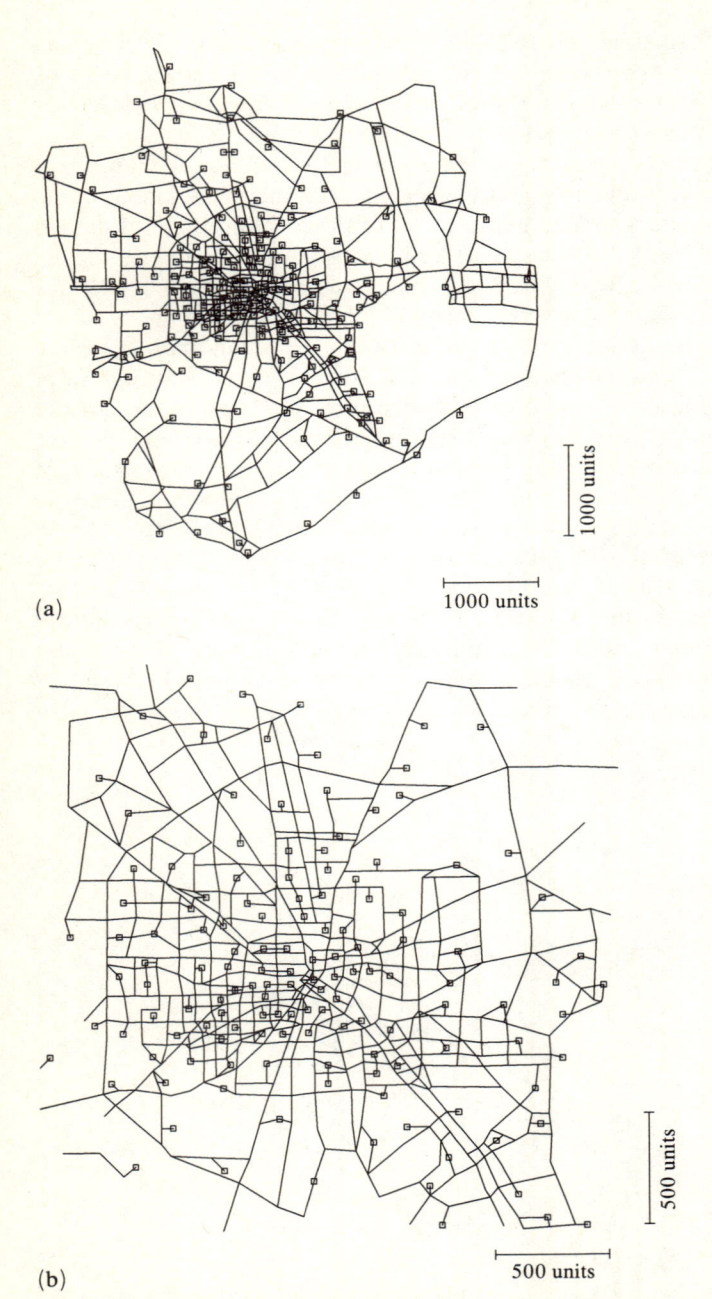

Figure 9.7. The 1980 highway network for (a) Houston region, and (b) central Houston.

quarter of the twentieth century, beginning perhaps in the early 1970s. This growth, coming at a time when extensive road systems already existed and where land-use controls were virtually nonexistent, produced extraordinary suburbanization as well as extensive subnucleation of both employment and population. Here, too, a predominantly rural agricultural landscape witnessed wholesale conversion to suburban development.

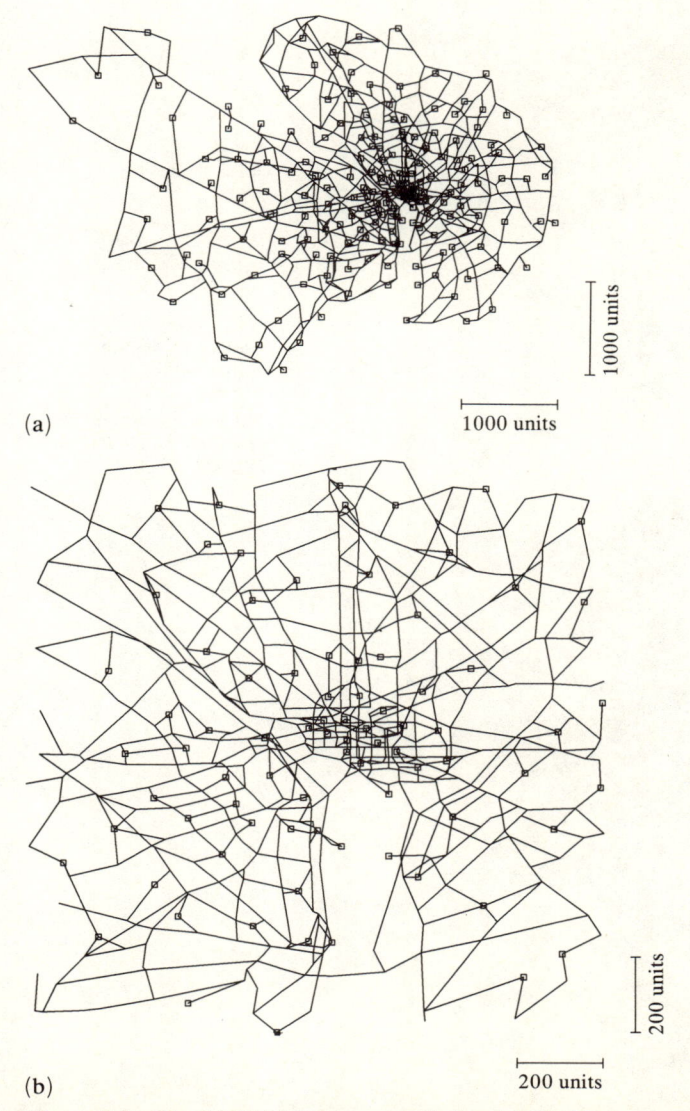

Figure 9.8. The 1980 highway network for (a) Washington region, and (b) central Washington.

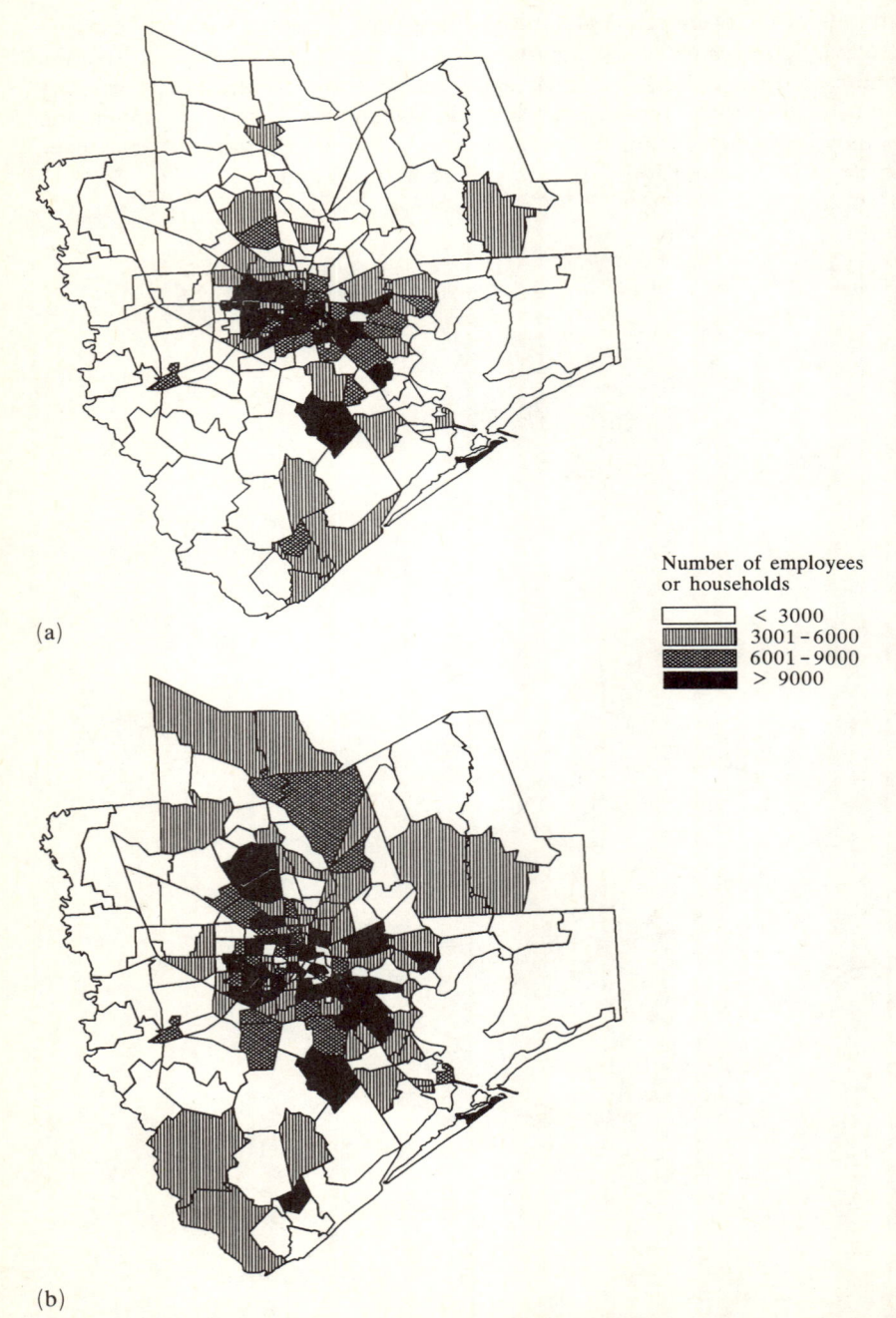

Figure 9.9. Houston, TX, region: 199-zone system. (a) The location of employment, and (b) the location of households, for 1980.

Number of employees or households
- < 3000
- 3001–6000
- 6001–9000
- > 9000

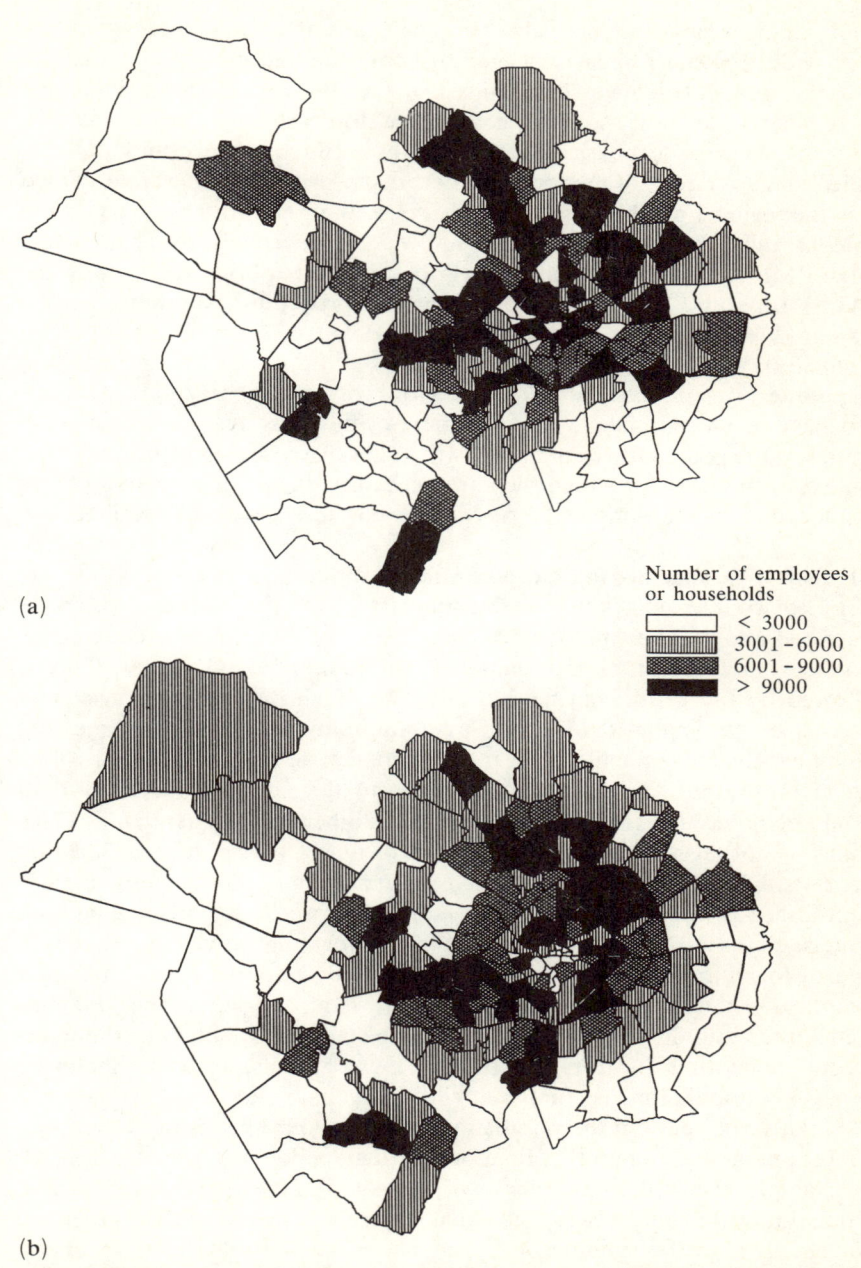

Simultaneous location–transportation models: 2

Figure 9.10. Washington, DC, region: 182-zone system. (a) The location of employment, and (b) the location of households, for 1980.

For all the series of numerical experiments described here it was necessary to provide regional forecasts of employment, population, trip making, and several regionwide conversion ratios. As the purpose of these numerical experiments was not to provide accurate forecasts for policy analysis, but rather to provide a basis for comparisons between model configurations, rather rough versions of the forecasts necessary for these runs were obtained from the regional planning agencies. The most important of these are given in table 9.2 and 9.3 in subsections 9.5.1 and 9.5.2, respectively. The parameters of the EMPAL and DRAM models were estimated for both regions with the CALIB program. The parameter-estimation procedure was described in chapter 3. Both models were discussed in detail in Putman (1983a). The parameters are given in the Appendix, and show the expected signs and magnitudes. As compared with other regions for which these model parameters have been estimated, the goodness-of-fit measures were, for both regions, quite good. Other data items will be discussed, as appropriate, in the context of the tests for which they are necessary. In the next section of this chapter the first, and simplest, series of test runs, test 1, will be described.

9.4 Test 1. Variable activity location with fixed transport cost

Test 1, of course, makes use of the simplest of the three model structures described above. The employment model, EMPAL, begins the sequence by producing a forecast of the spatial distribution of employment. This is followed by the residential model, DRAM, which produces a consequent forecast of the spatial distribution of households. The output of the first model has thus become the input to the second. The next recursion involves use of the output of DRAM as the input to the next run of EMPAL. It should be noted that this is a structure that is inherently sequential. EMPAL makes use of lagged employment data to produce its employment location forecasts. DRAM makes use of lagged household data plus current employment forecasts to produce its residence location forecasts. The fact that both models are disaggregated into several types of locating activity and that each type of location is determined by a multivariate, multiparametric function complicates the situation considerably. By comparison, in the standard spatial interaction model there are neither the lagged variables nor the interactions of multiple locators. In some of these models, an X% regionwide increase in population, as compared with some 'base run', will give the same X% relative increase in the population of every zone. Yet even in the more complex models being used here, if only a single model is run once, DRAM, for example, then the same effect will be seen. A 20% regional increase in population will produce twice the zonal change produced by a 10% regional increase. It is in the running of a sequence of such models, one recursion after another, that the nonlinear interactions begin to appear. The test-1 series of results described here are produced by such recursive sequences.

One further set of points must be made regarding the connection of employment forecast outputs from EMPAL to population forecast outputs from DRAM. First, the core of locational forecasting in DRAM is done in terms of households, by household type. In general the household types are specified in terms of socioeconomic classes or groups. In the specific cases of Houston and Washington the household types are specified by household-income quartiles. EMPAL operates in terms of employment specified in terms of industry type, such as manufacturing or retail. Thus the employment forecasts from EMPAL, which are spatial distributions of employment (at place of work) by employment type, must be converted to households by income quartile (at place of work). This conversion is accomplished by multiplying the employment by an employment-to-household matrix which is usually derived from regional statistics. This matrix can also be modified to include the effects, where known, of employee-per-household rates, unemployment rates, and the like. Given all this, the major point here is that regional employment forecasts become, after multiplication by the conversion matrix, the regional household forecasts. Thus the actual spatial allocation process accomplished in DRAM is of households at place of work to their respective place of residence. The only externally supplied regional input to DRAM is the total population which acts as a multiplier of the calculated distribution of employees at place of residence to yield the total population in each zone. By linking the employment, household, and population forecasts in this way, consistency between the models is ensured.

9.4.1 Houston

The regional forecasts used for the Houston runs were produced by HGAC in the mid-1980s, and reflected an appreciation of the substantial economic downturn in the region which began in the early 1980s. No doubt by the time this is published new regional forecasts will have been prepared. Yet the purpose here is not to produce spatial forecasts for agency use, but to discuss in a realistic context the behavior of the models. The forecasts of *regional* percentage changes in total employment were: 1980–85, 4.3%; 1985–90, 13.9%; 1990–95, 12.2%; and 1995–2000, 10.9%.

Figure 9.11 shows the changing spatial pattern of employment in the Houston region. In 1980, as was also shown in figure 9.9, employment was quite centralized, with only a few of the zones outside the center of the region having over 9000 employees. The region retains its centralized employment pattern through all four recursions of the models, out to the year 2000. Yet, by 2000, the employment forecasts clearly have spread out the employment in the region. There are fewer than three-dozen zones with under 3000 employees, even though such zones were in the great majority in 1980. Note, too, that although some of the zones in which there was new employment growth are spatially separated, much of the growth seems to be in an increasing number of contiguous zones. Last, note that the mean absolute percentage change from each time period to the next, for total *zonal* employment, was: 1980–85, 34.4%; 1985–90, 20.2%; 1990–95, 23.5%; and 1995–2000, 19.2%.

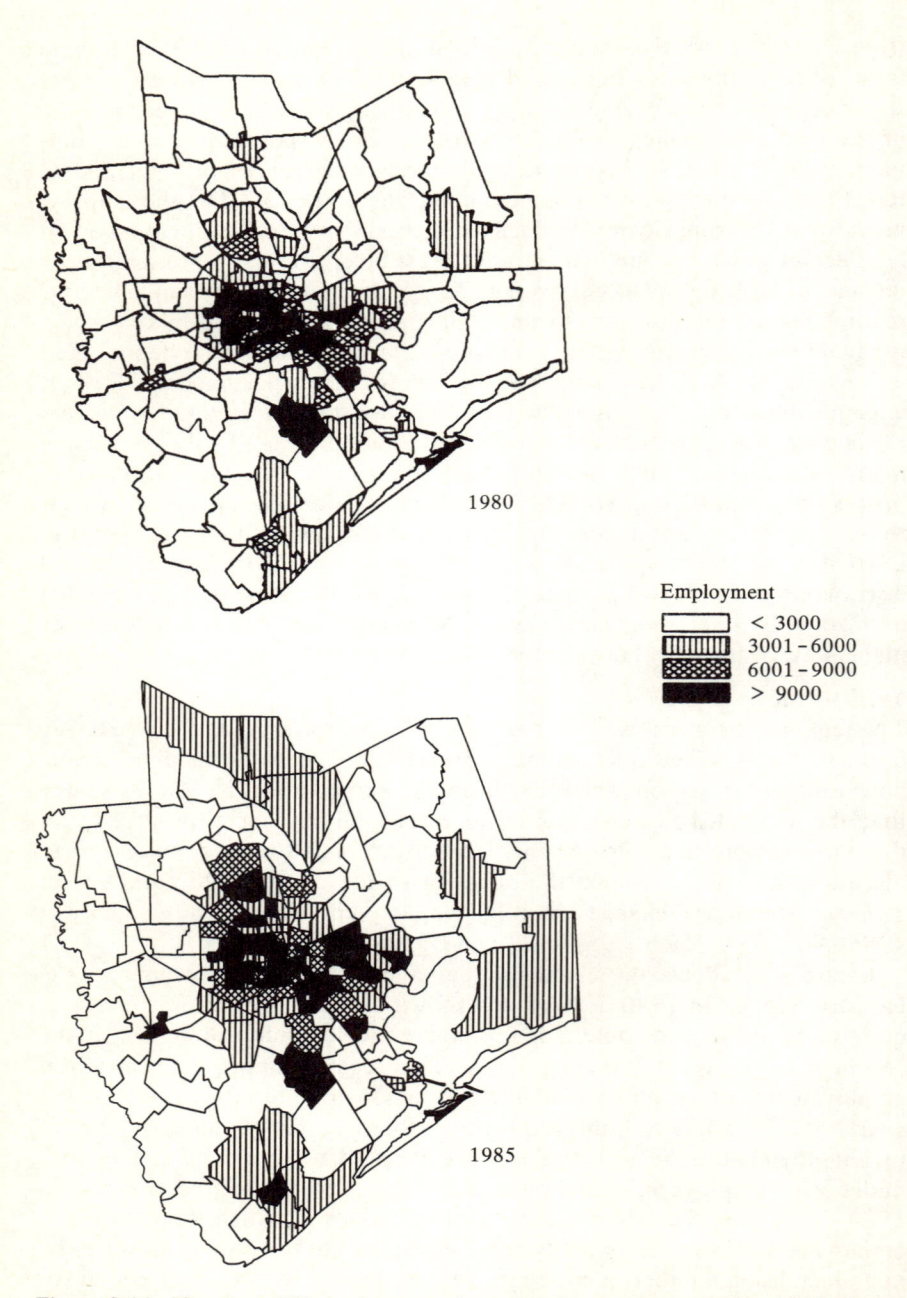

Figure 9.11. Houston, TX, region: test-1 results for employment for 1980, 1985, 1990, and 2000.

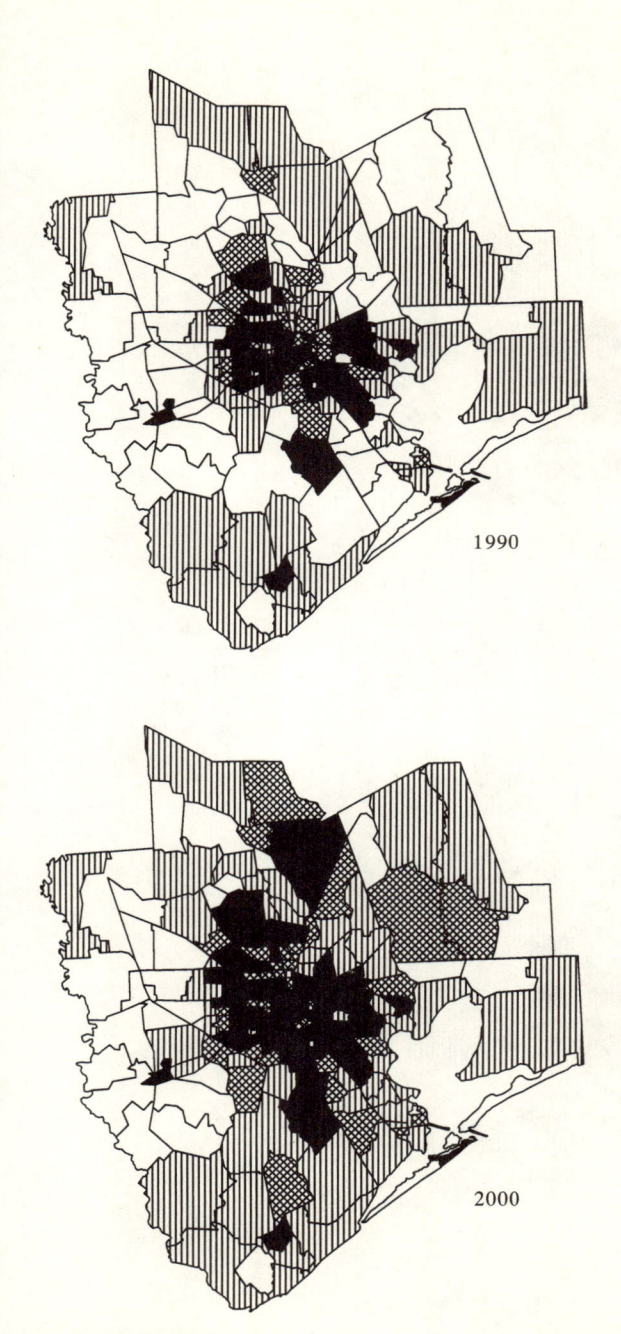

Simultaneous location–transportation models: 2

1990

2000

Figure 9.11. (continued).

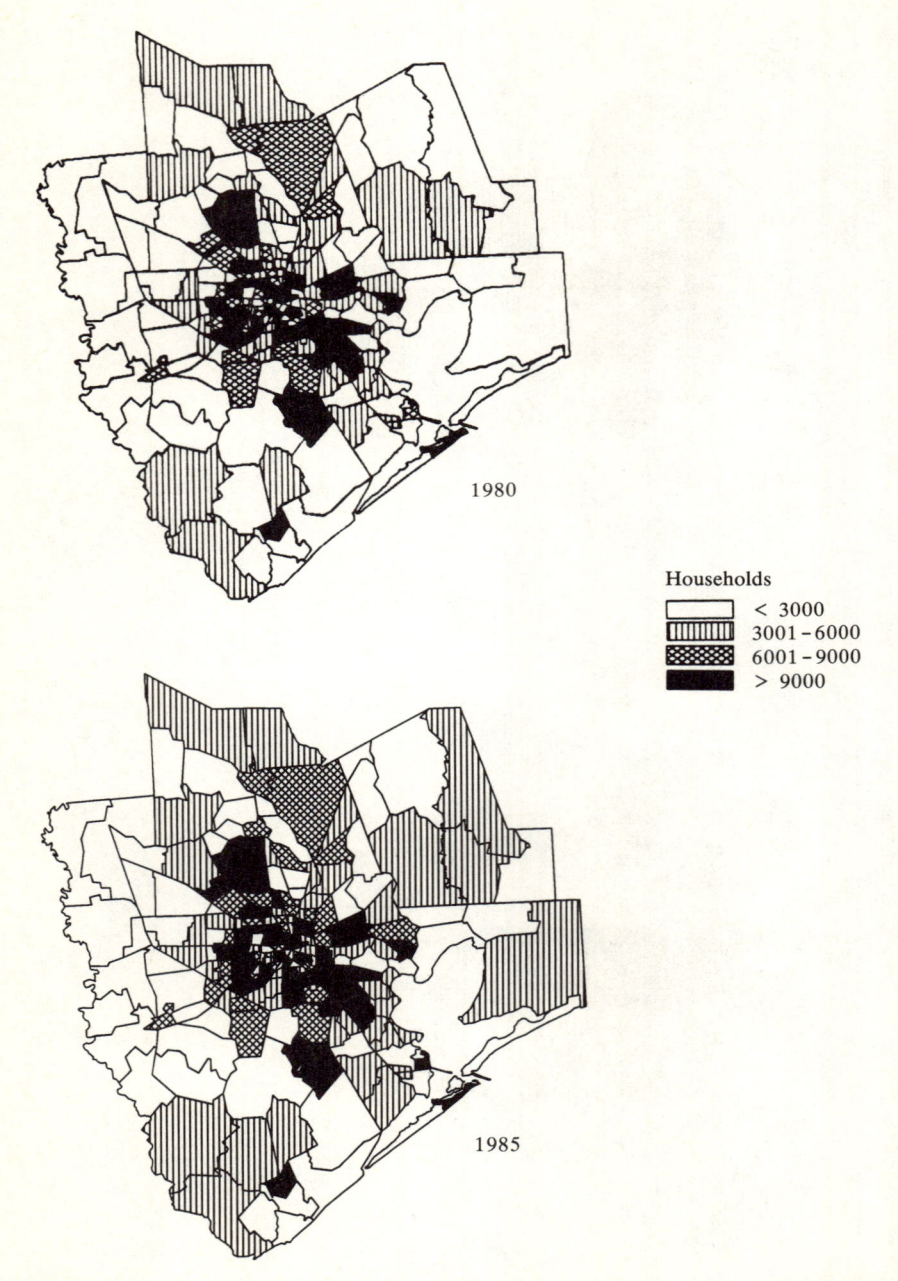

Figure 9.12. Houston, TX, region: test-1 results for population for 1980, 1985, 1990, and 2000.

1990

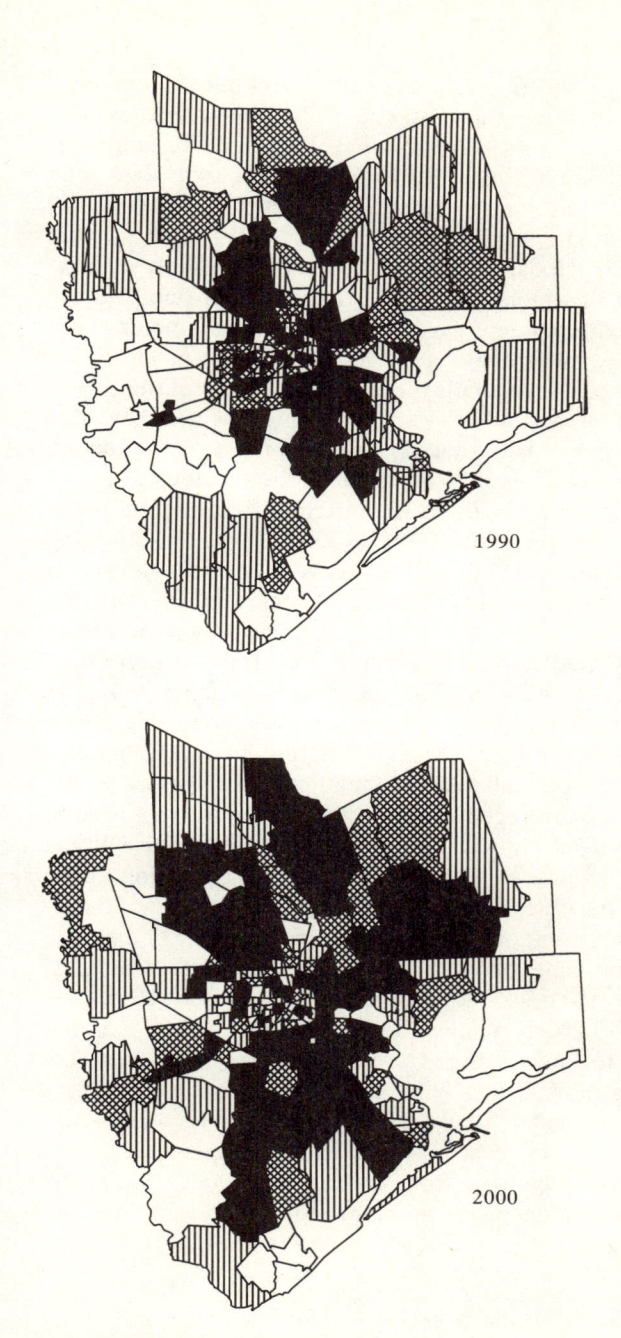

2000

Figure 9.12. (continued).

Before any discussion of the population results, a comment on these maps is needed. First, the reason for using the maps for the presentation of these results is simply that they are a very efficient way of presenting an enormous amount of information. Figure 9.11 for example, of the 199-zone Houston region at four different times, contains 796 items of zonal information, plus the thousands more items of information necessary to describe the spatial patterns of employment location. There is, however, one disadvantage. Consider two zones, each of which has 10 000 employees. If one of these zones is near the center of the region it may cover 1 square millimeter of the map. If the other zone is near the edge of the region, it may cover 10 square millimeters of the map. The 10 000 employees near the center of the region will have much less visual impact than the 10 000 employees near the edge of the region. This visual impact bias should be kept in mind by the reader, as maps of this sort will be used frequently in this and subsequent chapters.

The *regional* percentage changes in total households were: 1980–85, 12.3%; 1985–90, 7.8%; 1990–95, 3.9%; 1995–2000, 11.0%. The changing spatial patterns of population in the Houston region are shown in figure 9.12. The 1980 population in the region (note that the variable actually displayed on this and subsequent 'population' maps is total households), although generally centralized, is actually a centralized pattern of separate clusters. This can be seen a bit more clearly in figure 9.9. The pattern is basically the same in the 1985 'forecast', but with some dispersion becoming evident. By 1990, however, a very different pattern begins to dominate. Relatively high concentrations of population are dispersed to the extreme north and south of the region, and central zones which formerly had populations of 9000 or more, now have less. The forecast for 2000 is more extreme. Even after allowance is made for the visual distortion of the physically larger zones at the outer edges of the region, the pattern of population location is quite different. In terms of mean absolute percentage change *by zone* from each time period to the next, the results for total households per zone are: 1980–85, 21.5%; 1985–90, 40.8%; 1990–95, 36.5%; and 1995–2000, 43.2%. Note now the visual distortion effect, as the year 2000 map appears to have changed even more than these numbers suggest. This is because a more centrally located zone with 20 000 households looks considerably smaller than a large fringe zone with 9 005 households. Thus, in general, this first series of Houston tests shows both employment and population dispersing from the center of the region to the edges, with the movement being much more pronounced for population than for employment.

9.4.2 Washington
The regional forecasts used for the Washington runs were produced by WASHCOG in the mid-1980s. As was the case with the Houston forecasts, these will no doubt be superceded, but they are perfectly adequate for purposes of exposition here. The regional percent increases in total

employment are: 1980–85, 8.7%; 1985–90, 9.6%; 1990–95, 8.3%; and 1995–2000, 7.7%. The base-year (1980) spatial patterns of both employment and population were shown in figure 9.10.

The changing spatial patterns of employment are shown in figure 9.13. As with the test-1 results for Houston, the general tendency here, too, is towards dispersion, but with a considerable degree of persistent centralization. Note the clear delineation of the District of Columbia on the 1980 employment map (the diamond shaped set of zones at the center of the region). Obviously there must be some artificially (legislatively as distinct from economically) induced advantage to employers of being near but not inside the District of Columbia. This factor, although not explicitly included in EMPAL, can still be seen to have an effect, via the lag structure of the model, in the year 2000 forecasts. In addition, the growth of employment at the urban fringes of the region can be seen to have developed in and around zones having prior concentrations of employment. The mean absolute percent changes in total zonal employment were: 1980–85, 21.2%; 1985–90, 17.5%; 1990–95, 14.0%; and 1995–2000, 12.2%.

The percent changes in regional total households in the Washington region were: 1980–85, 14.8%; 1985–90, 9.6%; 1990–95, 8.3%; and 1995–2000, 7.7%. The spatial patterns of total households which are produced by the test-1 forecasts are shown in figure 9.14. The most obvious aspect of these maps is the very considerable growth of population in the suburban and rural fringe zones of the region. Note that although the central, District of Columbia, zones do not 'empty out', they do experience a significant decline. The major effect is the growth in the outer zones of the region. Note also the persistent effect of the boundaries of the District of Columbia or what they represent economically; the DC zones maintain their historical difference, even though, as mentioned above, there is no explicit inclusion of such a variable in the models. Last, the mean absolute percent change in the total number of households in the zones was, 1980–85, 20.1%; 1985–90, 53.9%; 1990–95, 27.4%; and 1995–2000, 20.3%.

9.4.3 Summary

The main point about test 1 is that in each region the transport costs (actually expressed in terms of travel times for both regions) are constant over the entire period from 1980 to 2000. Thus it is to be expected that the resultant spatial patterns of activities will be somewhat more dispersed than would be the case if an attempt were made to include the effects of network congestion (as will be done in the test-2 series of experiments).

In both regions, Houston and Washington, both employment and households do show significant decentralization over the forecast period. In both regions, as one would expect, the households disperse more than the employment. Yet, as mentioned above, decentralization is not a simple spreading out. There is marked subnucleation, as the dispersed locators cluster around dispersed centers of activity. Further, even though the

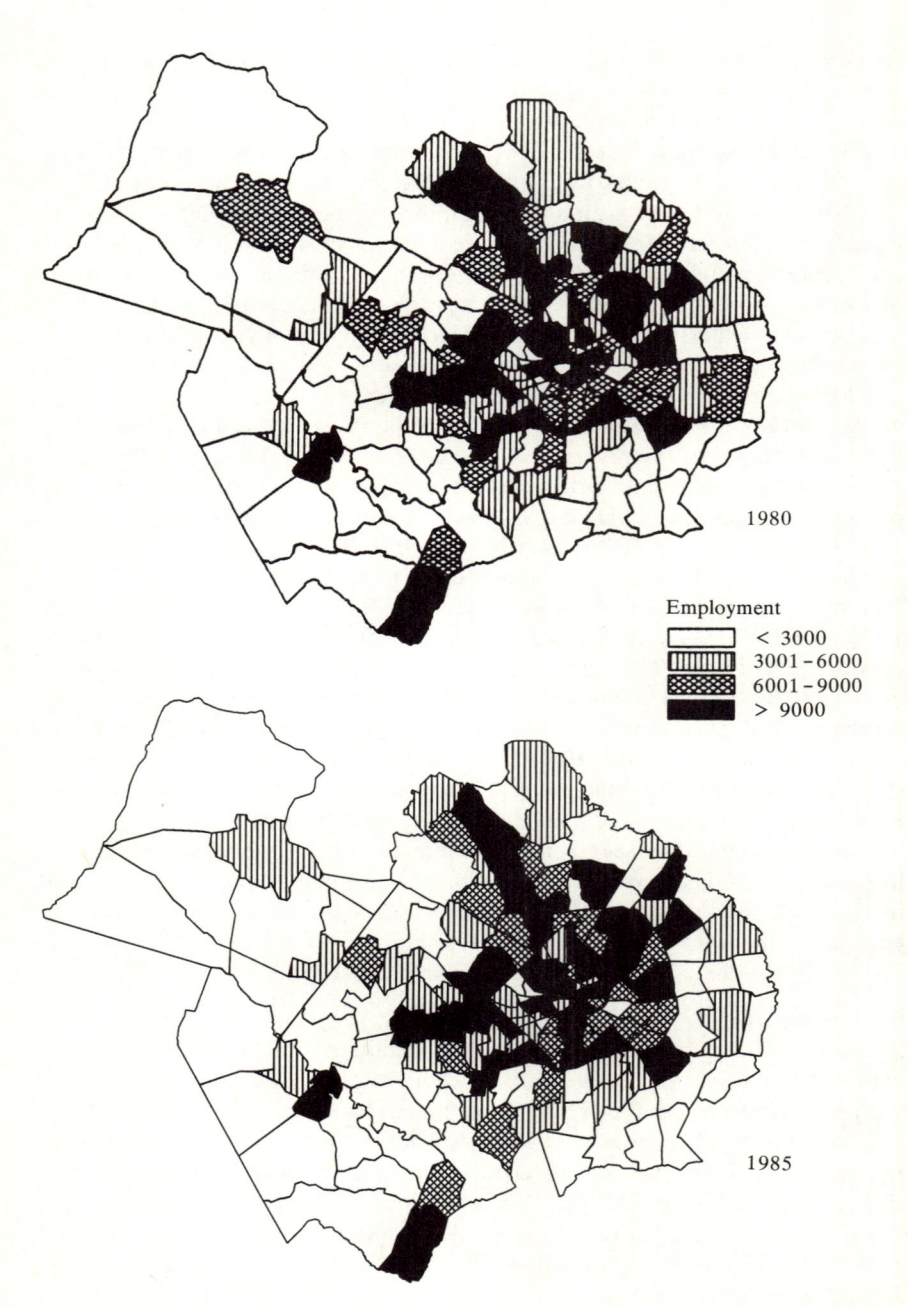

Figure 9.13. Washington, DC, region: test-1 results for employment for 1980, 1985, 1990, and 2000.

1990

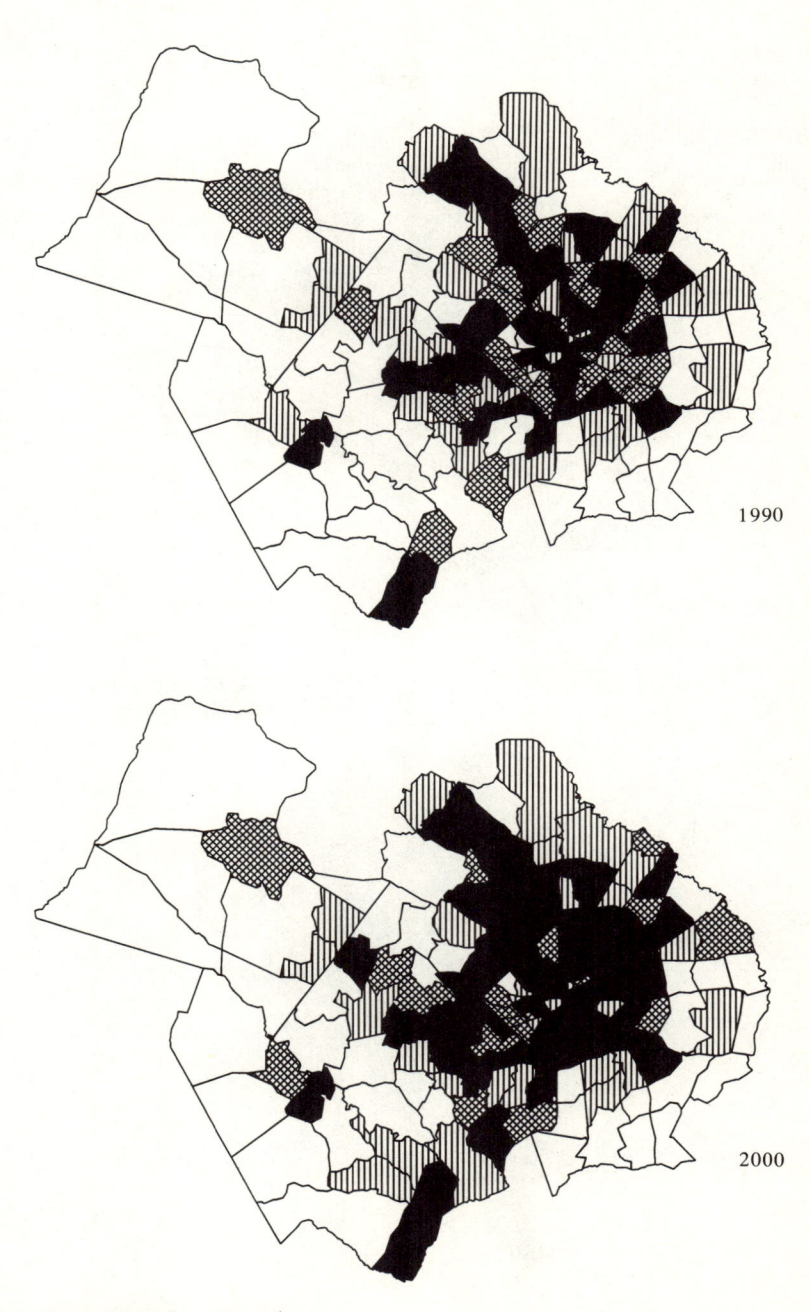

2000

Figure 9.13. (continued).

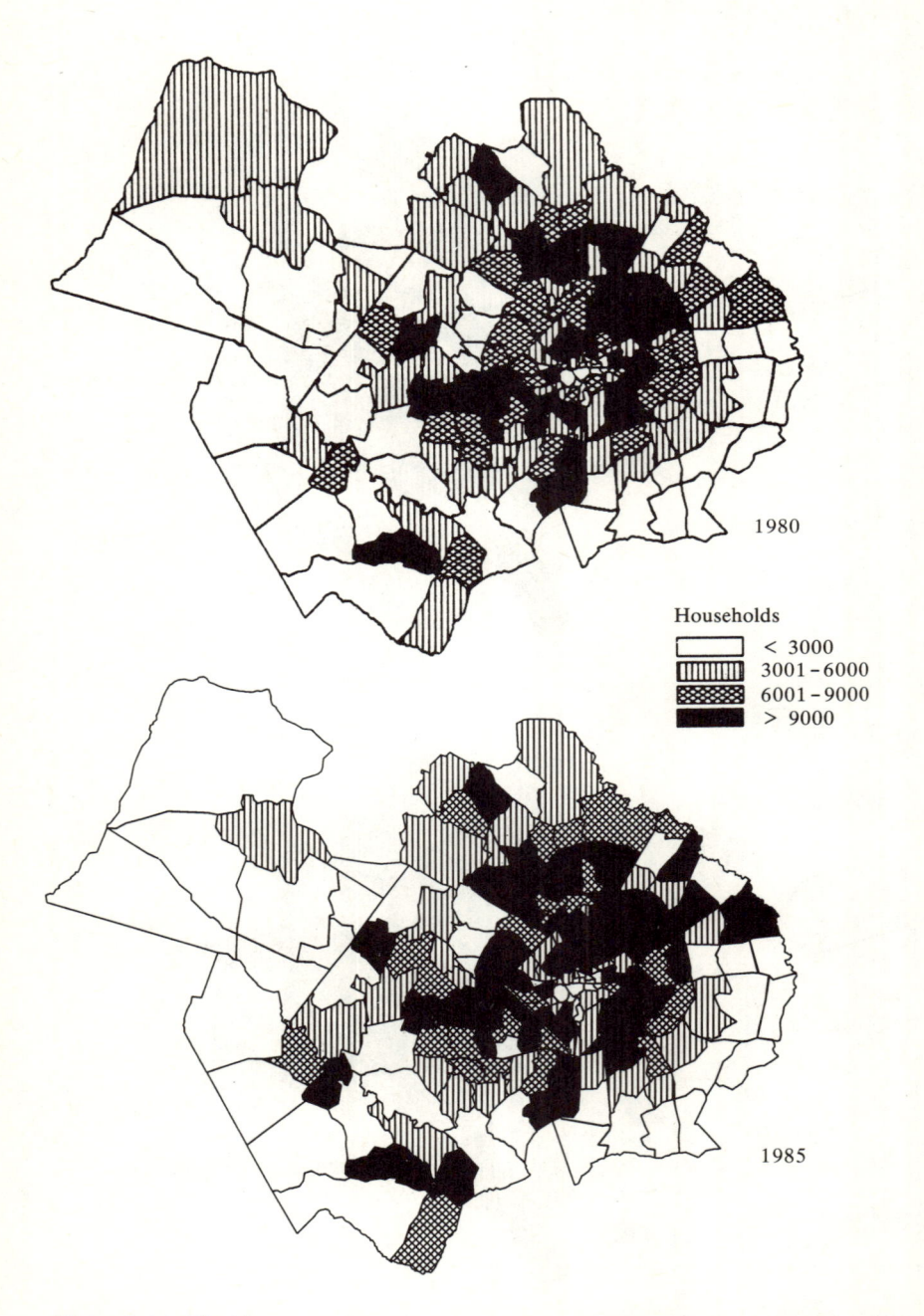

Figure 9.14. Washington, DC, region: test-1 results for population for 1980, 1985, 1990, and 2000.

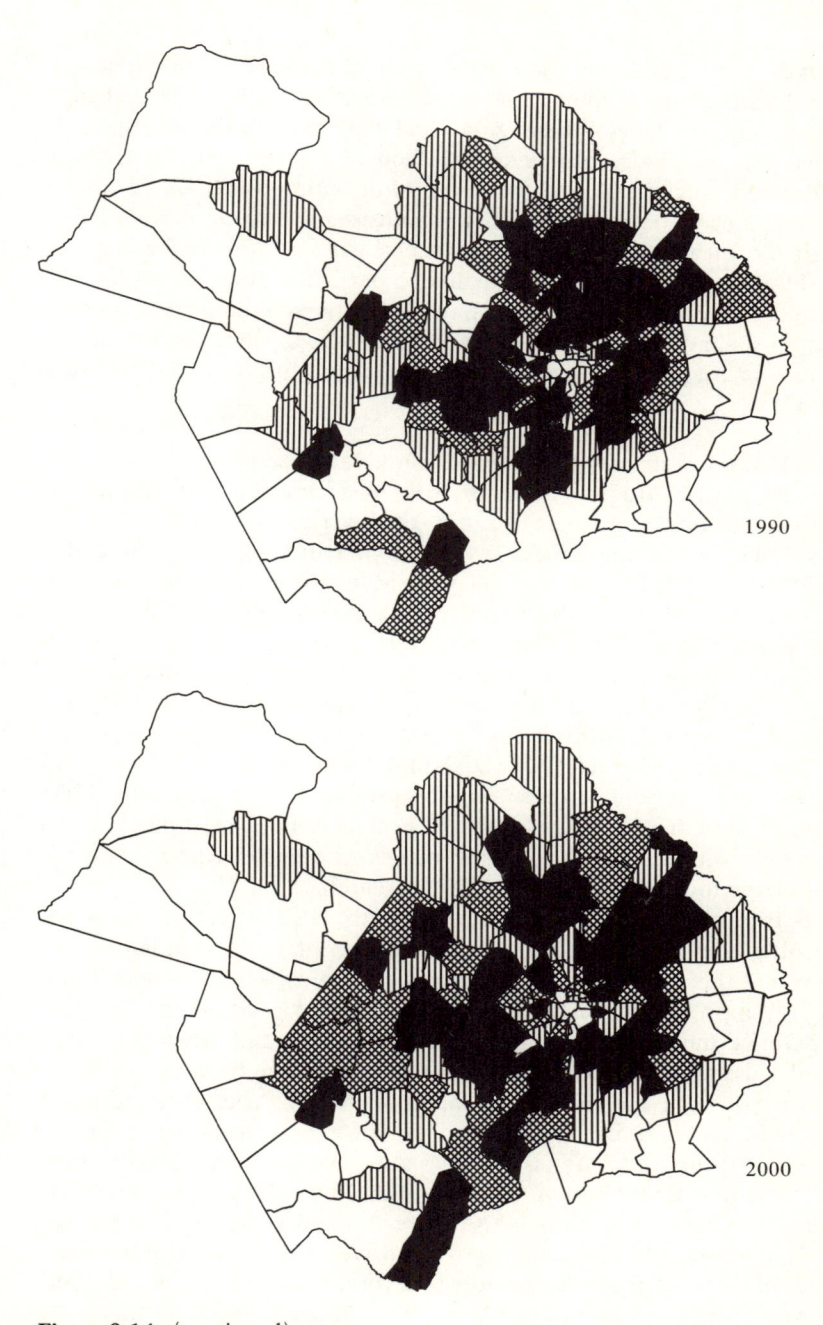

Simultaneous location–transportation models: 2

1990

2000

Figure 9.14. (continued).

transport costs are fixed over the forecast period, the lags and nonlinearities in the model structures result in different employment and household changes by zone, varying by locator type, zone, and time period. For example, the mean absolute percentage change in total households by zone, for Houston from 1990-95, was 50.0%, and the standard deviation was 0.42. Compared with the mean of 0.50, there was clearly a considerable amount of variation. Although the mean absolute percentage changes are similar for the four household types, the actual changes can, in a particular zone, be positive for one household type and negative for another. Thus even though the constant transport costs undoubtably produce a forecast bias towards dispersion, the overall results are probably fairly reliable indicators of the spatial evolution of the regions.

9.5 Test 2. Variable activity location with variable transport cost: one mode

The second model structure used in these tests involved, in addition to what was done for test 1, the utilization of the trip-distribution capability in DRAM, and the linking of a network assignment program to the end of the EMPAL-DRAM location model pair. Many complications arise from this augmentation of the model set. First, the trip-distribution capability of DRAM yields trip-probability matrices for three types of trip: work to home, work to shop, and home to shop. It is necessary to provide the total numbers of trips by which these probability matrices are multiplied to yield actual trips. These should probably be zone specific and trip-type specific person-trip generation rates, which should be followed by trip-purpose specific persons-per-vehicle rates. For the present tests a regional vehicle-trip total is used for each trip purpose, which is then multiplied by the proper probability matrix to yield the matrix of vehicle trips by purpose. There is an obvious need here for further research.

These trip matrices then become input to NETWRK, the network-assignment model. There are a substantial number of unresolved issues clustering around this component of the model system as well. The selection of the general approach to be followed, that is, which assignment algorithm, single path or multipath, iterative incremental, or equilibrium, is always open to discussion. The current series of tests were done by using two increments of a stochastic multipath assignment algorithm. After each increment the link volumes were used to recalculate the travel times for congested links. Even putting aside the issue of assignment-algorithm selection there remain other questions. Which volume-delay function is to be used? The traditional functions in common use were derived (several decades ago) from observations of actual on-the-road travel times and trip volumes. Virtually all network modelling involves several (often many) actual roads or highways being represented as one mathematically defined 'link'. The travel time and vehicle capacity of this abstract 'link' are then modified during the trip-assignment process, according to the volume-delay function.

First, how well do abstract 'links' represent actual physical facilities, that is, what consistent methods should be used to transform the times and capacities of the actual facilities into abstract link descriptions? Second, are the traditional volume-delay functions too steeply sloping for these aggregated abstract links? Further, what, if anything, is to be done about the issue of generalized cost of travel, which deals with the problems of aggregating several measures of travel difficulty (for example, time, costs) into some single measure for each zone-to-zone pair?

Other issues are associated with the actual model-to-model linkages as well as the procedure to be used in solving the models for numerical answers. A particularly troublesome problem is that of calibration, in a crude sense, of the overall model system. For the test-1 series the zone-to-zone impedances used in calibration were provided by the government agency from whom the initial data set was obtained. They were obtained by using selected nodes from a more complex network then in use by the agency, and by creating a matrix of the travel times between these selected nodes to represent zone-to-zone times for the aggregate zone system used in these tests. These impedances were specified as being representative of afternoon peak-hour travel times and were used in the calibration of EMPAL and DRAM. A similarly prepared future-year impedance matrix was used in the test-1 simulations. The same model parameters prepared for test 1 were used for the test-2 computer runs. Yet, when the DRAM-produced trips were loaded on the aggregated-link highway network, the congested travel times for individual zone-to-zone pairs were four to ten times greater than the original impedances used in test 1. Examination showed the vehicular capacity of the aggregated network to be substantially less than that of a detailed abstracted network for the region, and probably even more substantially less, by comparison, than that of the actual road network. Thus it was necessary to scale the numbers of trips to be loaded, by a process of trial and error, down to the point where the mean minimum path times on the loaded (congested) abstract network were approximately equal to the zone-to-zone times of the original impedance matrix used in the calibrations and for test 1. This was just the smallest step towards unravelling the enormously complex problem of system calibration for linked models.

All these issues and problems not withstanding, a series of model runs was made, called test 2, and analagous to test 1.

9.5.1 Houston

The regional forecasts used for the test-2 runs are identical to those for the test-1 runs described above. The regional trip-total forecasts are given in table 9.2 (see page 220). These totals were multiplied by a factor of 0.2 to compensate for the difference in total capacity of the coded highway network used in these runs and the capacity of the actual 'on-the-ground' road network. This scaling value was selected by a trial-and-error procedure, with the criterion being that the mean value of zone-to-zone travel time on

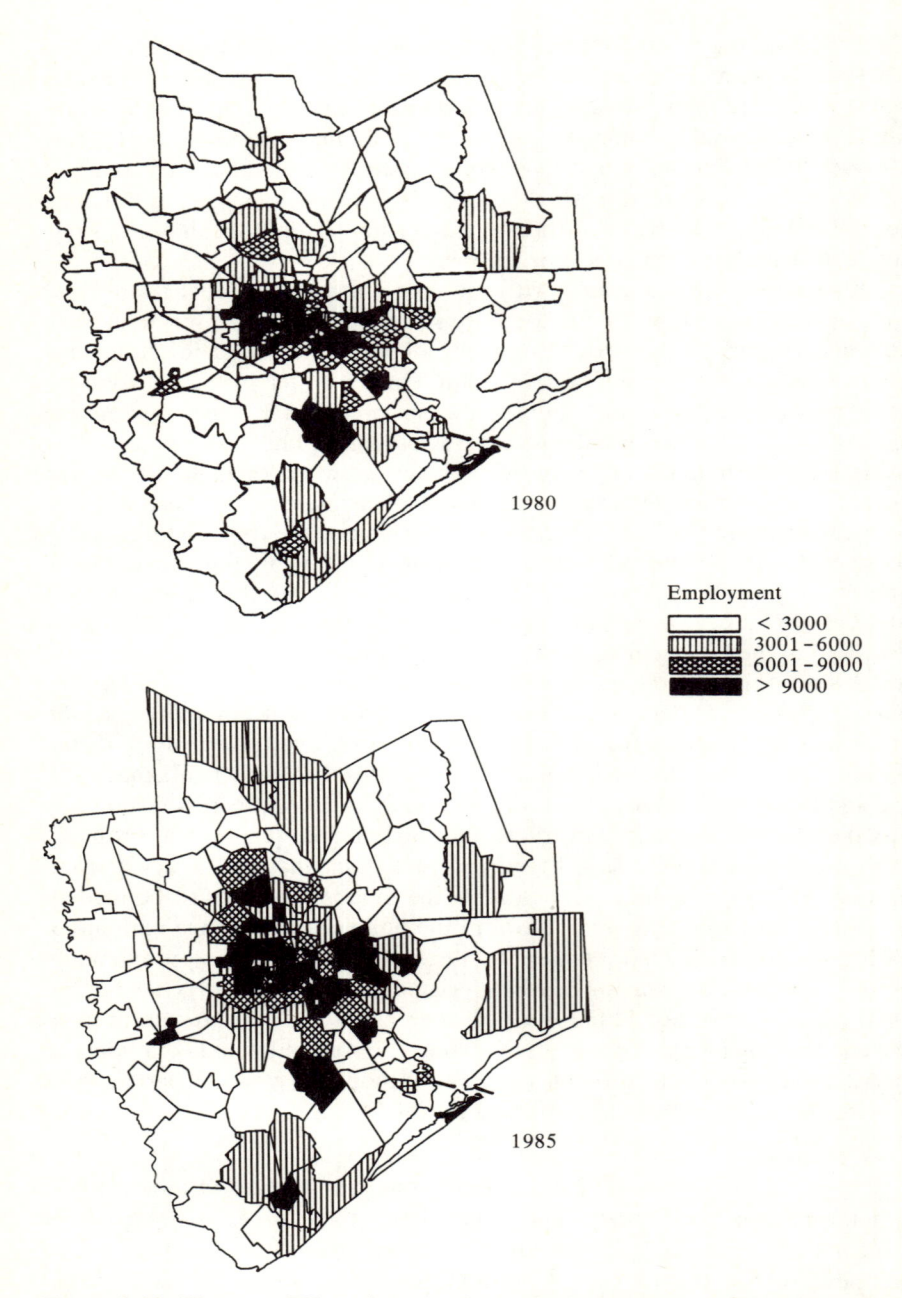

Figure 9.15. Houston, TX, region: test-2 results for employment for 1980, 1985, 1990, and 2000.

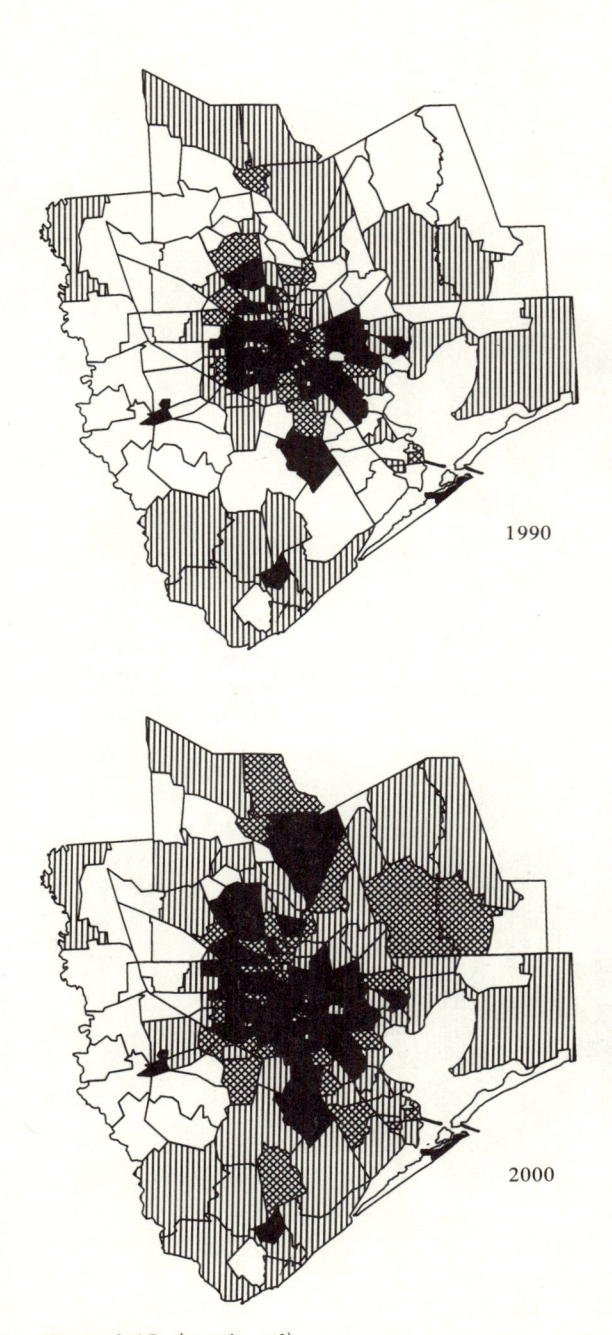

Simultaneous location–transportation models: 2

1990

2000

Figure 9.15. (continued).

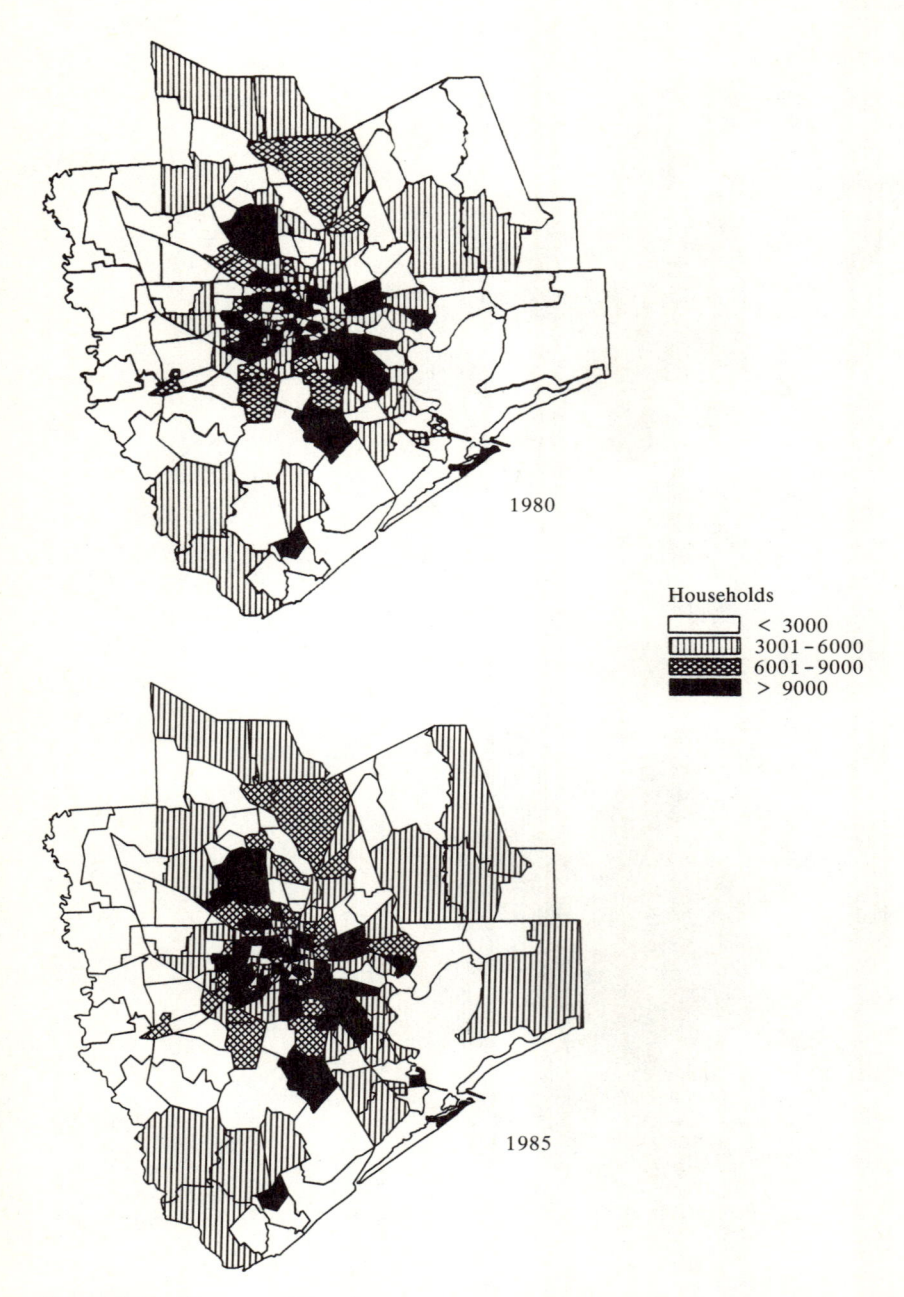

Figure 9.16. Houston, TX, region: test-2 results for population for 1980, 1985, 1990, and 2000.

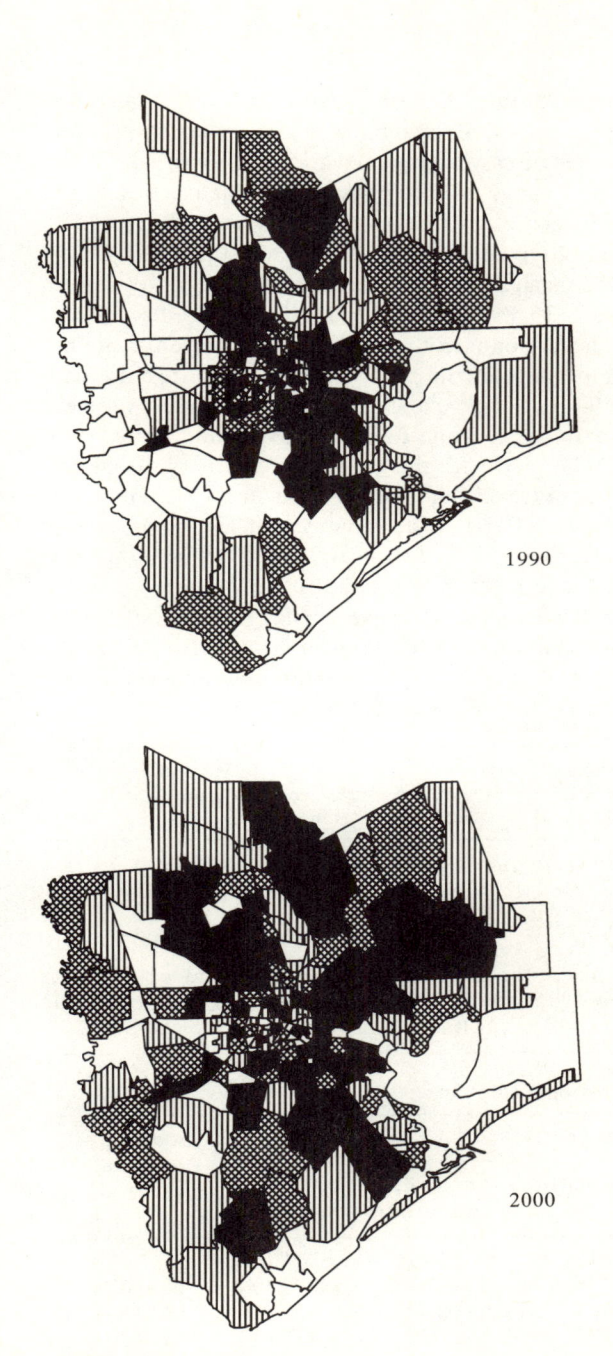

Simultaneous location – transportation models: 2   219

1990

2000

Figure 9.16. (continued).

the assigned (congested) network must be comparable with the mean value of zone-to-zone travel times provided by HGAC. The mean travel time increased over the forecast period, as the regional trip totals increased; but, for these tests, there was no change in the 'physical' characteristics of the highway network (for example, no links were added, no links were improved).

The spatial patterns of employment which resulted from these test-2 runs are shown in figure 9.15. The mean absolute percentage changes in total *zonal* employment were: 1980–85, 35.4%; 1985–90, 19.4%; 1990–95, 25.8%; and 1995–2000, 24.9%. These numbers are only slightly different from the test-1 values. A visual comparison of the test-1 results in figure 9.11 with the test-2 results in figure 9.15 shows virtually no difference until the year 2000. The actual increase in mean zone-to-zone travel time from 1980 to 2000 is just under 18%. Coupled with the rather lower sensitivity of employment location to changes in travel time these results are what one would expect. A comparison of the test-1 spatial distributions of employment with those for test 2 gives the following results for mean absolute percentage difference: 1985, 0.8%; 1990, 1.2%; 1995, 3.4%; and 2000, 5.7%.

Household location is somewhat more sensitive to changes in zone-to-zone travel times. The spatial distributions of total households resulting from the test-2 runs are shown in figure 9.16. The mean absolute percentage changes

Table 9.2. Houston, TX: regional forecasts.

Year	Employment forecasts				Regional population forecasts
	industrial	institutional	office	retail	
1985	507 000	158 000	544 000	346 000	3 623 956
1990	531 800	165 729	543 271	381 400	3 650 000
1995	556 600	173 457	700 543	416 800	3 969 000
2000	581 400	181 186	857 814	452 200	4 288 000

Note: unemployment was assumed the same for all four sectors, and was 7.8% for 1985, 8.0% for 1990, 6.5% for 1995, and 5.5% for 2000; the employee-to-household ratio was assumed to be 1.31 for 1985, 1.27 for 1990, and 1.37 for 1995 and 2000.

Table 9.3. Washington, DC: regional forecasts.

Year	Employment forecasts				Regional population forecasts
	industrial	government	wholesale and services	retail	
1985	188 470	596 817	522 953	344 768	3 118 325
1990	204 919	631 536	602 323	373 204	3 287 078
1995	217 558	663 961	681 288	400 218	3 439 501
2000	232 622	697 783	756 204	427 149	3 574 701

Note: unemployment was assumed to be 5% in the first three sectors and 6% in retail for the entire forecast run; the employee-to-household ratio was assumed to be constant at 1.4010 for the entire forecast run.

in total *zonal* households were: 1980-85, 14.7%; 1985-90, 56.8%; 1990-95, 50.3%; and 1995-2000, 41.3%. Here, visible differences between test 1 and test 2 begin to show up (by comparing figure 9.12 with figure 9.16) in 1990. Although the year-2000 maps are generally similar, there are some clear differences, with the test-2 results being perhaps a bit less evenly spread over the region. A comparison of the two sets of spatial distributions shows the mean absolute percentage difference between test 1 and test 2 for total zonal households as: 1985, 3.4%; 1990, 9.8%; 1995, 15.6%; and 2000, 22.5%.

9.5.2 Washington

As for Houston, the test-2 regional forecasts were identical to the test-1 regional forecasts. The regional trip totals were scaled to adjust for the coded network capacity as compared with the on-the-ground capacity. The actual values were given in table 9.3. Rather than belabor the point, suffice it to say that similar results to those of Houston were obtained for Washington. The spatial distributions of total employment and total households are given on figures 9.17 and 9.18. Again, the test-1 employment distribution is not very different from that of test 2. The test-2 household distribution is similar, but shows significant differences. Again, the principal difference in the

Table 9.2. (continued).

Regional trip forecasts

work-to-home	work-to-shop	home-to-shop
484 218	215 208	349 713
505 974	224 877	365 426
527 730	234 547	381 138
549 486	244 216	396 851

Table 9.3. (continued).

Regional trip forecasts

work-to-home	work-to-shop	home-to-shop
326 160	58 560	117 120
369 539	65 810	130 401
418 688	73 957	145 189
474 374	83 113	161 653

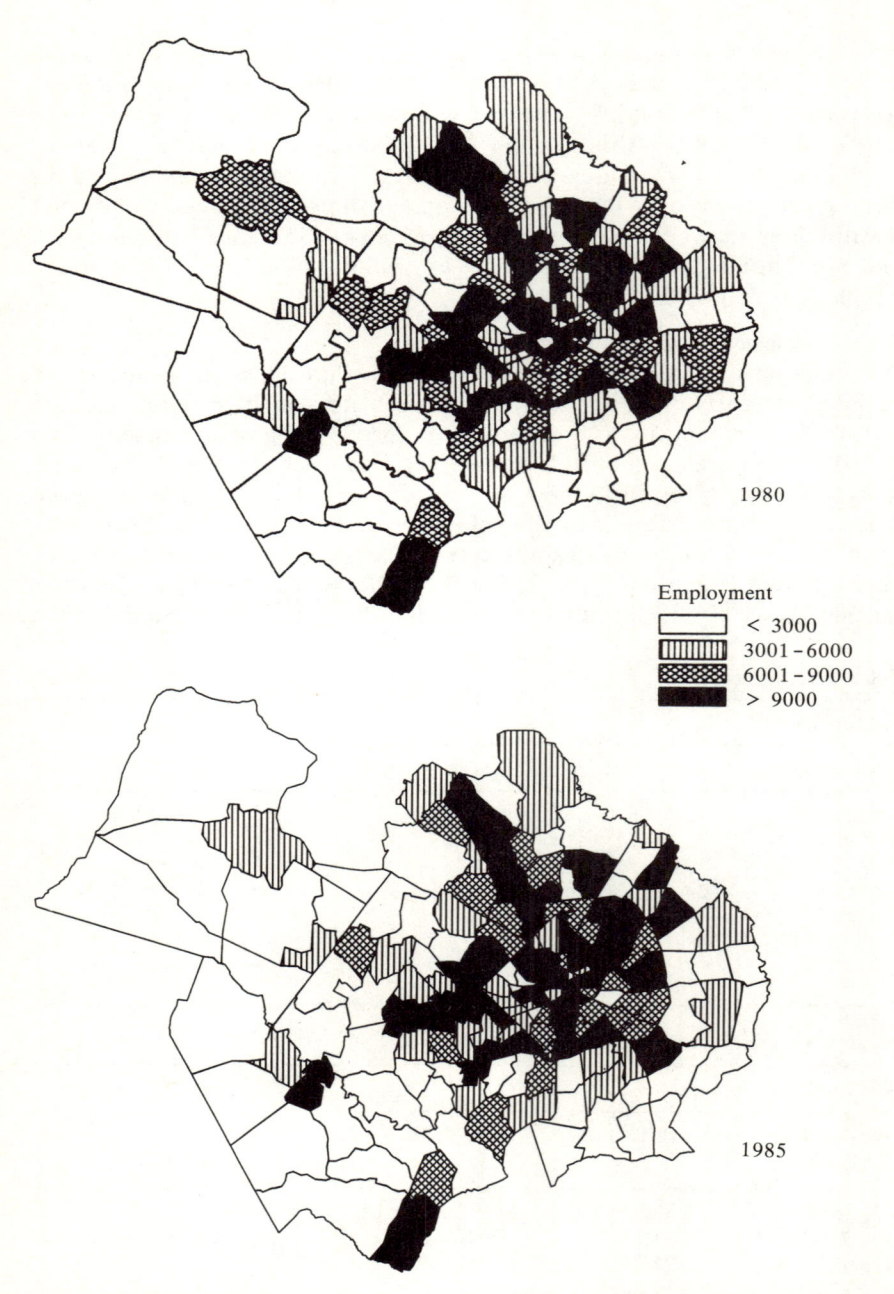

Figure 9.17. Washington, DC, region: test-2 results for employment for 1980, 1985, 1990, and 2000.

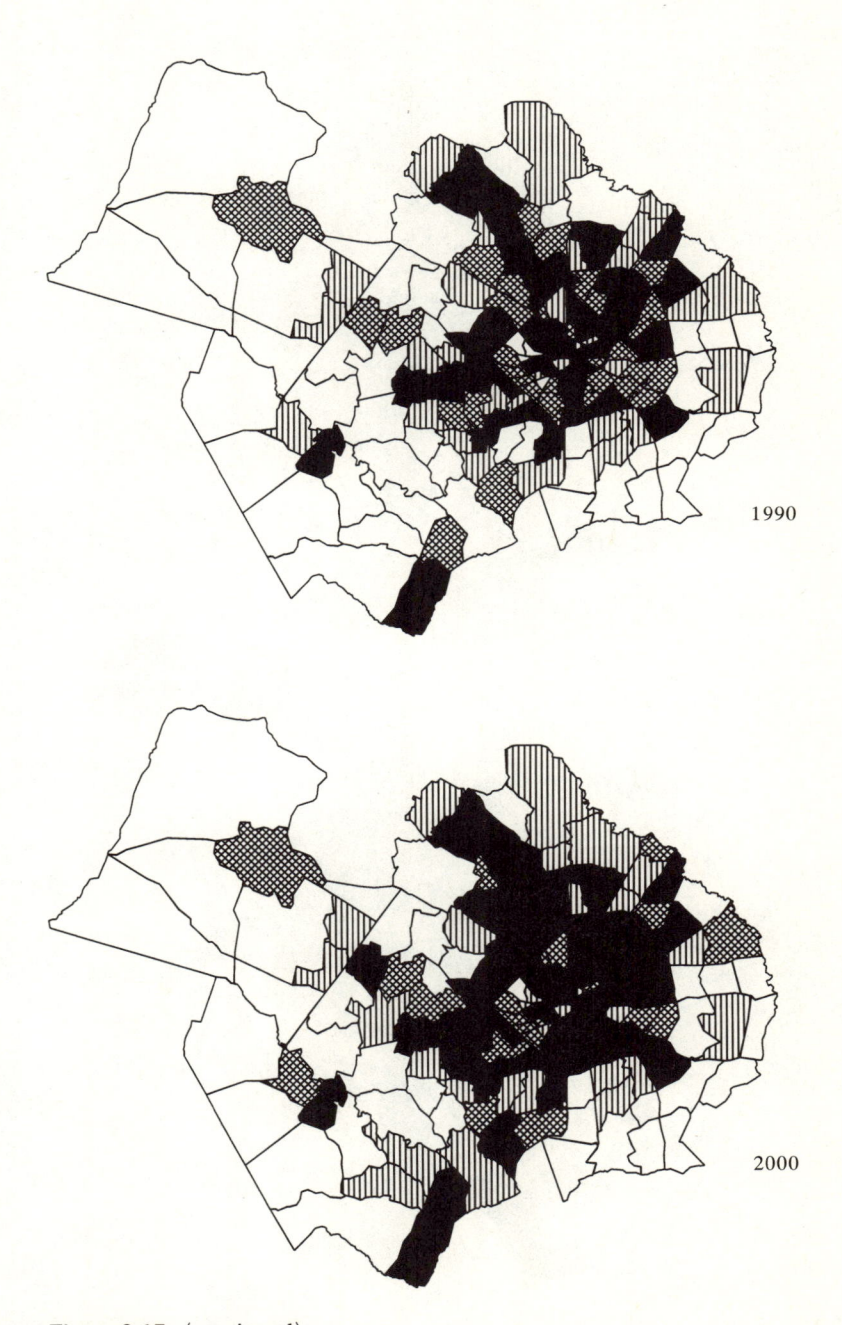

Simultaneous location–transportation models: 2   223

1990

2000

Figure 9.17. (continued).

224   Chapter 9

Households
☐ < 3000
▦ 3001–6000
▩ 6001–9000
■ > 9000

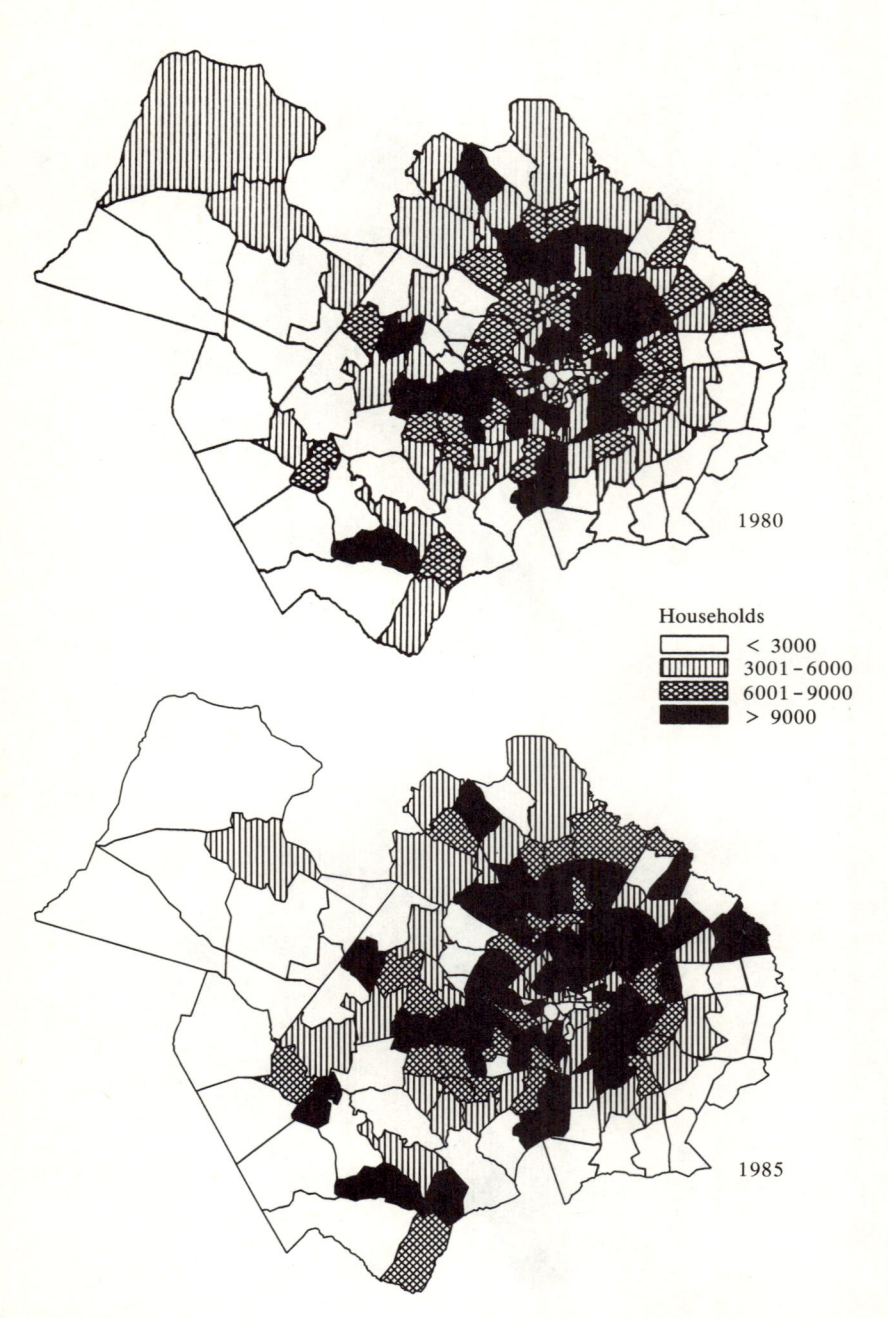

1980

1985

Figure 9.18. Washington, DC, region: test-2 results for population for 1980, 1985, 1990, and 2000.

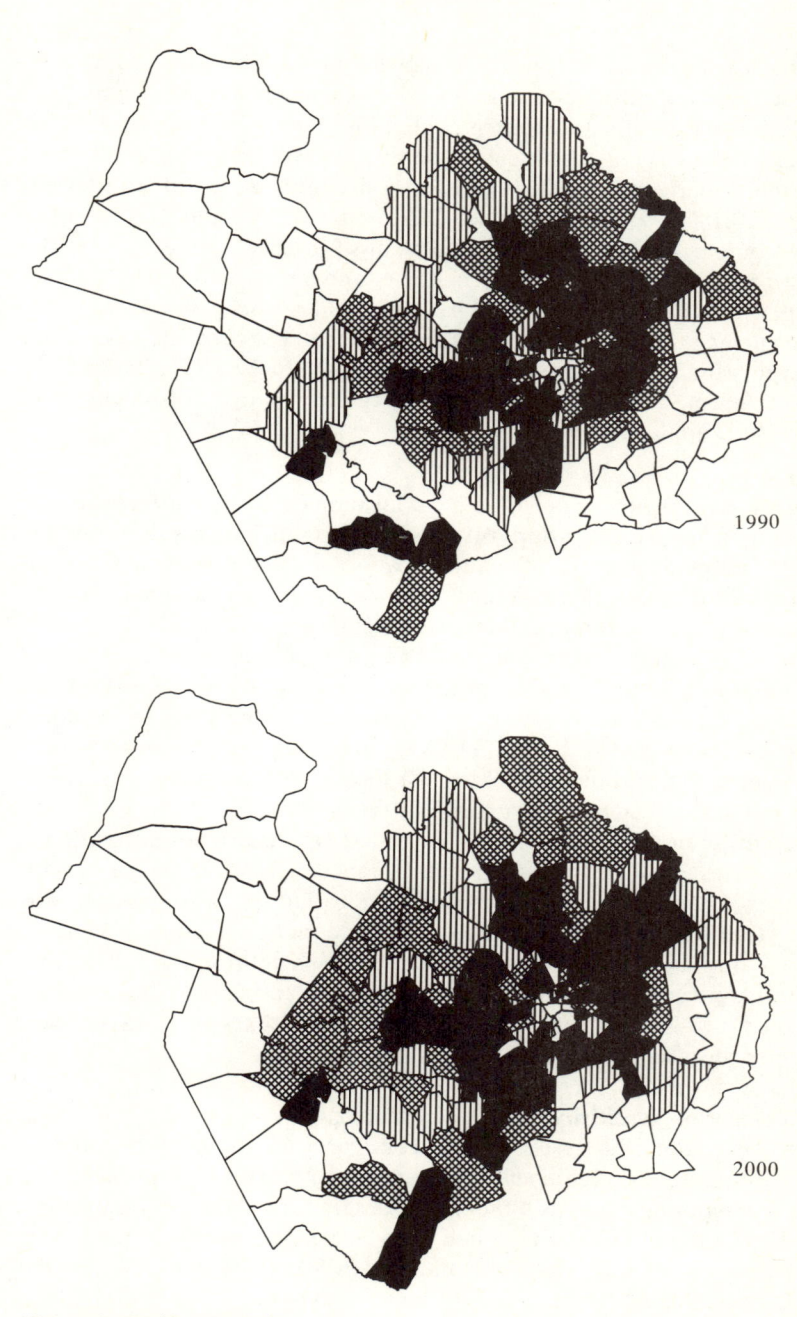

Simultaneous location–transportation models: 2   225

1990

2000

Figure 9.18. (continued).

household distribution is, it appears, somewhat more subnucleation (that is, dispersion), but concentrated around preexisting concentrations of activity.

The *zonal* mean absolute percentage differences for employment in Washington were virtually identical to the test-1 results, but the test-2 *zonal* mean absolute percentage differences for households were much different: 1980-85, 20.1%; 1985-90, 67.1%; 1990-95, 32.2%; and 1995-2000, 22.6%. A comparison of the Washington results for test 1 and test 2 gave the following *zonal* percentage differences for employment: 1985, 0.0%; 1990, 0.9%; 1995, 1.7%; and 2000, 2.5%. The household results gave considerably larger *zonal* mean absolute percentage differences between the test-1 and test-2 outputs for Washington: 1985, 0.0%; 1990, 12.8%; 1995, 24.6%; and 2000, 36.6%. Yet, overall, the larger scale regional patterns of results from test 1 and test 2 appear to be rather similar.

9.5.3 Summary
Test 2 introduces the effects on subsequent location of the network congestion which results from prior location. Perhaps the most striking results from this test is that the broad patterns of spatial locations of employment and households were not dramatically different from the test-1 results. This is not to say that there was no difference, there were significant differences, but rather that, to a considerable degree, the patterns were similar in both test 1 and test 2. This is as it should be. One of the worst failings of early location and land-use models was that they were overly sensitive to changes in transport times or costs. In EMPAL and DRAM the multivariate attractiveness measures serve the purpose of making the location of activities sensitive to many more factors than transport costs (or times) alone.

Yet, the differences between test-1 and test-2 results are not insignificant. The test-1 to test-2 mean absolute percent zonal difference in the year-2000 total household location in Houston was 22.5%, and for Washington it was 36.6%. To the extent that the EMPAL and DRAM models and their estimated parameters may be considered to be reliable, this represents a clear and important response to congestion-induced changes in highway travel times. Clearly then, the issue is the extent to which these changes in travel times are a proper representation of the effects of congestion.

9.6 A comment on sensitivity to data and assumptions
In progressing from the test-1 model configuration to the test-2 model configuration there is a significant increase in complexity of approach. The increase in computing expense, although modest in today's terms, is not insignificant. EMPAL and DRAM together, for a single time period or recursion, require about 4 s of CPU time on an IBM 3090/180E computer. If the NETWRK model is added in, it brings the total to almost 40 s of CPU time, an order of magnitude increase. The differences between the test-1 and test-2 results are significant but not astonishing. As will be shown here, seemingly minor assumptions about data can have effects which are almost as great.

One matter to be discussed here concerns the diagonal elements of the zone-to-zone travel times (or costs). As such, it is necessary to step all the way back to the preparation (coding) of the highway network for use in modelling. It is customary that each of the zones (analysis areas) in the model have a point, or centroid, at which the zone, in effect, is connected to the network. From this centroid one or more connector links are defined. These connector links join the centroid to the links of the network. It is customary that these links be very much abstract, in that they represent many 'on-the-ground' roads, and are often not at all representative of any one such real bit of road. As such, the capacities of the connector links are usually set at an arbitrarily high value.

In the computer-program representation of most traffic-assignment algorithms the traffic is assigned from the origin zone to all destination zones. Trips from the origin which terminate in the same (origin) zone are not usually assigned to the networks. Yet these intrazonal trips can be quite substantial in numbers. The principal reason that these trips go unassigned is that there is not an adequate representation of the intrazonal network to which the trips could be assigned. The intrazonal network is represented only in terms of the above-mentioned connector links. The immediate consequence of this situation is that intrazonal travel times (or costs) are likely to be poorly estimated.

A series of additional tests were done with the Houston data to explore the sensitivity of model outputs to alternative assumptions and procedures for dealing with intrazonal travel times. As a point of reference, the exogenously provided zone-to-zone travel-time matrix used for the test-1 forecasts had a mean zone-to-zone travel time of 56.6 minutes. This matrix was used to start the test-2 forecasts also.

As mentioned above, there is a crude system-calibration process which must be done with the trip-assignment portion of the model package. This involves scaling back the actual vehicle-trip estimates to compensate for the substantial difference in capacity of the coded network and the actual on-the-ground road network in the model. For the Houston network used in these tests, a 50% trip scaling gave a mean zone-to-zone travel time of 96.4 minutes. Scaling by 30% gave 68.7 minutes, and 20% gave 61.7 minutes. As the exogenously provided travel-time matrix was known to understate peak-hour travel times, the 20% scaling was taken as the value to be used in the test-2 series of experiments.

In the test-2 experiments the travel time for each network link is a function of the flow (volume) on the link. The procedures used for these calculations were discussed in detail in chapters 6 and 7. These procedures do not, however, provide a mechanism for dealing with intrazonal times or costs, as the coded networks do not include intrazonal network links. One way around this difficulty is to take some exogenously estimated value for the intrazonal travel time. This approach, however, does not provide for a congestion-related increase in intrazonal travel time for successive model runs. A second approach, used in previous work with ITLUP, involved 'congesting' the intrazonal time as a function of congestion on the highway links that originate or terminate at the load node in the zone. Unfortunately, the Houston and the Washington networks connect the load nodes to the network via very high-capacity 'connector' links. These never show any congestion. As a consequence, the only possibility available was to take the diagonals of the exogenously provided travel-time matrix and use them as the intrazonal times in the congested matrix. This is what was done for the test-2 series of forecasts.

To test the implications of this procedure a second test-2 series of forecasts was done, where the intrazonal times were increased by 25% (that is, multiplied by 1.25). In the original test-2 runs the mean zone-to-zone travel time started at 56.6 minutes for 1985, and became 61.5 minutes for 1990, 63.7 minutes for 1995, and 67.0 minutes for 2000. In the revised test-2 runs the values were 56.6 minutes, 62.1 minutes, 63.9 minutes, and 67.8 minutes respectively. By 2000, the zonal average absolute percent difference was 13.8% for total households. Comparisons of the spatial patterns of households for 2000 are given in figure 9.19. For total households there were only a few zones that differed, mostly located to the east of the center of the region. Overall there were numerous zonal percentage changes, but few were of sufficient absolute magnitude to change the categories of the zones on the maps. For comparison a second set of runs was done, with the intrazonal times increased by 50%. A comparable set of maps is given in figure 9.20.

Yet another set of runs was done to examine the response of the system to trip scaling and intrazonal travel times. In this case the regional trip total was scaled by 30% (as compared with 20% in the original test-2 runs). This is equivalent to a 50% increase in vehicle trips on the highway network in the region. The exogenously supplied intrazonal times were here increased by 35% (that is, multiplied by 1.35) as compared with the 25% increase of the prior set. The mean zone-to-zone travel times for this set of runs were 56.6 minutes, 70.7 minutes, 73.2 minutes, and 83.3 minutes for 1985, 1990, 1995, and 2000, respectively. For 2000, the zonal mean absolute percentage difference from the original test-2 results was 18.3%, and from the first of these special tests, 15.0%. Figures 9.21 and 9.22 show comparisons of the year-2000 spatial pattern forecasts of this run with those of the original test-2 run. Again, there are a few employment changes, and somewhat more household changes.

Taken together these results are, at the same time, both reassuring and sobering. The gross spatial patterns show relatively modest changes, suggesting that, at such a scale, small errors in the intrazonal travel times or in the regional trip scaling will not result in dramatic differences in forecast results. At the zone-by-zone level of examination there are numerous significant percentage differences. The zonal mean absolute percent difference between the alternative test-2 runs for the year 2000 ranged from over one half to over three quarters of the difference between test 1 and test 2. To the

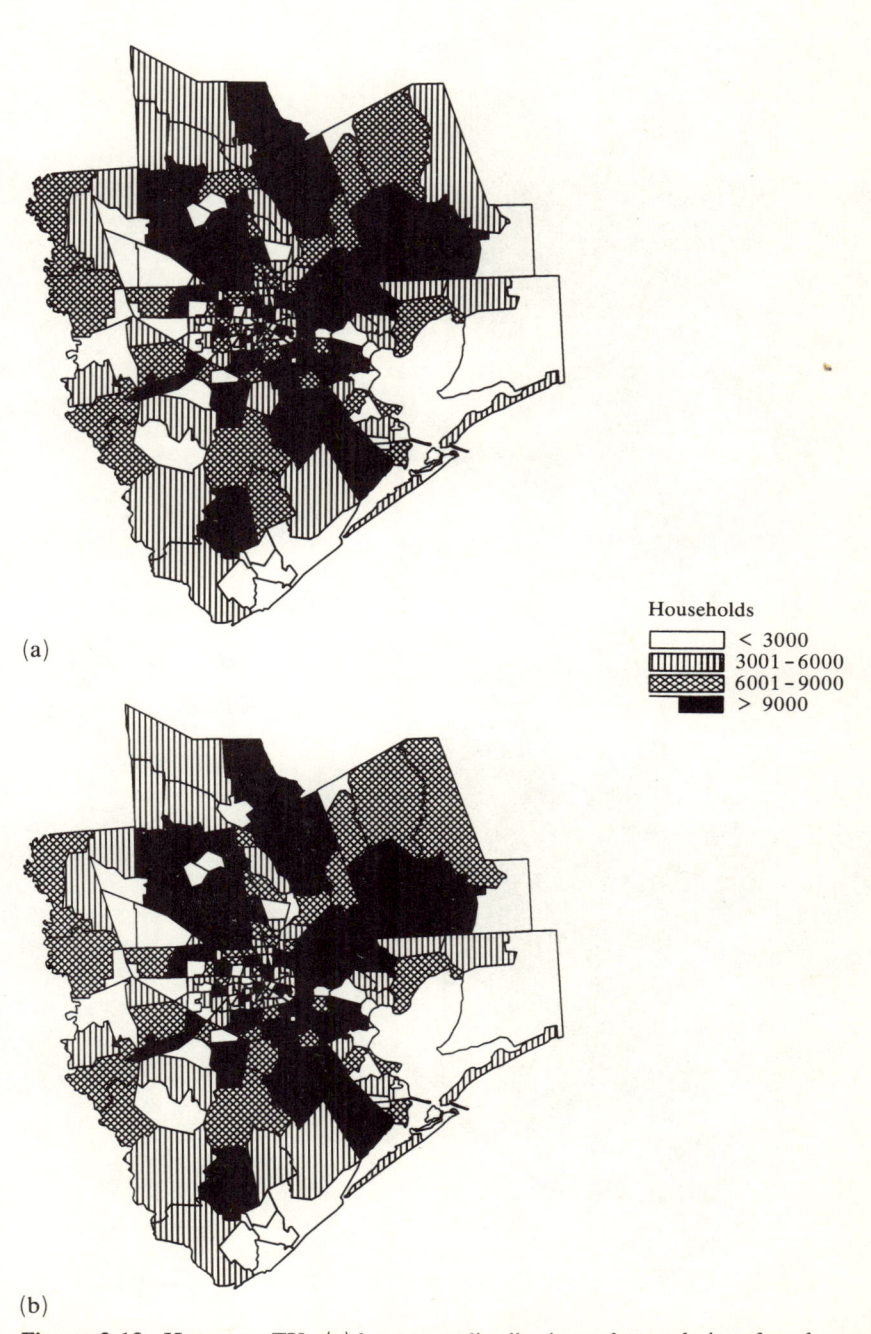

Simultaneous location–transportation models: 2

Figure 9.19. Houston, TX: (a) base-run distribution of population for the year 2000 and (b) the distribution of population with intrazonal costs 25% greater than in the base run, for the year 2000.

(a)

Households
☐ < 3000
▦ 3001–6000
▨ 6001–9000
■ > 9000

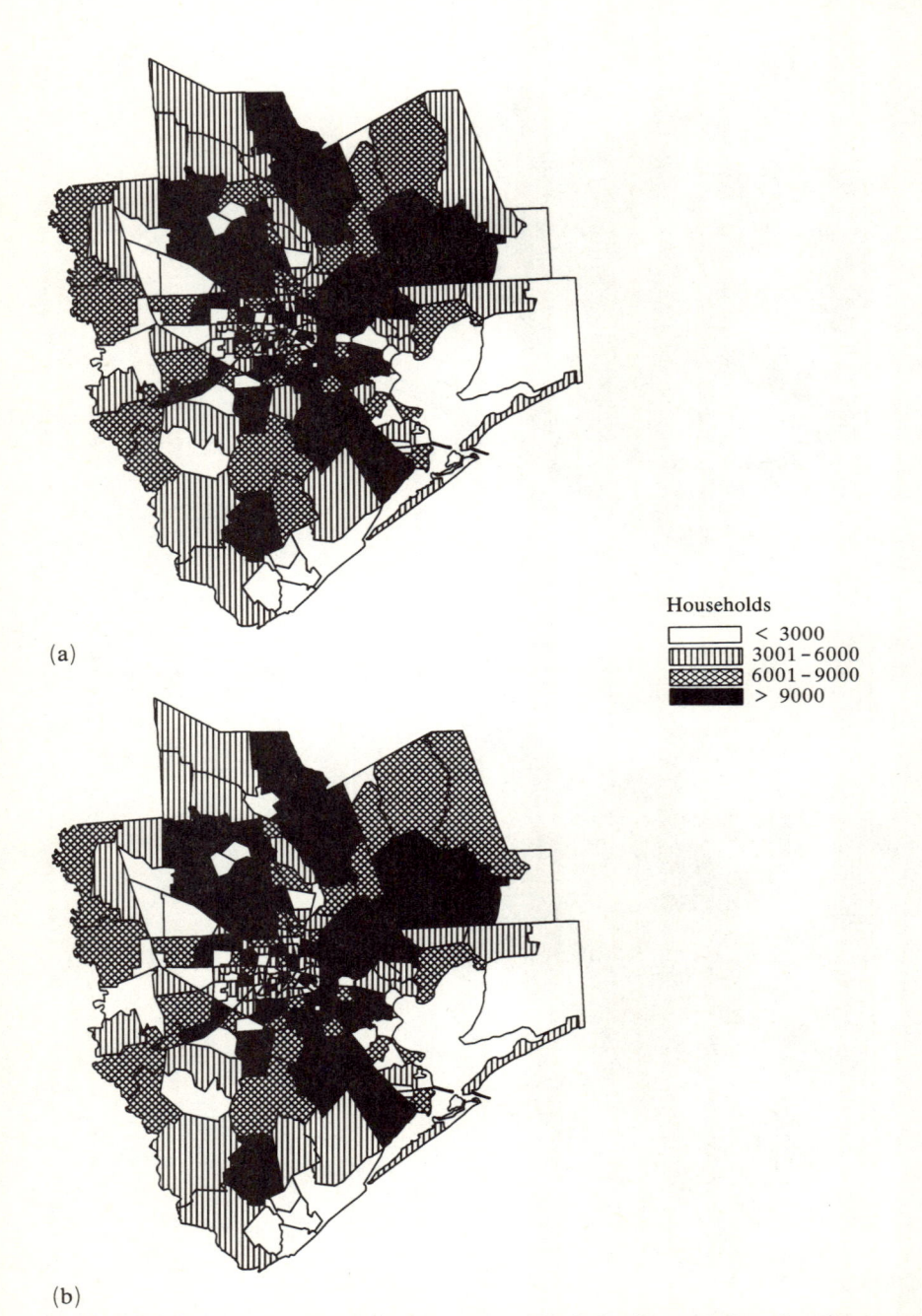

(b)

Figure 9.20. Houston, TX: (a) base-run distribution of population for the year 2000 and (b) the distribution of population with intrazonal costs 50% greater than in the base run, for the year 2000.

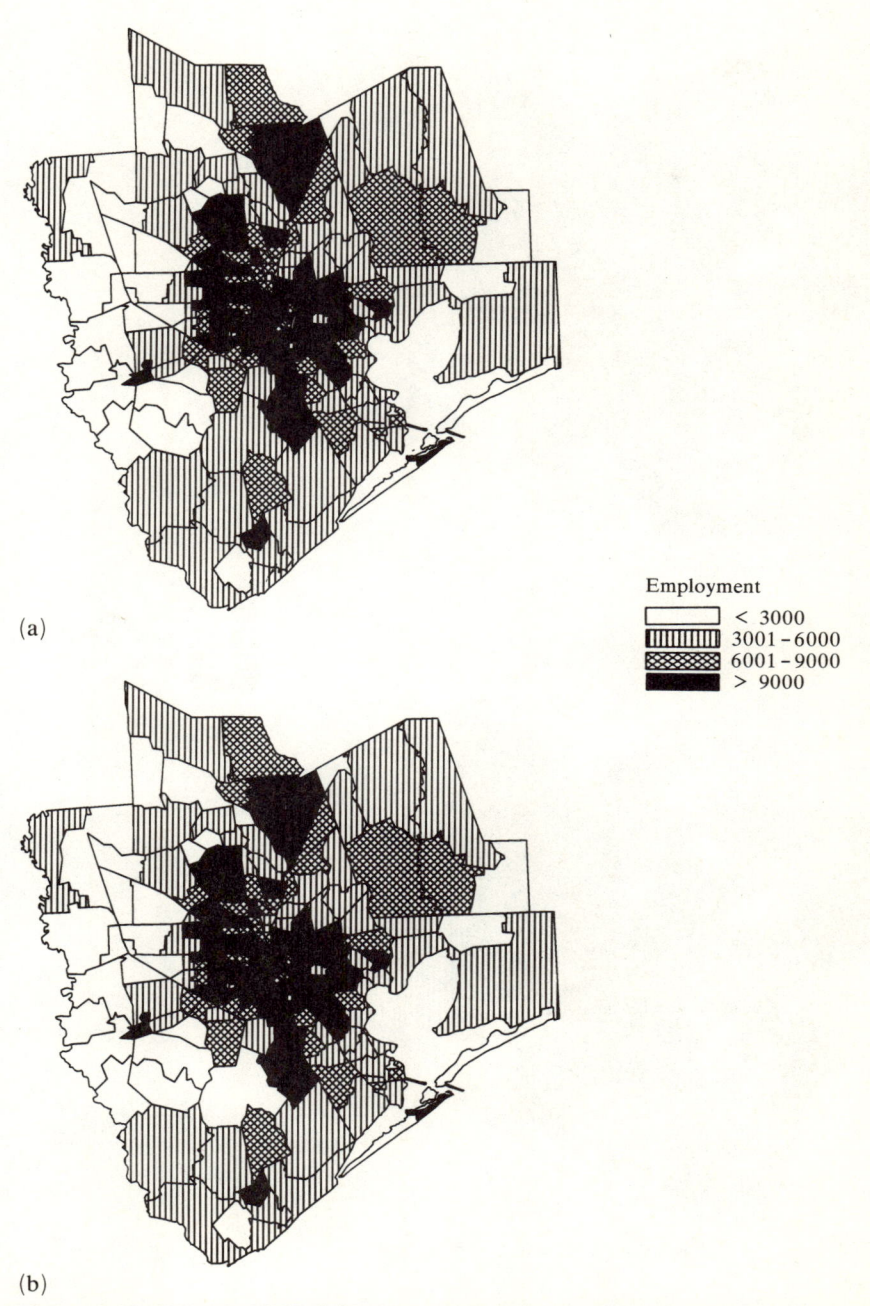

Figure 9.21. Houston, TX: (a) base-run distribution of employment for the year 2000 and (b) the distribution of employment with the regional trip generation increased by 50% compared with the base run, for the year 2000.

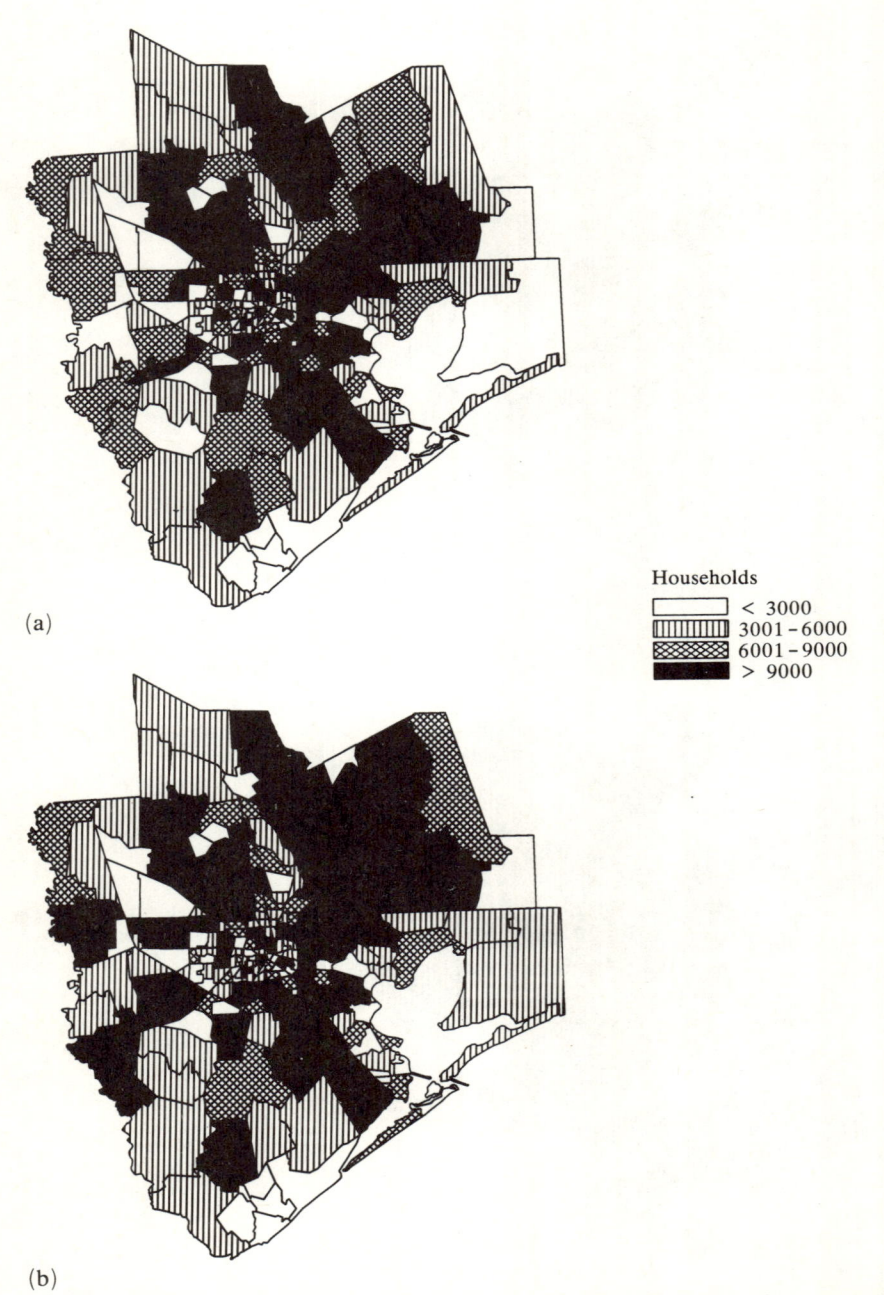

Figure 9.22. Houston, TX: (a) base-run distribution of population for the year 2000 and (b) the distribution of population with the regional trip generation increased by 50% compared with the base run, for the year 2000.

extent that one is concerned with the zone-by-zone accuracy of the forecasts, the regional-trip scaling factor must be rather carefully determined, and pains should be taken to estimate the intrazonal travel times correctly. Indeed, it is here made clear that under certain circumstances these rather basic assumptions can be as important as the decision as to which model package configuration is used to produce the forecasts.

9.7 Test 3. Variable activity location with variable transport costs: two modes

In order to make any practical use of urban models it is necessary to accept, at face value, a substantial number of assumptions. In the previous discussion of test 1 and test 2 the model structures were shown to rely upon many such assumptions. The addition of modal choice to the system for the test-3 series requires that yet another series of assumptions be made. First there is consideration of the general structure of the mode-split procedure, for example, predistribution or postdistribution mode split? Then there is the problem of choosing a particular model to use. In the model selected (in this case a multinomial logit formulation) there is the question of variables and, as a subquestion, the matter of selecting a way of calculating generalized costs, or times, given the two data sets being used in this chapter.

Even when these questions are dealt with, there is the further question of the composite cost (or time) calculation. Here again, the question of system calibration appears. The use of composite times results in rather different impedance matrices as compared with those used in test 1 and test 2. Should some form of composite-cost impedances have been used in the initial calibration (or a recalibration) of EMPAL and DRAM for test 3? Here the question of zone size also becomes important, as many transit trips are shorter than automobile trips and may take place entirely within a single analysis zone. This implies the need for rather careful consideration of intrazonal trips and intrazonal mode split, as these determine the values of the diagonals of the impedance matrices later input to EMPAL and DRAM, and to which both models, as shown above, are rather sensitive.

By way of further exploring the composite cost or time question, consider a simple spatial interaction model of the form

$$T_{ij} = E_j B_j W_i \exp\beta c_{ij} \,, \tag{9.1}$$

where

T_{ij} is the number of trips between zone i and zone j,
E_j is the number of employees (at place of work) in zone j,
W_i is some measure of the attractiveness of zone i,
c_{ij} is the time (cost) of travel between zone i and zone j,
B_j is a balancing factor, described by the equation

$$B_j = \left[\sum_i W_i \exp(\beta c_{ij}) \right]^{-1} \,,$$

β is an empirically determined parameter (expected to be negative).

Note that travel cost and travel time are used interchangeably in the following discussion. In many cases such a model is summed over trip ends to produce estimates of residential location, which gives

$$N_i = \sum_j T_{ij} = \sum_j E_j B_j W_i \exp(\beta c_{ij}) , \qquad (9.2)$$

where
N_i is the number of residents in zone i.

Subject to the availability of data the actual values of c_{ij} may be travel time, travel cost, or in some cases distance. The question of what, in theory, makes the best measure here is not resolved in the current literature. The purpose of the measure is to quantify the 'difficulty of interaction'. Clearly the measure must, when possible, take into account the circumstances of the tripmakers, their income, their value of time, etc. In practice, this is such a complex and poorly defined question that rather simple expedients are taken. In principle, one would like some *generalized measure of the cost of interaction* which would combine properly all the appropriate factors into a single value. Again, in practice this is often not possible and a simple travel-time or travel-cost measure must be used.

A second aspect of this problem appears with regard to models which represent multimodal transport facilities. Consider, for example, the case of a system which deals with two modes: private automobile and public transit. To skip over the question of generalized cost, assume that for each mode only travel time is known. How should these two times, for each $i-j$ zone pair, be used?

One possibility is that the two costs should be used as separate variables yielding the following model form:

$$N_i = \sum_j E_j B_j W_i \exp(\beta_1 c_{ij}^1) \exp(\beta_2 c_{ij}^2) , \qquad (9.3)$$

where

$$B_j = \left[\sum_i W_i \exp(\beta_1 c_{ij}^1) \exp(\beta_2 c_{ij}^2) \right]^{-1} , \qquad (9.4)$$

and where there are now two empirically determined parameters, β_1 and β_2, associated with the costs c_{ij}^1 and c_{ij}^2 for each of the two modes. These two parameters would be estimated as part of the calibration of the model. In practice it is virtually certain that there would be a very high correlation between c_{ij}^1 and c_{ij}^2. Recent research on calibration of models such as these has clearly shown that high correlation between independent variables can result in serious difficulties in determining the correct values of their parameters (Putman and Kim, 1984). In models such as DRAM, where there are several parameters in an expanded formulation of the attractiveness variable W_i, this problem is further exacerbated.

Another alternative is to develop a procedure for combining the modal costs into a *composite-cost* variable prior to their use in the model.

The problem then becomes one of determining the precise formulation for such a composite-cost calculation. The taking of a simple weighted average was often used in early multimodal analyses. The weighting used was the modal shares, which resulted in a formulation such as

$$c'_{ij} = \sum_k a^k_{ij} c^k_{ij} , \qquad (9.5)$$

where
a^k_{ij} is the percent of trips taking mode k between zones i and j, and where it is required that

$$\sum_k a^k_{ij} = 1.0 . \qquad (9.6)$$

Unfortunately this formulation has a major drawback. If a new mode is added between zones i and j, and if that new mode is more costly than the prior composite cost (or if there was only one mode to begin with, and if the new mode is more costly than the first mode) then the composite cost increases. This is counterintuitive. Any new mode added to an existing system should result in a improvement in zone-to-zone impedance.

An alternative formulation follows from an analogy to electronic circuit theory, when several resistances are connected in parallel. In such a case the addition of a parallel path, no matter how high its resistance, cannot increase the resulting path resistance. In the case of travel costs, the formula would be

$$\frac{1}{c'_{ij}} = \sum_k \frac{1}{c^k_{ij}} , \qquad (9.7)$$

which can be transformed to

$$c'_{ij} = \left[\sum_k (c^k_{ij})^{-1} \right]^{-1} . \qquad (9.8)$$

Of course, the restriction that c^k_{ij} is positive must be imposed to avoid dividing by zero

Recent work on the ties between random utility theory, probabilistic choice models, and spatial interaction models suggests yet another composite-cost function (Wilson et al, 1981). In particular, the following form is suggested:

$$c'_{ij} = -\frac{1}{\lambda} \ln \sum_k \exp(-\lambda c^k_{ij}) , \qquad (9.9)$$

where
λ is an empirically derived parameter.

The rationale for this formulation is discussed in considerable detail in several articles from the late 1970s (Senior and Williams, 1977; Williams, 1977; Williams and Senior, 1977). Some of what follows derives from that discussion.

It is well known that several model system configurations for travel-demand modelling have been developed and tested in practice. Three of the most popular of these are:
structure 1   G – MS – D – A,
structure 2   G – D – MS – A,
structure 3   G – (D + MS) – A,
where
G   is trip generation,
MS is modal split,
D   is trip distribution,
A   is trip assignment,
and any of these joined with a '+' sign means that they are done by some simultaneous procedure. The structure of ITLUP used in test 3 is a variant of structure 2, being

(G + D) – MS – A ,

because trip generation and trip distribution, for the work trips and the shopping trips, are done simultaneously with the location of households and employment. In the following discussion the ITLUP test-3 structure will be taken as given.

It is useful to further consideration to write this model structure (in part) in what is known as the 'product-share' form (Manheim, 1973). For structure 2 this becomes

$$T_{ij}^{nk} = G_i^n M_{ij}^n M_{ij}^{nk} , \qquad (9.10)$$

where
T_{ij}^{nk} is the total number of trips taken by persons of type n on mode k between zones i and j,
G_i^n is the total number of trips by persons of type n originating from zone i,
M_{ij}^n is the proportion of G_i^n which go to zone j,
M_{ij}^{nk} is the proportion of $G_i^n M_{ij}^n$ which go on mode k.
It should be noted that

$$T_{ij}^n = G_i^n M_{ij}^n , \qquad (9.11)$$

where
T_{ij}^n is the total number of trips made by persons of type n between zones i and j.

Now it should be noted that in justifying the form of equation (9.9) for composite costs, it was assumed that the equations for both M_{ij}^n and M_{ij}^{nk} were of logit form. Thus generation and distribution were accomplished by

$$T_{ij}^n = G_i^n B_i^n A_j D_j \exp(-\beta^n c_{ij}^n) , \qquad (9.12)$$

where
D_j is the total number of trips attracted to (terminating in) zone j,

B_i^n and A_j are balancing factors such that

$$\sum_j T_{ij}^n = B_i^n \,, \tag{9.13}$$

$$\sum_{i,n} T_{ij}^n = A_j \,, \tag{9.14}$$

c_{ij}^n is the composite cost (difficulty of interaction) which represents all available modes between zones i and j,
β^n is an empirically derived parameter.
The mode split is then calculated by the following

$$M_{ij}^{nk} = \frac{\exp[-\lambda^n(c_{ij}^k + \delta^{nk})]}{\sum_k \exp[-\lambda^n(c_{ij}^k + \delta^{nk})]} \,, \tag{9.15}$$

where
c_{ij}^k is the generalized cost (difficulty of interaction) between zones i and j on mode k,
λ^n and δ^{nk} are empirically derived parameters.

Note that the summation in the denominator of equation (9.15) must be over only the modes which are available to persons of type n. The generalized cost, c_{ij}^k, is sometimes represented as a complex function of out-of-pocket costs, prorated fixed costs of vehicle ownership (for automobile modes), in-vehicle time, out-of-vehicle time, etc. There may well be additional parameters in this function which also must be estimated. The parameter δ^{nk} may be referred to as a 'modal penalty' and, in essence, can be considered to represent otherwise unobservable modal attributes and to adjust the modal split proportions accordingly. The parameter λ^n represents the sensitivity of the mode-choice decision to the generalized cost of interaction, and also acts as a dispersion parameter to distribute trips over modes. Last, to refer back to equation (9.12), the parameter β^n represents the sensitivity of trip-making, and thus of location, to composite costs.

DRAM, the residential submodel of the ITLUP package is a singly constrained model and thus, in form, would resemble the following revised version of equation (9.12).

$$T_{ij}^n = W_i^n A_j^n D_j \exp(-\beta^n c_{ij}^n) \,, \tag{9.16}$$

where, now

$$D_j = \sum_h a_{hn} E_j^h \,, \tag{9.17}$$

where
E_j^h is the amount of employment of type h in zone j,
a_{hn} is a regional ratio of the number of type-n residents per type-h employee,
and
W_i^n is a surrogate measure of the residential utility of zone i to type-n residents,

A_j^n is a balancing factor, expressed by

$$A_j^n = \left[\sum_i W_i^n \exp(-\beta^n c_{ij}) \right]^{-1}. \tag{9.18}$$

In addition, the surrogate residential utility measure W_i^n is normally formulated as a multivariate multiparametric function of specific attributes of zone i such as vacant land, percentage developed (urbanized) land, and the income distribution of the resident population. All of this has been described elsewhere (Putman, 1983a). The important point to note here is that although β^n is the only parameter in equation (9.12), it is but one of many (as many as nine in some applications) in equation (9.16).

If one takes equations (9.11) and (9.12) together, then

$$T_{ij}^{nk} = T_{ij}^n M_{ij}^{nk}, \tag{9.19}$$

and, since DRAM uses a version of equations (9.16), (9.17), and (9.18), all that is left is the M_{ij}^{nk} which may be estimated by using any of a variety of postdistribution multinomial logit mode-split models.

The logit model has been implemented for numerous urban areas for mode-choice analysis. There have been two principal types of logit model applied—aggregate and disaggregate. In the aggregate type, zone-to-zone data on modal shares are used as dependent variables in calibration. The independent variables include zone-specific variables such as average income of residents, and interchange-specific variables such as zone-to-zone travel times or costs. The calibration of disaggregate-type models is based on survey data describing the mode-choice behavior of individuals. The dependent variables are actual observed modal choices of individuals. The independent variables are attributes of the individuals, such as income or automobile availability, and interchange-specific (for the individual's trip origin and destination zones) variables such as travel times and costs, perhaps divided into in-vehicle and out-of-vehicle times and costs. The calibrated disaggregated model equation is then applied to the forecasted population in each zone to produce the aggregate zone level mode-split forecasts. Use of the disaggregate mode-split model is favored in current practice.

Regardless of whether the aggregate or disaggregate logit model is used, virtually all contemporary applications will involve multiparametric formulations. In nearly all cases there are at least three and as many as five parameters involved. A typical model of this, multinomial logit, form, would be

$$M_{ij}^{nk} = \frac{\exp[-\lambda_n^k + \gamma_n^k X_i^1 + \eta_n^k X_{ij}^2 + \phi_n^k X_{ij}^3 + ...]}{\sum_k \exp[-\lambda_n^k + \gamma_n^k X_i^1 + \eta_n^k X_{ij}^2 + \phi_n^k X_{ij}^3 + ...]} \tag{9.20}$$

where
X_i^1 is a zone-i specific variable,
X_{ij}^2, X_{ij}^3 are zone-i zone-j interchange-specific variables,
$\lambda_n^k, \gamma_n^k, \eta_n^k, \phi_n^k$ are empirically derived parameters specific to mode k and type-n persons (travellers).

Note, of course, that for the disaggregate model it would be necessary to go through two steps; first to calculate the probability that a particular person type would choose mode k for each $i-j$ interchange, and then, based on the zonal attributes (for example, X_i^1), to calculate the market share for mode k for that interchange. Even so, the point of multiparametric functions in this discussion remains valid. It is this point [that is, that here we deal with several parameters for equation (9.15)] that causes difficulties. In particular, the composite-cost function of equation (9.9) makes use of the λ parameter from equation (9.15) [with the understanding that equation (9.9) would normally be expected to have an n superscript throughout]. The question is, given equation (9.20) as the modal share equation, what value of λ is to be used in the composite-cost function of equation (9.9)? It is not clear how the mathematics that produced equation (9.9), given equation (9.15) (Williams, 1977), can be made to produce a revised equation (9.9), given equation (9.20).

In lieu of the anlaytics, a series of numerical experiments was conducted to gain an understanding of the shape of equation (9.9), and to learn about its sensitivity to changes in λ and c_{ij}. These experiments took fixed values of λ (a different value for each of the experiments in the series). Two travel modes were used—automobile and public transit. In each experiment the automobile cost, c_{ij}^1, was fixed, whereas transit cost, c_{ij}^2, varied from a value less than automobile cost to a value more than automobile cost. The resulting composite cost from equation (9.9) was then plotted. The graphs shown in figure (9.23) show the results of these experiments. It should be noted that on each graph, in addition to the family of curves for equation (9.9), the curve for equation (9.7) is also shown. For figure 9.23(a) the fixed value of c_{ij}^1 is 50, for figure 9.23(b) it is 75, for figure 9.23(c) it is 100, and for figure 9.23(d) it is 200. In all figures, the topmost curve, the solid line, is for λ equal to 0.20, the descending order of curves, with successively shorter dashes, correspond to λ equal to 0.10, 0.07, 0.05, 0.03, and 0.01, respectively. The line and double-dashed curve on each figure is the result from equation (9.7).

Williams (1977) posits four conditions which should be satisfied by a composite cost, for all $i-j$ pairs:
1 it must always be nonnegative,
2 it must always be monotonically increasing,
3 it must always be less than the minimum of its components, that is, less than the cost of the minimum-cost mode for the $i-j$ pair,
4 at the limit, as the dispersion (mode-switching) parameter λ tends to infinity, the composite cost must tend to the minimum-cost mode for the $i-j$ pair.

The fourth condition bears some comment, as it acts through both the mode-split calculation and the composite-cost calculation. When λ becomes extremely large there is no mode splitting, and all trips use the least-cost mode. Thus the composite cost should, in this circumstance, equal the minimum cost.

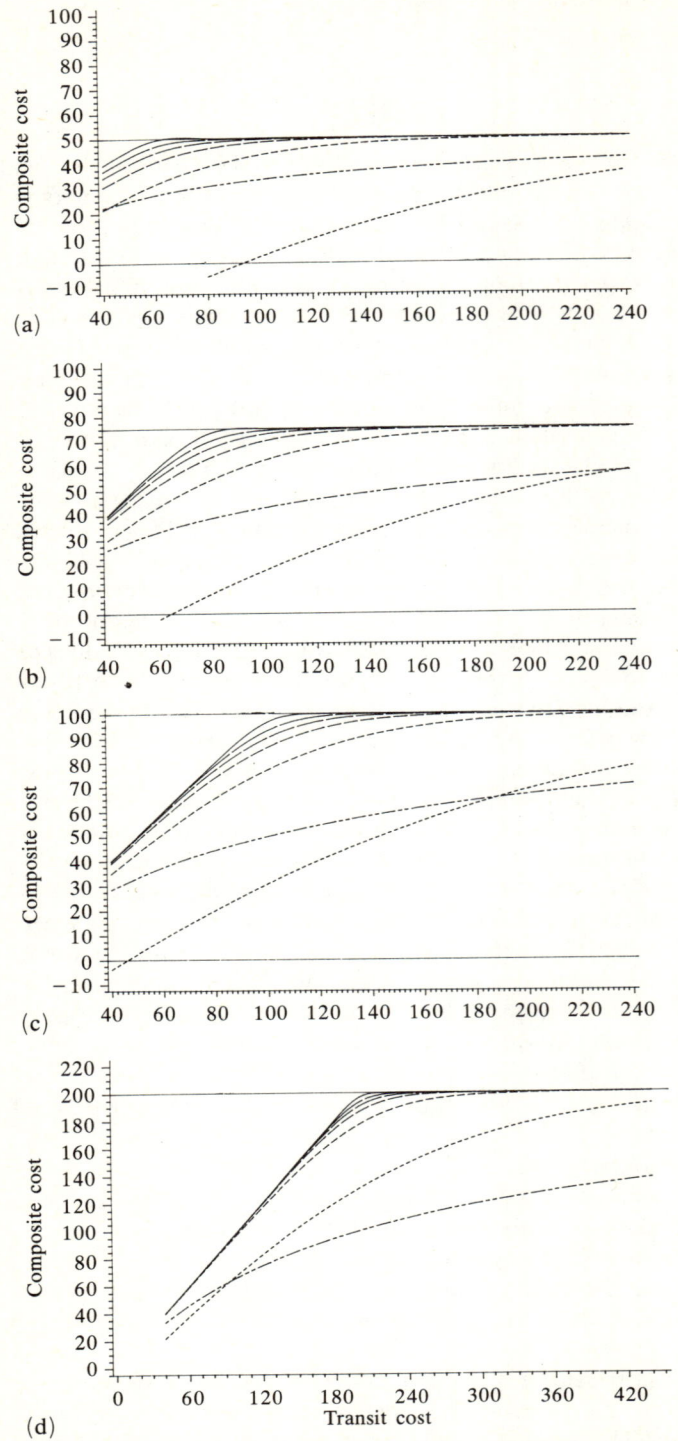

Figure 9.23. Composite-cost functions for a fixed automobile cost of (a) 50 units, (b) 75 units, (c) 100 units, and (d) 200 units.

By inspection of figure 9.23, it can be seen that the composite-cost function of equation (9.9) sometimes violates condition one. Its performance with respect to condition four is also somewhat problematic as it can be seen that simple substitution of extremely large values of λ into equation (9.9) will result in c_{ij}' approaching zero. The composite-cost function of equation (9.7) suffers from neither of these problems. In reasonable ranges of λ and c_{ij}^k, both functions seem to perform adequately. Because of its presumed theoretical superiority, the test-3 results were conducted by using equation (9.9) for composite cost, but further work is clearly necessary here.

The choice of equation (9.9) required the selection of a value of λ, as the multinomial logit model used for mode split is multiparametric and does not use travel costs directly, but rather uses differences in travel costs between modes. Thus for test 3 an arbitrary value of 0.05 was taken for λ in the composite-cost function. This was done in order to produce some, but not too much, sensitivity of composite cost to addition of modes. This choice clearly implies an additional system-calibration problem as discussed above. Thus the resolution of the composite-cost question, itself a subquestion in the model-integration problem, is somewhat elusive. Some structuring of the situation is possible. Elimination of some clearly incorrect alternatives is also possible. But the selection of a 'guaranteed correct' answer does not seem to be possible. As an exemplar of urban systems modelling problems this is an unfortunately too common outcome.

Having said all this, the actual results from the test-3 run are somewhat anticlimactic. The actual percentage of transit usage in the Houston region in 1980 was about 4%. It did not seem likely that there would be any observable difference in spatial pattern forecasts as a result of such a low level of transit usage, so the test-3 runs were not done for the Houston data set. In Washington, the base-year percentage of transit usage was of the order of 11%, perhaps large enough to show some change in model outputs as a result of its inclusion in the modelling process. Even so, the maps of spatial patterns resulting from test 3 did not show enough differences from those of test 2 to warrant their being included here.

The zonal percentage differences, for total employment, between test 2 and test 3 for Washington were: 1985, 0.0%; 1990, 0.4%; 1995, 0.7%; and 2000, 1.0%. For total households the zonal percentage differences were somewhat greater: 1985, 0.0%; 1990, 4.5%; 1995, 6.8%; and 2000, 9.1%. Clearly some of the individual differences are smaller than the differences which might be observed from alternate assumptions of the sort discussed in the preceding section of this chapter. From a policy perspective this implies that in an already developed region transit will not cause significant region-wide changes in spatial patterns. This *does not* mean that the introduction of a new transit facility to a portion of the region where none had existed before, or where highway access was poor, would not result in a significant local change. In such cases transit may well produce significant effects, though, in reality, it will always be difficult to know whether the results were produced by the transit improvement, or whether the transit improvement, located by presumably knowledgeable planners, was placed in a growth corridor of the region to begin with.

With the conclusion of the runs of tests 1, 2, and 3, there remains the question of equilibrium versus nonequilibrium solutions to these models. In all these tests the models were run in sequence, within each time period, or recursion. In the next section of this chapter the question is posed as to whether some attempt should be made to have the models converge to a stable solution for each time period by iterating the model sequence within each recursion.

9.8 The equilibrium solution of the variable activity location with variable transport cost

In the test-2 runs discussed above, a single pass through EMPAL, DRAM, and NETWRK is done for each time period. Each of these recursions is used to calculate a five-year step in the forecast process. Ever since the first experimental version of this model package was developed in 1972, the issues of stability and convergence of solutions have been problematic. Two principal approaches were explored in early work with the model package (Putman, 1973). The first of these involved simply reducing the forecast increment from the fifteen-year increment of the first runs in 1972 to the five-year increments used in test 1 and test 2 in this chapter. This was some help as the extreme fluctuations seen in the first experiments were reduced, but they were not eliminated. A second approach involved averaging the work-trip matrices from successive iterations, prior to doing trip assignment. Again, an improvement was obtained, but the general problem was not solved. Over the years other schemes were considered—for example, very small time increments (equivalent to one year), additional iterations within each recursion, etc. None of these provided a definitive solution.

The last section of chapter 8 described the use of the MSA algorithm to solve a combined location–distribution–assignment problem in which a simple spatial interaction model was used for the location–distribution calculations. A modest expansion of this approach allows the calculation of stable solutions with the model package. In each recursion, instead of a single pass through EMPAL, DRAM, and NETWRK, several iterations are made until an equilibrium solution is reached within each recursion. The resulting equilibrium spatial patterns and network congestion are used as the starting point for the next recursion which, itself, is then iterated to equilibrium. The application of the MSA algorithm is to the network link flows, the combined values of which are used to calculate new link travel times which are, in turn, used as input to the next iteration. A number of very interesting conceptual and computational issues arise from consideration of this approach. These will be skipped over here and discussed in some detail in the next two chapters. Suffice it to say that (virtually) complete convergence for all employment types, all household types, and link flows (and travel times) is achieved in three to five iterations. What follows here is a discussion of the results of applying this approach to the Washington data.

Consider first the recursion from the base year, 1980, to the first forecast year, 1985. Using the uncongested 1980 network times (for all the numerical experiments reported here there were no forecast-year networks available, thereby implying no highway construction after 1980), the 1980 spatial patterns of activity, and the 1985 regional forecasts, a preliminary forecast of 1985 spatial patterns and trips was calculated. These trips were then assigned to the network, to yield a set of congested-network link times and minimum-path trees. The entire procedure was then repeated, to yield a second 1985 estimate of spatial patterns and trips. These trips were then assigned to the empty (1980) network and a second set of congested-link times was calculated. The link times were then combined, by using the MSA algorithm, and a set of minimum-path trees was calculated based on the combined link times. The iterative procedure continued in this fashion, each time using MSA to calculate combined link times. Convergence was usually achieved after four iterations (beyond the initial estimates), but for these runs five additional iterations were calculated in each recursion. The solution thus obtained is the equilibrium solution to the combined employment location and residence location, land consumption, trip generation and distribution, and trip-assignment problem. Mode split could have been included, but was not in this case, in order to keep computation costs as low as possible. Even so, it should be obvious that doing an additional five iterations within each recursion increases the cost of solving the models by 500%. Each recursion takes approximately 180 s of CPU time on the IBM 3090/180E computer. This includes the use of VS Fortran Version 2 vectorization. Thus to calculate the equilibrium solutions for 1985, 1990, 1995, and 2000 takes a total of about 720 s of CPU time.

The statistics regarding employment and household convergence for the 1980-85 recursion are given in table 9.4(a). The largest change (note that the numbers in the table are percentages (for example, 1.0 means 1.0%) is the change from the starting point, the 1980 base year, to the first iteration, called 1985A. After that first change, the change from one iteration to the next (for example, from 1985A to 1985B, or from 1985B to 1985C) becomes progressively smaller for all activity types. If one were to take a 1% change as being the standard for convergence, then employment converged after three iterations and households after four. To standardize the procedure for these preliminary tests, five iterations (beyond the first forecast estimate, that is, 1985A) were calculated. As will be seen, this gave iteration-to-iteration mean absolute percentage changes less than 0.1% in all cases. The magnitude of the percentage difference between the first iteration, 1985A, and the last iteration, 1985F, is given in the AF column of table 9.4(a), where it can be seen that the households change substantially more than employment in reaching an equilibrium. Last, the column headed 1985N-1985F shows the difference between a normal forecast of 1985

Table 9.4. Washington: summary statistics—the mean absolute percentage (MAP) change from iteration to iteration for (a) 1980–85 and (b) 1985–90.

Locator[a]	MAP change							
	1980–85A	AB	BC	CD	DE	EF	AF	1985N–1985F
(a) 1980–85[b]								
EMP1	41.20	3.49	1.14	0.30	0.14	0.04	2.92	2.92
EMP2	35.68	3.91	1.07	0.58	0.26	0.04	4.05	4.05
EMP3	42.88	2.02	0.64	0.29	0.12	0.01	1.97	1.97
EMP4	22.61	1.85	0.87	0.31	0.12	0.02	1.97	1.97
TOTEMP	21.20	1.25	0.56	0.27	0.07	0.01	1.23	1.23
LIHH	19.39	11.96	2.65	0.91	0.14	0.06	11.57	11.57
LMIHH	18.56	13.05	3.06	1.03	0.15	0.06	12.37	12.37
UMIHH	20.21	13.46	3.31	1.13	0.16	0.06	12.66	12.66
UIHH	26.73	16.40	3.54	1.22	0.17	0.07	15.47	15.47
TOTHH	20.08	13.02	3.05	1.05	0.15	0.06	12.24	12.24
	1985F–90A	AB	BC	CD	DE	EF	AF	1990N–1990F
(b) 1985–90[c]								
EMP1	30.10	0.91	0.41	0.10	0.03	0.01	0.64	1.97
EMP2	25.28	1.72	0.36	0.13	0.05	0.01	1.49	3.36
EMP3	23.67	1.10	0.19	0.07	0.03	0.01	0.98	1.81
EMP4	13.53	1.63	0.24	0.09	0.03	0.01	1.50	2.05
TOTEMP	17.60	0.91	0.18	0.07	0.02	0.01	0.79	1.18
LIHH	51.80	2.96	0.93	0.27	0.09	0.04	2.27	16.50
LMIHH	53.76	3.23	1.03	0.29	0.09	0.05	2.45	13.54
UMIHH	60.02	3.33	1.11	0.31	0.10	0.05	2.51	12.08
UIHH	72.61	3.74	1.33	0.37	0.10	0.06	2.78	12.90
TOTHH	62.42	3.21	1.07	0.30	0.09	0.05	2.43	12.55

[a] EMP1, EMP2, EMP3, EMP4, employment types are defined in table 9.3; TOTEMP, total employment; LIHH, LMIHH, UMIHH, UIHH, household types low income, low middle income, upper middle income, upper income, respectively; TOTHH, total households.

[b] 1985–85A is the difference of the base-year (1980) value from the first-iteration (1985A) value; AB is the change in value from iteration 1985A to 1985B, and similarly for BC, CD, DE, EF, and AF; 1985N–1985F is the difference between the 1985F iteration and the 1985 test-2 forecast (1985N); AF and 1985N–1985F values are identical, because 1985A is equal to the 1985 test-2 forecast 1985N.

[c] 1985F–90A is the change from the 1985 equilibrium solution to the first-iteration value, 1990A; AB is the difference of the 1990A iteration value from that of 1990B, and so on for BC, CD, DE, EF, and AF; 1990N–1990F is the difference between the 1990 test-2 forecast (1990N) and the 1990F equilibrium forecast.

and the equilibrium forecast. In the case of 1985, AF and 1985N–1985F are identical. This is because 1985A, the first-iteration forecast of 1985, is identical to the normal forecast, 1985N, of 1985. In subsequent recursions (time periods) this will not be true, as the equilibrium forecasts will continue to diverge from the normal forecasts.

Table 9.5. Washington: summary statistics—the mean absolute percentage (MAP) change from iteration to iteration for (a) 1990–95 and (b) 1995–2000.

Locator[a]	MAP change							
	1990–95A	AB	BC	CD	DE	EF	AF	1995N–1995F
(a) 1990–95[b]								
EMP1	20.19	0.70	0.54	0.18	0.06	0.01	0.43	2.42
EMP2	17.33	0.87	0.29	0.16	0.11	0.01	0.77	3.70
EMP3	18.82	0.56	0.15	0.09	0.06	0.01	0.50	2.49
EMP4	10.68	1.01	0.21	0.12	0.06	0.01	0.95	3.73
TOTEMP	14.09	0.51	0.18	0.11	0.04	0.01	0.44	1.82
LIHH	33.51	2.73	1.02	0.51	0.08	0.04	2.06	18.77
LMIHH	28.28	3.14	1.35	0.64	0.08	0.05	2.28	13.84
UMIHH	25.74	3.32	1.66	0.77	0.09	0.05	2.31	11.08
UIHH	27.49	3.61	1.91	0.86	0.10	0.06	2.45	10.54
TOTHH	29.40	3.08	1.47	0.69	0.09	0.05	2.18	12.11
	1995F–2000A	AB	BC	CD	DE	EF	AF	2000N–2000F
(b) 1995–2000[c]								
EMP1	16.59	0.86	0.94	0.43	0.14	0.03	0.45	2.18
EMP2	14.22	0.92	0.41	0.38	0.28	0.03	0.75	3.63
EMP3	15.53	0.55	0.24	0.23	0.14	0.01	0.49	2.55
EMP4	9.27	0.27	0.27	0.24	0.15	0.02	0.83	4.19
TOTEMP	12.20	0.49	0.34	0.27	0.11	0.01	0.40	1.86
LIHH	29.21	3.01	2.46	1.58	0.14	0.08	2.03	22.42
LMIHH	22.58	3.82	2.99	1.85	0.15	0.09	2.49	16.26
UMIHH	18.73	4.44	3.52	2.10	0.16	0.09	2.74	12.25
UIHH	17.66	5.13	3.93	2.25	0.17	0.10	3.10	10.49
TOTHH	21.41	3.91	3.17	1.95	0.16	0.09	2.44	12.83

[a] See table 9.4.
[b] 1990F–95A is the change from the 1990 equilibrium solution to the first-iteration (1995A) value; AB, is the change from iteration 1995A to iteration 1995B, and so on for BC, CD, DE, EF, and AF; 1995N–1995F is the difference between the 1995 test-2 forecast (1995N) and the 1995 equilibrium forecast (1995F).
[c] 1995F–2000A is the change from the 1995 equilibrium solution to the first-iteration (2000A) value; AB is the change from iteration 2000A to iteration 2000B, and so on for BC, CD, DE, EF, and AF; 2000N–2000F is the difference between the 2000 test-2 forecast (2000N) and the 2000 equilibrium forecast (2000F).

After solving for the 1985 equilibrium, the entire process was repeated. The 1985F results, the equilibrium solution, were used as input to the 1990 runs. In table 9.4(b), which contains the results of the 1990 equilibrium run it can be seen that, again, the biggest percentage changes are between the 1985 equilibrium and the 1990 first iteration. Again, successive iterations show fairly rapid convergence, the 1% tolerance being met after the second iteration for employment and after the third iteration for households. Note, however, that for 1990 the differences between the first iteration, 1990A, and the sixth iteration, 1990F are much smaller than those between 1985A and 1985F. This is likely to be a result of the commonly occuring but rarely discussed phenomenon of dissipation of the unexplained variation from the base year. No model for which equation coefficients are estimated based on observed data is likely to be a perfect fit to that data. To the extent that such a model is not a perfect fit, there is unexplained variation in the base-year data. When such a model is used in a recursive forecasting scheme (the output of one forecast becoming the input to the next) this unexplained variation is not carried forward from the base year. Thus a significant portion of the change which takes place in the first forecast period is because the model 'reshapes the world' to match the equations and their parameters. As in the 1985 tabulations, the 1990N – 1990F column gives the changes between the normal forecast and the equilibrium forecast. Note that the differences for 1990 are about the same size as those for 1985.

Tables 9.5(a) and 9.5(b) give the same kinds of tabulated results for the 1995 and 2000 equilibrium runs. The results are similar, in the pattern of changes, to the 1985 and 1990 equilibrium runs. Convergence is achieved after two to four iterations (with the use of the 1% standard). The magnitudes of the differences between the normal forecasts and the equilibrium forecasts are about the same. This suggests a possibility, which will be explored further in the next chapter, of a set of equilibrium solutions running from one recursion to the next, in a general sense, parallel to the normal, one-iteration solutions.

The mean zone-to-zone impedance is a crude indicator of the situation on the region's highway network during these equilibrium tests. Table 9.6 gives the mean impedance values for each iteration of each recursion. In each

Table 9.6. Washington: mean zone-to-zone impedance (in minutes) for equilibrium-run iterations.

Run	Imp.	Run	Imp.	Run	Imp.	Run	Imp.
1985A	52.58	1990A	55.47	1995A	59.95	2000A	65.66
1985B	50.85	1990B	54.14	1995B	57.61	2000B	61.90
1985C	51.43	1990C	54.35	1995C	58.17	2000C	63.08
1985D	51.43	1990D	54.32	1995D	58.13	2000D	63.02
1985E	51.46	1990E	54.33	1995E	58.14	2000E	63.02

Note: Imp. Impedance.

recursion the mean impedance after the first iteration is the highest, that after the second iteration is the lowest, and convergence is reached after the third iteration. The changes in link flows, though not tabulated here, also converge to fluctuations of less than 1%.

Figures 9.24 and 9.25 show the spatial patterns of total households from the equilibrium forecasts compared with those from normal forecasts. Again, as with many of the earlier maps in this chapter, the differences are minor. Although some zones do change, the overall patterns are, for each time period, rather similar. As was seen in tables 9.4 and 9.5, the difference between equilibrium and nonequilibrium is on average less than 2% for employment and about 12% for households. In general, this is less than the differences found between the test-1 (fixed transport time) and test-2 (variable transport time) runs. The differences between equilibrium and nonequilibrium tend to be about the same for all four time periods individually. These differences are, however, cumulative, beginning at 0% for 1985 (for both employment and households) and ending at 2.5% for employment and 37% for households in 2000, when comparing the equilibrium results with the test-2 results. The question of which results give the better forecasts of reality will be taken up in the next two chapters.

9.8.1 Summary

Perhaps the principal point to be made here is that it is clearly possible to compute equilibrium solutions to these integrated location, land-use, and transportation model systems. In and of itself, this is not 'new', Boyce has been working on combined model formulations for some time (for example, Boyce and Lundquist, 1987). One thing that is interesting here, however, is that the equilibrium solution is obtained via some rather modest changes in the existing configuration of the integrated model system. The computational costs are higher, but not impossible, and although file management becomes a bit tricky, it is not overwhelming.

Perhaps of more interest, substantively, are the results of comparing the equilibrium solutions with the nonequilibrium solutions. As mentioned above, after four recursions the differences between fixed travel-time and variable travel-time model configurations are substantially greater than the differences between equilibrium and nonequilibrium variable travel time model configuration results. It is especially noteworthy that from the first through the fourth recursions, taken individually, the differences between the equilibrium and nonequilibrium results remain about the same.

The development of the solution procedure which was used here to achieve equilibrium solutions began with the discussion of SUE trip assignment in chapter 7, where the MSA algorithm was introduced. The notion of solving the combined trip-distribution and assignment problem was introduced in chapter 8. This was followed in the same chapter by discussion and numerical examples of solving the combined location, distribution, and assignment problem using the MSA algorithm embedded in the overall

(a)

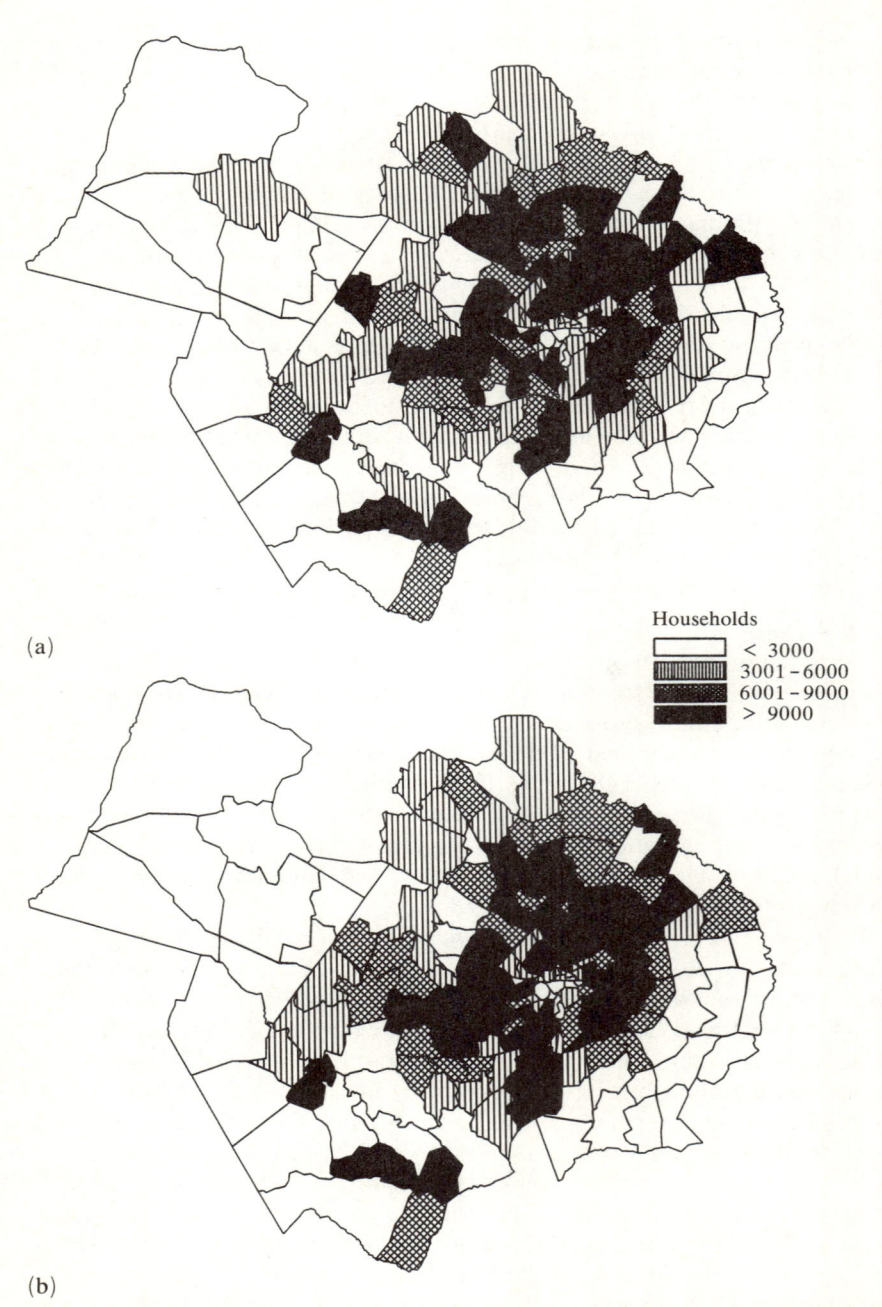

Households
- < 3000
- 3001–6000
- 6001–9000
- > 9000

(b)

Figure 9.24. Washington, DC, region: population equilibrium compared with test-2 results. (a) 1985 test-2 results, (b) 1990 test-2 results, (c) 1985 equilibrium results and (d) 1990 equilibrium results.

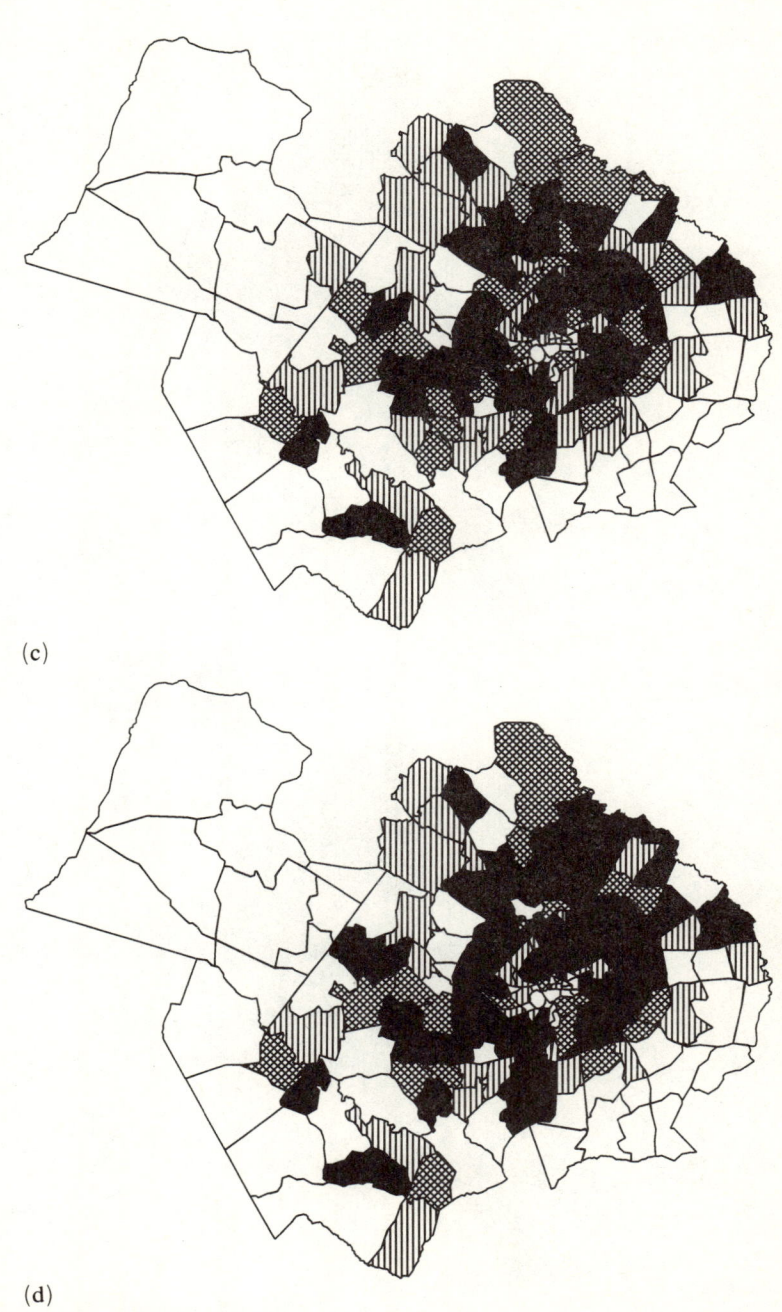

Simultaneous location–transportation models: 2

(c)

(d)

Figure 9.24. (continued).

(a)

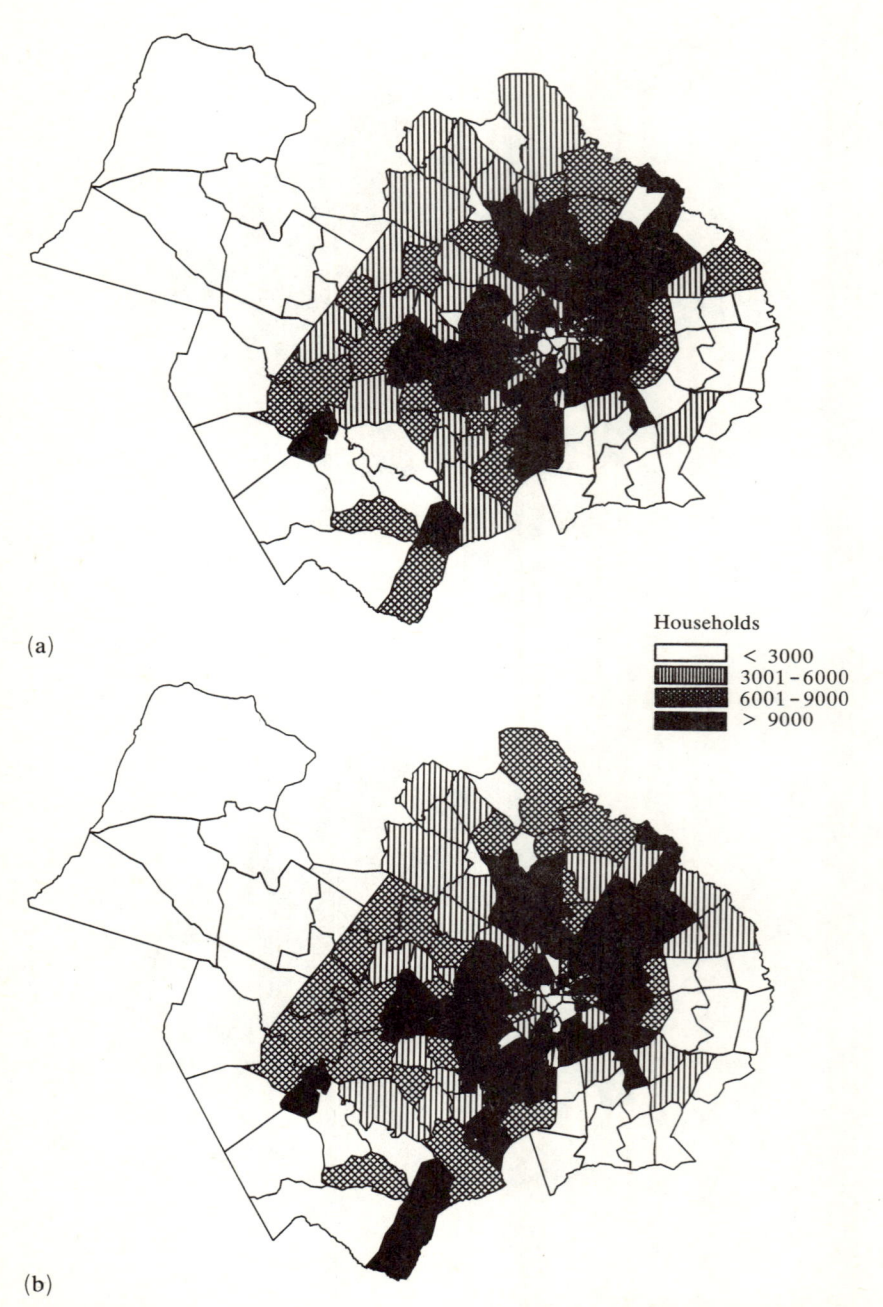

(b)

Figure 9.25. Washington, DC, region: population equilibrium compared with test-2 results. (a) 1995 test-2 results, (b) 2000 test-2 results, (c) 1995 equilibrium results, (d) 2000 equilibrium results.

Households
- < 3000
- 3001–6000
- 6001–9000
- > 9000

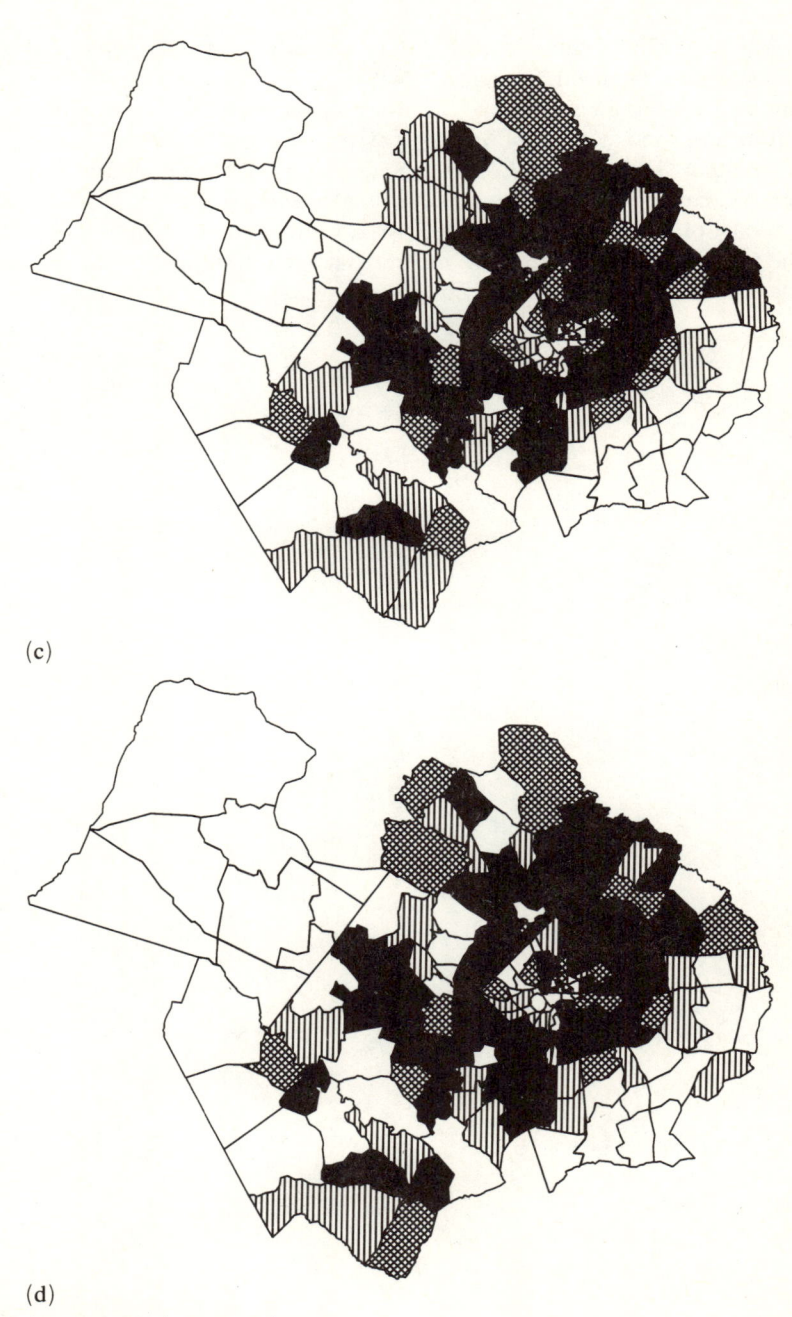

Simultaneous location–transportation models: 2   251

(c)

(d)

Figure 9.25. (continued).

Evans approach. In this chapter it was shown, for full-scale data sets, how this approach could be used to modify the ITLUP package to calculate equilibrium solutions to an even more extensive problem, including employment location and land consumption along with the other portions of the combined problem. To have shown that it can be done is not the end of the issue. It is not clear that an equilibrium solution to these models will produce the best forecast of reality. Wegener (1986) argues this point too. In the next chapter the focus will be on the relationships between nonequilibrium and equilibrium solutions and dynamic models.

10

Equilibrium solutions to location models

10.1 Introduction

In chapter 8 several small numerical examples were given of alternative approaches to the solution of the combined activity-location and trip-assignment problem. In chapter 9 one of these approaches was used to produce a series of equilibrium solutions to a complete combined problem, based on the full-scale Washington data set. In the description of these efforts a number of questions arose, the discussion of which was postponed to this chapter. These questions cluster around the issue of the meaning, for these models, of an equilibrium solution.

The second section of this chapter, contains a discussion of the structures of the individual models in ITLUP and what is implied by equilibrium solutions to each of them. In the third section, the models are taken in their interconnected configuration and the meaning of an equilibrium solution to the entire model system is considered. Finally, there is a fourth section in which the discovery and analysis of chaotic behavior in the attempt to solve for equilibrium with certain model system configurations is described.

10.2 Solution methods for individual models

10.2.1 EMPAL

The employment location model, EMPAL, has a relatively simple equation structure:

$$E_{m,j,t} = \lambda^m Q_j + (1-\lambda^m) E_{m,j,t-1} , \qquad (10.1)$$

where

$$Q_j = \sum_i P_{i,t} \frac{E_{m,j,t-1}^{\gamma^m} L_j^{\delta^m} c_{i,j,t}^{\alpha^m} \exp \beta^m c_{i,j,t}}{\sum_k E_{m,k,t-1}^{\gamma^m} L_k^{\delta^m} c_{i,k,t}^{\alpha^m} \exp \beta^m c_{i,k,t}} , \qquad (10.2)$$

$P_{i,t}$ is the total number of employed residents residing in zone i at time t,

$E_{m,j,t}$ is the number of employees of type m working in zone j at time t,

$c_{i,j,t}$ is the travel time (or cost) from zone i to zone j at time t,

L_j is the land area of zone j,

$\alpha^m, \beta^m, \gamma^m, \delta^m, \gamma^m$ are empirically estimated parameters for type-m employment.

A shorthand notation for this is useful to show the temporal relationships. Let equation (10.1) be represented as

$$E_t = f(E_{t-1}, P_t, c_t) . \qquad (10.3)$$

As was described in the previous chapter, the simplest configuration, called 'test 1', of EMPAL and DRAM involved running first one and then the next of these models. It is quite clear that equation (10.3) cannot be solved by the test-1 configuration, as P_t is not known when equation (10.3) is first solved. The test-1 configuration can only be solved if P_{t-1} is substituted for P_t in equation (10.3). The extent to which this substitution is of concern depends both on the length of the interval between $t-1$ and t, and on what hypotheses are being made about the speed of response of employment location to changes in population location. In virtually all EMPAL applications the length of the interval from $t-1$ to t has been five years. If P_t is used in equation (10.3), the corresponding hypothesis is that present employment location is a function of present population location (as well as of past employment location and present travel time). If P_{t-1} is used, then the corresponding hypothesis is that present employment is a function of the population distribution of five years past. In all likelihood, neither of these hypotheses is strictly correct and the 'true' hypotheses will vary from region to region and amongst employment types. What is more, it is quite possible that employment location depends upon *both* past and present population location as well as upon expectations of future population location. In point of fact, the modeller often will not be able to choose, owing to data limitations, and thus will not be likely to be able to conduct the statistical analyses necessary to decide the matter either.

Suffice it to say here that if P_{t-1} is used in equation (10.3), and if transport times are constant, then the equation gives the solution for E_t directly. If, however P_t is used, then it will be necessary to iterate between EMPAL and DRAM (the population location model) to get the solution for E_t. Implicitly, the test-1 configuration gives the solution for E_t by using P_{t-1}. This is true for the test-2 configuration as well, and may or may not be true for the equilibrium-test configuration, depending upon which variables are passed between models from one iteration to the next. Last, if P_{t-1} is used, because E_t is not found on both sides of equation (10.3), no 'self-iteration' is necessary for its solution.

10.2.2 DRAM

The residence location model, DRAM, has a standard singly-constrained spatial interaction model structure, with an expanded zonal attractiveness term, $N_{h,i,t}$:

$$N_{h,i,t} = \sum_j \tilde{E}_{j,t} Z_i^h , \qquad (10.4)$$

where

$$\tilde{E}_{j,t} = \sum_l a_{h,l} E_{l,j,t} , \qquad (10.5)$$

and

$$Z_i^h = \left[L_{v,i,t}^{q^h} X_{i,t}^{r^h} L_{r,i,t}^{s^h} \prod_h n_{h,i,t}^{b_h^h} c_{i,j,t}^{a^h} \right]$$

$$\times \left[\sum_k L_{v,k,t}^{q^h} X_{k,t}^{r^h} L_{r,k,t}^{s^h} \prod_h n_{h,k,t}^{b_h^h} c_{k,j,t}^{a^h} \right]^{-1} , \qquad (10.6)$$

where
$N_{h,i,t}$ is the number of households of type h residing in zone i at time t,
$a_{h,l}$ is the number of heads of household type h per employee of type l (employee-to-household conversion ratio),
$E_{l,j,t}$ is the number of employees of type l employed in zone j at time t,
$L_{v,i,t}$ is the amount of vacant developable land in zone i at time t,
$X_{i,t}$ is equal to 1.0 plus the ratio of developed to developable land in zone i at time t,
$L_{r,i,t}$ is the amount of residential land in zone i at time t,
$n_{h,i,t}$ is equal to 1.0 plus the ratio of type-h households to total households in zone i at time t,
$q^h, r^h, s^h, b_h^h, \alpha^h$ are empirically estimated parameters for type-h households.

As with EMPAL, a shorthand notation will be useful to point out the temporal relationships. Thus equation (10.4) may be rewritten as

$$N_t = f(E_t, L_t, N_t, c_t). \qquad (10.7)$$

Now it is quite clear that equation (10.7) describes a simultaneous equation model of residence location. To date, however, and thus in the tests described in chapter 9, the model has not been solved according to equation (10.7). It has been the practice to solve the following:

$$N_t = f(E_t, L_{t-1}, N_{t-1}, c_{t-1}). \qquad (10.8)$$

This was done in order to avoid the complications of the simultaneous solution. Again, the extent to which the difference between equation (10.7) and equation (10.8) is of concern depends upon the length of the interval between $t-1$ and t and upon what hypotheses the model is intended to represent. In all applications of DRAM to date the length of the interval has been five years. The theory on which models of this type are based is not sufficiently explicit to tell whether the difference between equations (10.7) and (10.8) is of importance. Further, this author has been unable to locate any data sets, prior to this writing, which would allow an empirical investigation to be done to decide the matter.

In and of themselves, neither equation (10.7) nor equation (10.8) require an iteration with EMPAL. If equation (10.8) is used then the hypothesis is that current residence location depends on current employment location and current transport costs and on lagged residence location and lagged land use. In this form, given employment location and transport cost, equation (10.8) gives current residence location directly. If equation (10.7) is used, then, given current employment and current transport cost, the equation must be solved simultaneously (probably by iteration) and a land-consumption calculation must be included in the model structure.

10.2.3 LANCON

This is not the place to give a lengthy discussion of land-consumption procedures, but a few words are in order. Most spatial interaction models deal directly with the location of activities such as population or service employment. After the numbers of these activities to be located in a particular zone are calculated, it is usually assumed that they will use land (for example, acres per household) at the prior average rate for that zone. Clearly, such a procedure will give inaccurate results. The problem is compounded by the fact that most such models completely reallocate all activities in each time period and thus recalculate the land consumption for all activities as well.

Although it is by no means a definitive solution to the problem of forecasting land consumption, the procedure used with EMPAL and DRAM does represent a significant improvement. First, a set of log–linear multivariate equations are used to estimate land consumption for each activity, by zone (Putman, 1983b). Second, even though, as in other models, all activities are reallocated for each time period, land consumption is calculated only for the change in numbers of activities in each zone. If there is a decrease, say, in households, then residential land is 'released' at the prior average rate of residential land per household. If there is an increase in households, the 'new' households consume land at a rate given by the land-consumption equation.

For purposes of the discussion of equilibrium solutions of DRAM, the important point regarding land consumption is that it complicates the problem. The land-consumption equations are cross-sectional in nature. They include variables which describe the type and amount of land use already in the zone, and variables describing the income distribution of the residents in the zone. In a sense, these variables taken all together, are a surrogate measure of land price. All the variables in the land-consumption equations, residential as just described, as well as basic and commercial which will be described below, are measured at the same time. Thus time-t residential land consumption depends upon time-t basic and commercial land use as well as upon time-t residential location. The residential land-consumption equation is:

$$\frac{L_{\mathrm{r},i,t}}{N_{\mathrm{T},i,t}} = k_0 \left(\frac{L_{\mathrm{d},i,t}}{L_{\mathrm{D},i,t}}\right)^{k_1} \left(\frac{L_{\mathrm{b},i,t}}{L_i}\right)^{k_2} \left(\frac{L_{\mathrm{c},i,t}}{L_i}\right)^{k_3} \left(\frac{N_{1,i,t}}{N_{\mathrm{T},i,t}}\right)^{k_4} \left(\frac{N_{4,i,t}}{N_{\mathrm{T},i,t}}\right)^{k_5} L_{\mathrm{D},i,t}^{k_6} \,, \tag{10.9}$$

where
- $N_{\mathrm{T},i,t}$ is the total number of households in zone i at time t,
- $L_{\mathrm{b},i,t}$ is the amount of 'basic' land use in zone i at time t,
- $L_{\mathrm{c},i,t}$ is the amount of 'commercial' land use in zone i at time t,
- $L_{\mathrm{d},i,t}$ is the amount of 'developed' land use in zone i at time t,
- $L_{\mathrm{D},i,t}$ is the amount of developable (developed plus vacant developable) land in zone i at time t,

Equilibrium solutions to location models 257

$N_{1, i, t}$ is the number of low-income households in zone i at time t,
$N_{4, i, t}$ is the number of high-income households in zone i at time t,
$k_0, k_1, k_2, k_3, k_4, k_5, k_6$ are empirically estimated parameters.

The 'basic' industry land-consumption equation is:

$$\frac{L_{b, i, t}}{E_{b, i, t}} = g_0 \left(\frac{L_{d, i, t}}{L_{D, i, t}}\right)^{g_1} \left(\frac{E_{b, i, t}}{E_{T, i, t}}\right)^{g_2} \left(\frac{L_{b, i, t}}{L_{D, i, t}}\right)^{g_3} \left(\frac{L_{r, i, t}}{L_{D, i, t}}\right)^{g_4} L_{D, i, t}^{g_5}, \qquad (10.10)$$

where
$E_{b, i, t}$ is the amount of 'basic' employment in zone i at time t,
$E_{T, i, t}$ is the total employment in zone i at time t,
$g_0, g_1, g_2, g_3, g_4, g_5$ are empirically estimated parameters.

Last, the 'commercial' land-consumption equation is

$$\frac{L_{c, i, t}}{E_{c, i, t}} = p_0 \left(\frac{L_{d, i, t}}{L_{D, i, t}}\right)^{p_1} \left(\frac{E_{c, i, t}}{E_{T, i, t}}\right)^{p_2} \left(\frac{L_{c, i, t}}{L_{D, i, t}}\right)^{p_3} \left(\frac{L_{r, i, t}}{L_{D, i, t}}\right)^{p_4} L_{D, i, t}^{p_5}, \qquad (10.11)$$

where
$E_{c, i, t}$ is the amount of 'commercial' employment in zone i at time t,
$p_0, p_1, p_2, p_3, p_4, p_5$ are empirically estimated parameters.

It is important to note that

$$L_{D, i, t} = L_{d, i, t} + L_{v, i, t}, \qquad (10.12)$$

and that

$$L_{d, i, t} = L_{r, i, t} + L_{b, i, t} + L_{c, i, t} + L_{M, i, t}, \qquad (10.13)$$

where $L_{M, i, t}$ represents 'other' or 'miscellaneous' developed land uses such as parks, streets, and highways. The specific definitions of employment types (for example, 'basic' and 'commercial') depend on the region being analyzed and its economy.

This model, too, can be written in a shorthand version, which gives

$$L_t = f(L_t, E_t, N_t). \qquad (10.14)$$

Again, this is a simultaneous-equation model. In practice it has been replaced by

$$L_t = f(L_{t-1}, E_t, N_t), \qquad (10.15)$$

in order to avoid having to solve the simultaneous equations. As was the case for EMPAL and DRAM, the question of the importance of any difference in land consumption estimated by equation (10.15) and that which would be estimated by equation (10.14) is a matter of the underlying hypotheses about how quickly land-consumption rates respond to the land market as represented by the variables in equation (10.9) and (10.11). Here, too, lack of data has precluded the appropriate empirical investigation of this question.

10.2.4 NETWRK

In both EMPAL and DRAM an important input variable is the zone-to-zone travel cost (or as it is often represented, travel time). In early use of spatial interaction models, as in the test-1 results described in chapter 9, travel cost is taken as being both exogenous and constant. In ITLUP usage (that is, in integrated location and transportation modelling), travel cost is endogenous and is thus variable. This was the model configuration used to produce the test-2 results of chapter 9. The explicit mechanism for varying the travel cost is by the use of a volume-delay function for the links of the highway network. As volume on the highway links changes, the travel time and/or cost to traverse the link changes as well. In chapters 6 and 7 the various techniques for determining link volumes and the consequent congested link times and/or costs were discussed.

In the same shorthand notation introduced above, the travel cost may be considered to be a function of the zone-to-zone trips matrix which, itself, is the result of the employment and residence locations in each zone. Thus,

$$c_t = f(E_t, N_t) \,. \tag{10.16}$$

As discussed in chapters 6 and 7, any of several algorithms can be used to solve this, with either user-equilibrium or stochastic user-equilibrium being preferred from the standpoint of theory. From the standpoint of practice, other algorithms may give as good, or better, numerical results.

The mode-split issue may be dealt with here as well. There is, of course, a substantial literature on mode split, and there remain many unresolved questions (Ben-Akiva and Lerman, 1985). For present purposes, the role of a mode-split model in the overall model system will be to reduce the number of trips being assigned to the highway network. This reduction will not be uniform, as there will be different mode-split proportions for different zone-to-zone pairs and different types of residents or employees. Even so, the net effect of the mode-split calculation will be to change c_t. Mode-split proportions are, however, a function of zone-to-zone travel costs. In the shorthand notation which has been used to describe the various models, the effect of including mode split in the transport-cost calculation would be to add c_t to the right-hand side of equation (10.16). Now, in fact, c_t is already there indirectly, as it is on the right-hand side in equations (10.3) and (10.7) for E_t and N_t. Thus the inclusion of mode split makes the presence of c_t on both sides of equation (10.16) more direct.

10.3 Solving the linked set of models

Consider the models as described in the shorthand notation of equation (10.3):

$$E_t = f(E_{t-1}, P_t, c_t) \,.$$

For all intents and purposes, P_t may be considered to be a scalar transform of N_t, so equation (10.3) may also be written as in equation (10.17).

Thus the model of employment location can be expressed as

$$E_t = f(E_{t-1}, N_t, c_t),\qquad(10.17)$$

the model of residence location is written as in equation (10.7), as

$$N_t = f(E_t, L_t, N_t, c_t),$$

the model of land consumption is represented by equation (10.14), as

$$L_t = f(L_t, E_t, N_t),$$

and the model of congested travel cost by equation (10.16), as

$$c_t = f(c_t, E_t, N_t).$$

In theory these could be solved as a single large nonlinear simultaneous-equation model. In practice it does not appear possible to solve these equations in such a manner. First, the number of equations involved would be quite substantial. For a 100-zone model with four employment and four household types there would be 400 expressions of the form of equation (10.17), 400 of equation (10.7), 300 of equation (10.14), and 10 000 of equation (10.16), to give a total of 11 000 equations, all highly nonlinear. For a 200-zone region there would be 42 200 equations to solve. Further, these figures take no account of the fact that the c_t are a result of the tracing of minimum paths through the congested highway network. In practice, the only way to even consider dealing with these equations is by use of a computer and some approximization technique such as Gauss–Seidel (Chapra and Canale, 1985). In the case of these integrated model systems even more extensive measures must be resorted to, as was discussed in chapter 9. An example is the use of convex-combinations approaches within an overall construct such as the one developed by Evans (1976), and used for the test-3 equilibrium runs described in chapter 9.

10.4 Experiments with model solution procedures: DRAM

As mentioned above, the customary solution method for DRAM, considered in isolation from other models, has been according to equation (10.8). With the customary data sets, the interval $t \to t+1$ has been approximately five years. (Often it has been impossible to obtain data for all variables with the same dates, thus the qualification 'approximately'.) From the point of view of locational behavior, and with t equal to, say, 1990, the substantive meaning of equation (10.8) may be stated as follows. The spatial distribution of households, by type, in 1990 is a function of the travel cost in 1990, of the spatial distribution of employment in 1990, and of the spatial distributions of households and land use in 1985. In a sense, such a formulation makes an assertion about the availability of information to residential locators, as well as about their rationality. In the parameter estimation for DRAM, the usual case is that all variables have the same time subscript.

This, of course, implies no more about the rationality of locators, but it does imply that instantaneous information is available and that it is used in making the residence location decision. It also implies a sufficiently rapid rate of adjustment to permit all households to use this information and locate accordingly within the specified time period.

Data were not available to test which of these hypotheses was the best representation of reality. However, it was possible to do a series of numerical experiments to see what differences in model forecasts resulted from solving the model according to these different hypotheses. In the first of these tests the N_{t-1} on the right-hand side of equation (10.8) was replaced by the unlagged N_t. This represents the alternate hypothesis that the spatial distribution of households of each type is jointly, and simultaneously, determined by the spatial distribution of households of all other types as well. In effect, this means that each household type is presumed to know, instantaneously, about the locational decisions of each other household type. In addition, it is assumed that each household type, in choosing a residence location, will act on this information in a manner consistent with its prior patterns of behavior (that is, its locational preferences) vis-à-vis each other household type.

The solution of the model in this form had the potential of being rather difficult. For the Washington data there were 182 zones and 4 household types, giving 728 equations. It is evident from equations (10.4)–(10.6) that the model is highly nonlinear. It was hoped that a simple successive substitution method would work to solve the modified version of the model (Pearson, 1986). For these equations the method involves, simply, the solving of the equations for a set of values of $N_{h,i,t}^1$ for the 4 types of household. These 1st iteration values, based on the $N_{h,i,t-1}$ values, are then substituted into the equations, and the equations are solved for $N_{h,i,t}^2$. The process is continued until the difference between the kth and $(k+1)$th values fall below some prespecified tolerance. There are situations when this method may fail to converge. In such cases more complex approaches, such as Newton's method, involving the calculation of derivatives of the equations, must be used (Pearson, 1986). Fortunately the successive substitution method did converge for this case.

The model was run for 50 iterations of the successive substitution method. The mean absolute percentage deviation was calculated, by zone, for each household type. The results between the 49th and 50th iterations were as follows: low-income households (LIHH), 0.32%; lower-middle-income households (LMIHH), 0.35%; upper-middle-income households (UMIHH), 0.39%; and upper-income households (UIHH), 0.14%. The starting values of households by type and zone were the values produced by the conventional solution procedure implied by equation (10.8). The final values, after 50 iterations, were quite different. The simple Pearson's correlations between the original solution procedure results and the 50 iteration results were: LIHH, 0.448; LMIHH, 0.566; UMIHH, 0.659; UIHH, 0.851; and TOTHH, 0.820.

The same test version of the model was then run for 150 iterations of successive substitution. The mean absolute percent deviation between the 149th and 150th iterations was then: LIHH, 0.016%; LMIHH, 0.036%; UMIHH, 0.10%; and UIHH, 0.008%. The results of 150 iterations were rather like the results of 50 iterations, with the simple Pearson's correlations between them being: LIHH, 0.994; LMIHH, 0.979; UMIHH, 0.961; UIHH, 0.993, and TOTHH 0.988. It is interesting to note that the actual minimum value of the mean absolute present deviation (from one iteration to the next) was not at 150 iterations. Each household type reached a minimum at a slightly different number of iterations: LIHH, 126; LMIHH, 136; UMIHH, 127, and UIHH, 127. The slight variations around these minima which occurred for succeeding iterations appeared to be the result of round-off errors in computation.

Figure 10.1(a) shows the spatial pattern which results from the conventional solution method produced by equation (10.8), as compared with the 50-iteration solution of the alternative method shown in figure 10.1(b). The spatial pattern of the 150-iteration solution is shown in figure 10.2(a), and is virtually identical to the 50-iteration spatial pattern [figure 10.1(b)]. The principal differences between the conventional and partly simultaneous solutions are the shifting of households out of the central (District of Columbia) area, to zones in the eastern part of the region, and a somewhat greater degree of concentration of households in the innermost ring of suburban zones. The mean absolute percentage deviations, by zone, between the conventional solution and the 50-iteration partly simultaneous solution were: LIHH, 174.2%; LMIHH, 82.0%; UMIHH, 60.4%; UIHH, 69.1%; and TOTHH, 60.8%. These are rather substantial differences. This equilibrium solution is achieved when each household type is assumed to know, instantly, the locational behavior of each other household, and to react to this knowledge instantaneously, while all other variables remain constant. It is also implicitly assumed here that all households are not only able to respond instantly, but that they adjust, fully, to the relocations of all other households during the forecast time period (five years, in this case).

A logical next step was to have the land-use variables change along with the household variables. The model which was solved for the first experiment, just described, was

$$N_t = f(E_t, L_{t-1}, N_t, c_{t-1}), \quad (10.18)$$

the model to be solved for this next experiment is

$$N_t = f(E_t, L_t, N_t, c_{t-1}). \quad (10.19)$$

This model can be solved in several ways. In general, first there are 'inner' iterative loops where the households are relocating with respect to each other's location decisions while other variables remain, temporarily, constant. Then, there are 'outer' iterative loops where land use is adjusted while all other variables remain constant. The inner iterations on households locating with respect to each other are called, here, HH iterations. Thus the

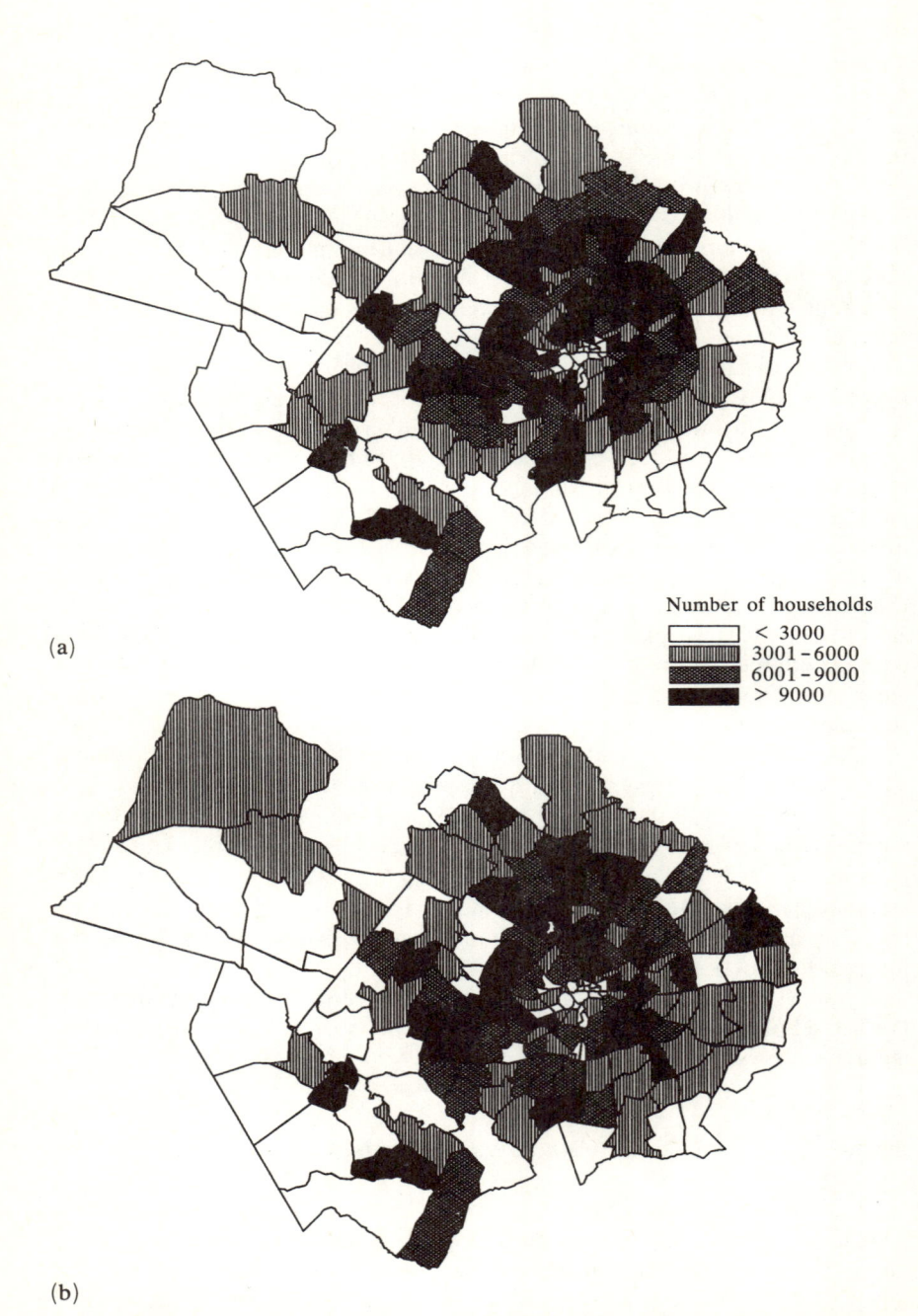

Figure 10.1. Washington, DC, region: results from (a) 1985 base-run distributions, (b) 1985 50-HH iteration. Note: HH iterations are inner iterations on households locating with respect to each other.

Equilibrium solutions to location models 263

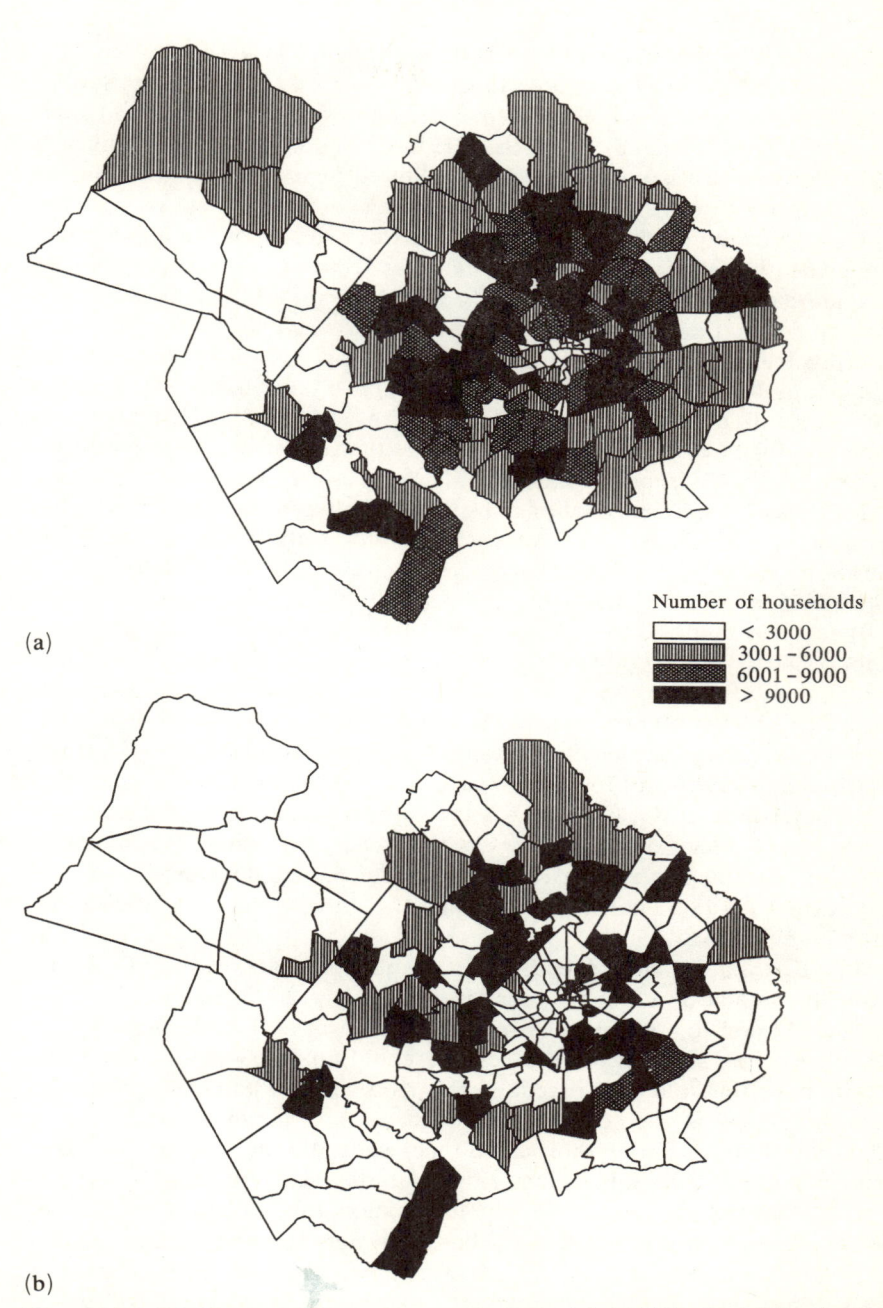

Figure 10.2. Washington, DC, region: results from (a) 1985 base-run distribution, and (b) 1985 3-HH iterations in 50-LU iterations. Note: HH iterations are inner iterations on households locating with respect to each other; LU iterations are land-use adjustment, outer iterations.

two previous model runs with 50-HH iterations and 150-HH iterations. The land-use adjustment, outer, iterations are called LU iterations. Several experiments were done. Here the first intimations of instabilities and solution difficulties appeared. When the HH iterations were run to equilibrium and then the LU iterations were done, followed by another equilibrium run of HH iterations, and so on, the system appeared to converge very slowly. In fact, no solution approach of this form was actually run to equilibrium. Only when the number of HH iterations was reduced below that necessary to achieve convergence (at least during the first several LU iterations) was it possible to run the entire system to convergence. A typical convergent solution involved 3-HH iterations followed by an LU iteration, the whole set being done 40 or 50 times. The spatial pattern of households which results from a solution involving 3-HH iterations in each of 50-LU iterations is shown in figure 10.2(b). It is clear from the figure that this solution differs quite significantly from the ones provided by 50-HH or 150-HH iterations. Indeed, the simple correlations, by zone, between household location estimates from 150-HH and the corresponding estimates from the 3-HH iterations in 50-LU iterations were: LIHH, 0.24; LMIHH, 0.52; UMIHH, 0.55; UIHH, 0.55; and TOTHH, 0.47.

It is worth considering the meaning of these results. In the 'standard' solution method each household type locates itself, basing its choice on a description of the region from the prior time period (usually five years earlier). The location decisions of each household type are made independently of the concurrent location decisions of other household types. A single calculation is performed for each household type, in which it is assumed that a full adjustment of location takes place in response to the variables of the prior period which describe zonal attractiveness and zone-to-zone travel time (or cost), and in response to estimates of the employment location of the current period. The 'standard' solution is the 1st iteration of all other solution methods.

The first, partly simultaneous, solution method that was tested differed from the standard solution in that it allowed for additional iterations of adjustment and for readjustment of the location of each household type in response to the location of all other household types. By performing many iterations of this solution procedure, this model ensures, in effect, that each household type is eventually located with respect to its attraction to, or repulsion from, each other household type, while all other relevant variables remain constant. The equilibrium solution in this case represents an expression of the preferences for household location with or apart from other household types as expressed in the base-year data from which the equation coefficients are estimated. This is obviously *not* a proper representation of reality inasmuch as the concomitant changes in other variables, which would, in reality, mitigate these household association preferences, are not included here.

In the second partly simultaneous solution experiments the land variables change along with the household locations. In this case, only travel time and employment location remain fixed. Here the solution converges under some circumstances but only approaches convergence under others. It appears, generally, that if the inner, HH, iterations are allowed to converge too far (that is, to get too close to an 'HH only' equilibrium) then the LU iteration cannot converge. By the same token, LU iterations with no HH iterations converge very slowly, if at all. Thus with 2-HH to 5-HH iterations within each LU iteration an equilibrium solution is achieved, though each configuration (in terms of numbers of HH and LU iterations) gives a somewhat different solution. The most likely interpretation of this fact is in terms of rate of adjustment. It is reasonable to expect that household location does indeed depend on household-to-household attraction or antipathy and on the land-use attributes of each zone. However, location most certainly depends on them differently, and changes in one or the other are met with varying degrees *and rates* of response.

10.5 Solution methods for linked models: DRAM and NETWRK

Next a series of several additional numerical experiments were done, incorporating travel time at time t into the solution. The equation thus becomes

$$N_t = f(E_t, L_t, N_t, c_t) \ . \tag{10.20}$$

In order to include c_t in the solution it was necessary to link the residential location calculations of DRAM to a network assignment procedure to calculate the resultant congested travel time. Any assignment algorithm might have been used, but to keep these results consistent with those of chapter 9 a stochastic multipath algorithm was selected. After each DRAM run the resulting implicit work-trip matrix was assigned to the Washington network. The resulting zone-to-zone travel times, calculated from the minimum paths through the congested network, were then available to use as input to the next DRAM run. Several different configurations of the DRAM solution procedure were thus linked to a network assignment procedure so that the zone-to-zone travel times would be a function of household locations which were themselves a function of zone-to-zone travel times. These tests may be considered as full simultaneous solution (FSS) tests and are now discussed.

In the first test, DRAM was solved by the 3-HH, 50-LU approach discussed above. The resulting trip matrices were assigned, by using the NETWRK model, to the Washington network. The resulting congested network travel times became input to the next solution of DRAM. Every iteration of DRAM began with the same forecast of employment location and the same inputs for base-year household location and land use. After 8 iterations of DRAM and trip assignment there were no indications of convergence to an equilibrium solution of household location. Within each DRAM-NETWRK iteration, DRAM was iterated to equilibrium by using the 3-HH,

(a)

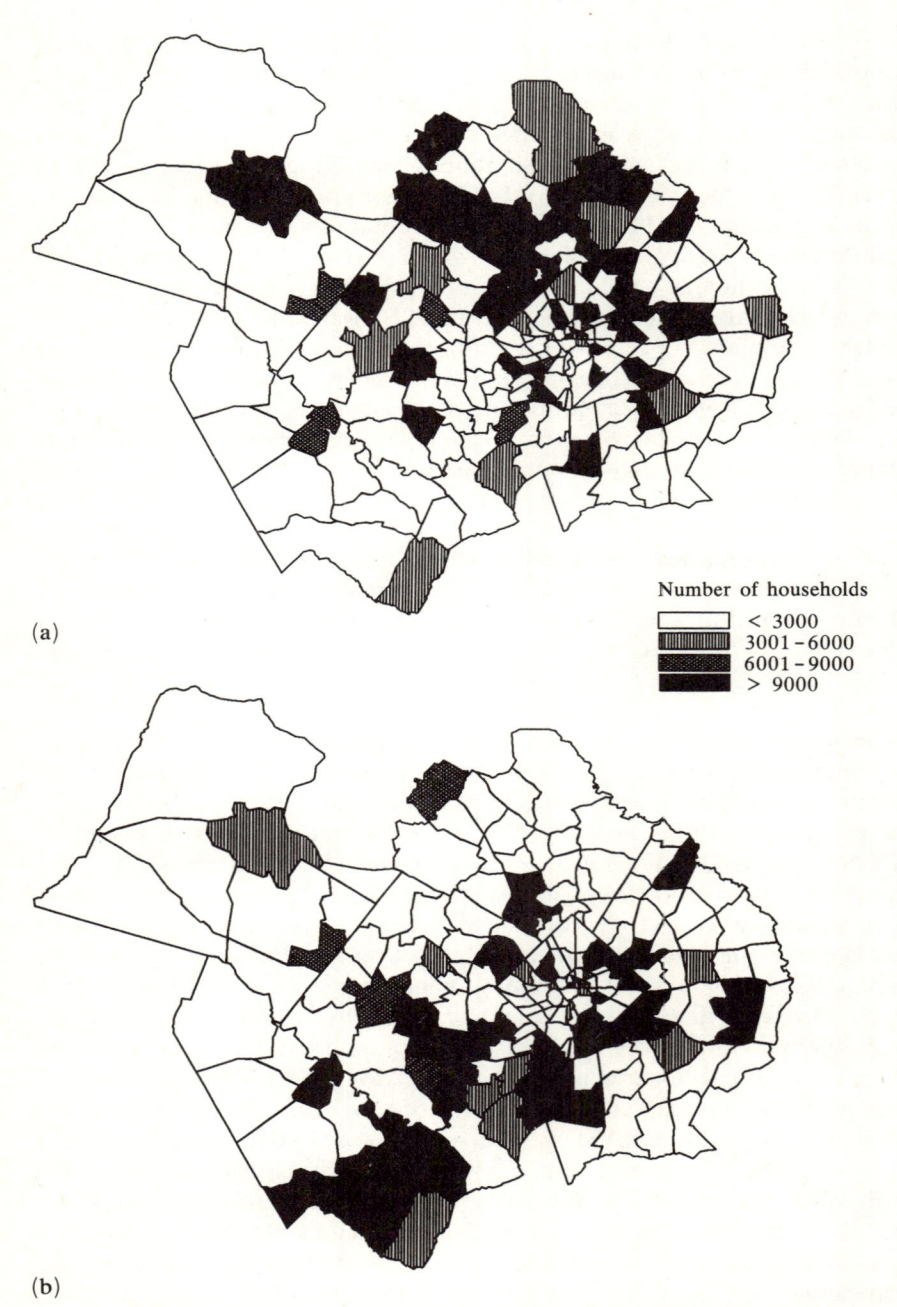

Number of households
- < 3000
- 3001–6000
- 6001–9000
- > 9000

(b)

Figure 10.3. Washington, DC, region: results from (a) the 7th iteration and (b) the 8th iteration of the first method of full simultaneous solution, for 1985.

50-LU approach. Yet, the entire system seemed to oscillate between a 'northern' and 'southern' location pattern. The spatial patterns of total households for the 7th and 8th iterations of this experiment are shown in figure 10.3.

In the second experiment of this series the only change to the above procedure was that the household and land-use outputs of each DRAM run, in each DRAM-NETWRK iteration, were used as the input to the next DRAM run. This procedure, too, showed no signs of convergence after 8 iterations. Here there was oscillation between an 'eastern' and 'western' pattern of location. The solutions to the 7th and 8th iterations of this approach are shown in figure 10.4. A glance at these figures will show clearly that there was virtually no correlation between the two sets of solutions nor, for that matter, between successive iterations in either solution procedure. For example, between the 7th and 8th iterations of the second of these two procedures (shown in figure 10.4) the simple correlations were: LIHH, -0.07; LMIHH, -0.01; UMIHH, -0.07; UIHH, -0.26; and TOTHH, -0.23. The simple correlations between the 8th iteration results for these two procedures were: LIHH, 0.19; LMIHH, 0.50, UMIHH, 0.44; UIHH, 0.41; and TOTHH, 0.44.

In chapter 9 the method of mean successive averages (MSA) was used to bring about convergence of sets of these linked location and trip-assignment models. Thus, as part of this new series of tests, several experiments were conducted where MSA was used with the hope of achieving convergence to an equilibrium solution. A third numerical experiment was performed. Here, as in the second experiment, the DRAM output of each iteration became input both to NETWRK and to the next iteration of DRAM. The MSA procedure in this application calculated weighted-average link flows (on the network) and then traced paths through the network congested by these weighted-average flows (trips). The zone-to-zone travel times for use in the 'next' iteration of DRAM were the lengths of these paths. The addition of the MSA procedure did serve to damp the oscillations somewhat. Figure 10.5 shows the spatial patterns of total households on the 7th and 8th iterations. Yet, even though these spatial patterns are more alike than the analogous results in the previous experiments, no equilibrium was reached. The rate of convergence was low enough to make it seem doubtful that an equilibrium would be reached. The simple correlations between the 7th and 8th iterations were: LIHH, 0.22; LMIHH, 0.65; UMIHH, 0.17; UIHH, 0.17; and TOTHH, 0.24.

A fourth experiment was tried. As it seemed that the problem in achieving convergence might be due to the 3-HH, 50-LU process being used to solve DRAM to an 'internal' equilibrium, in this experiment the original solution procedure for DRAM was substituted. Here again the output of each DRAM run was input to NETWRK and the next iteration of DRAM. If the 'self-iterative' form of DRAM produced spatial patterns which were too different from their starting points, it was reasoned, then perhaps the

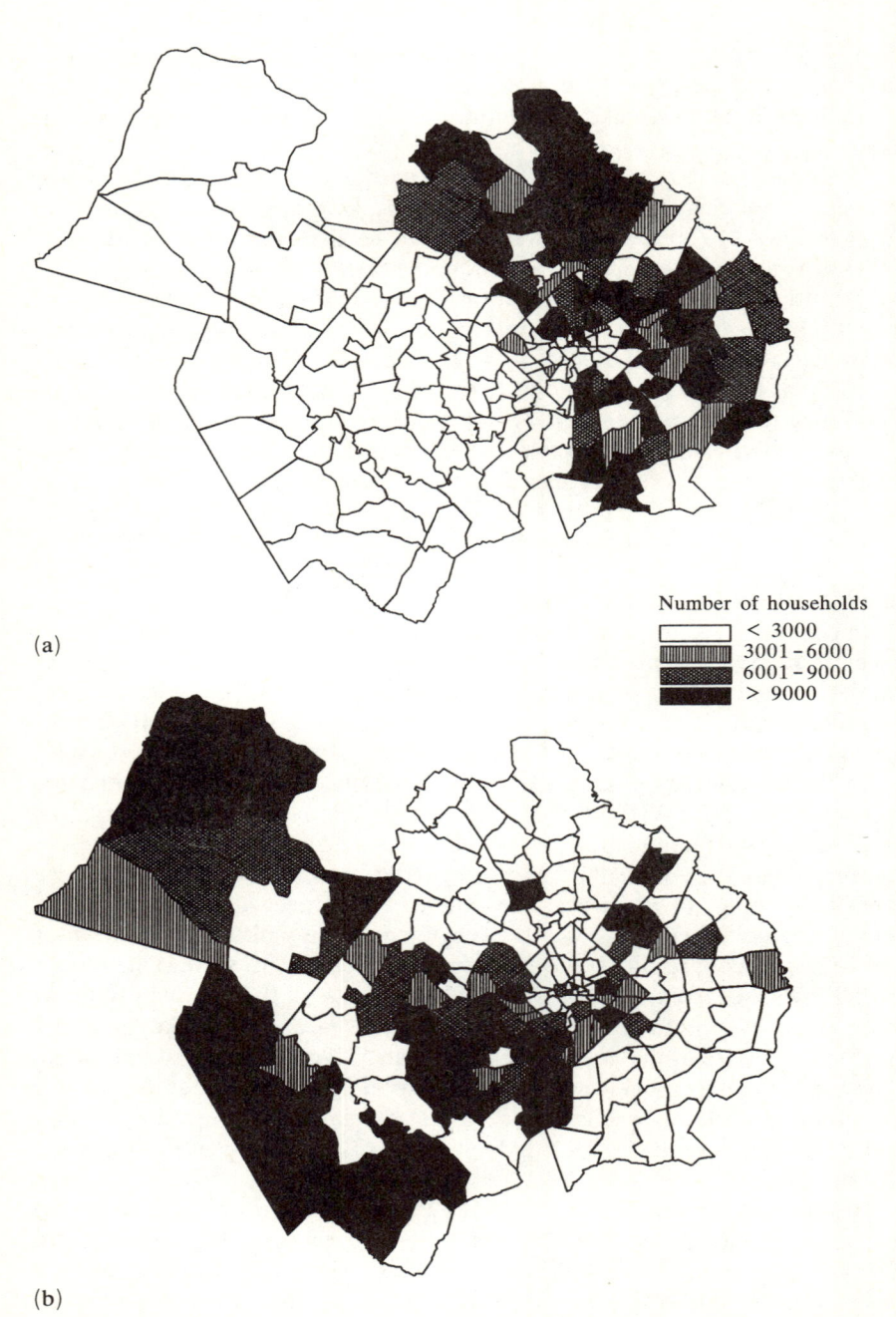

Figure 10.4. Washington, DC, region: results from (a) the 7th iteration and (b) the 8th iteration of the second method of full simultaneous solution, for 1985.

Equilibrium solutions to location models

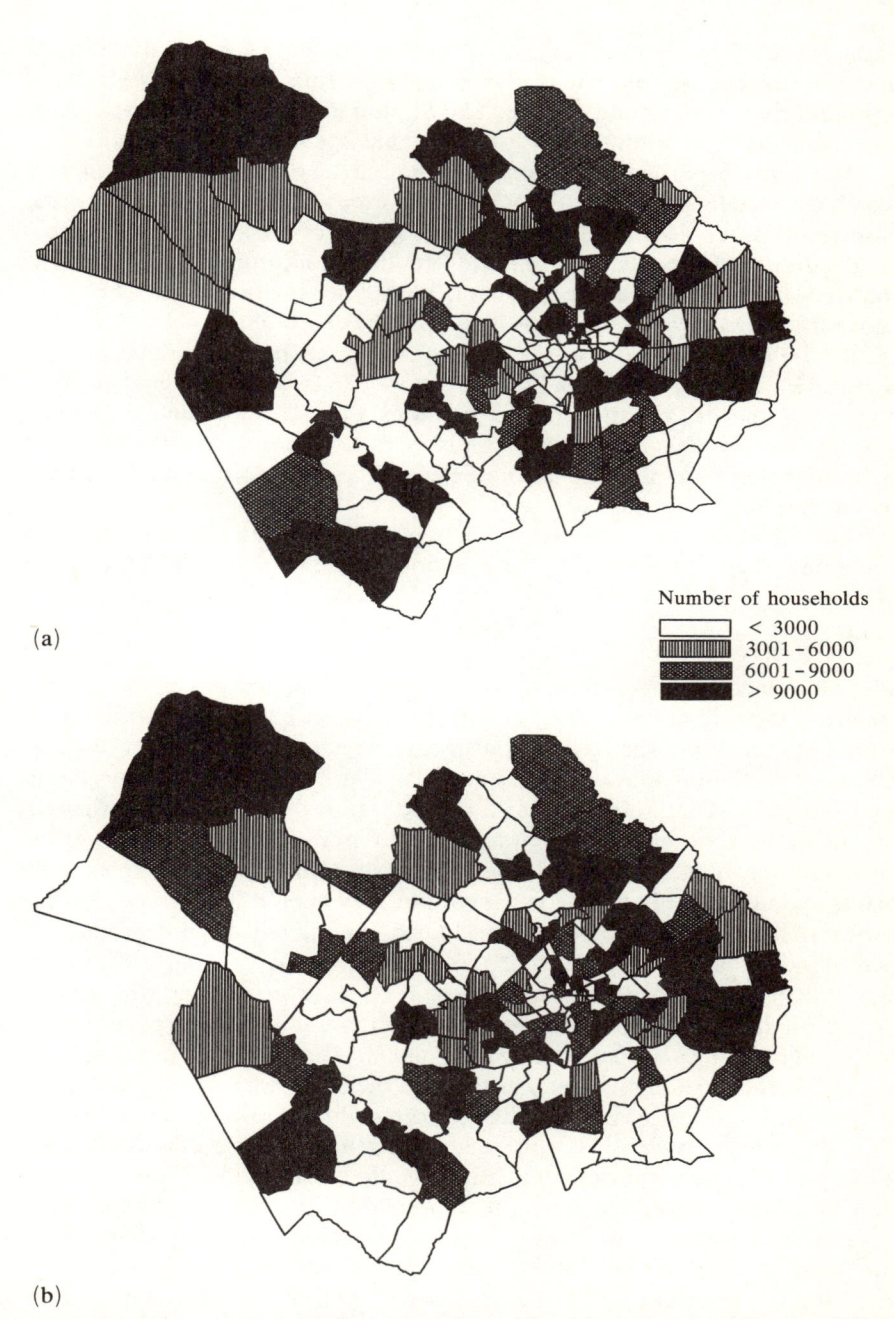

(a)

Number of households
- < 3000
- 3001–6000
- 6001–9000
- > 9000

(b)

Figure 10.5. Washington, DC, region: results from (a) the 7th iteration and (b) the 8th iteration of the third method of full simultaneous solution, for 1985.

traditional DRAM solution, going only one step (iteration) beyond the starting point in each case, would allow an overall equilibrium to be achieved. Once again, the system did not converge. The 8th iteration of this approach gives the spatial pattern for total households which is shown in figure 10.6(a).

Then a fifth experiment was done. In this case the only difference from the fourth experiment was that DRAM was 'restarted' in each iteration from the base year, rather than from the output of the prior iteration. This experiment converged quickly and smoothly to the equilibrium spatial pattern shown in figure 10.6(b). Here the simple correlations between iterations one and two were: LIHH, 0.99; LMIHH, 0.98; UMIHH, 0.98; UIHH, 0.98, and TOTHH, 0.98. The mean absolute percent deviations for these same groups were: LIHH, 12.4%; LMIHH, 13.7%; UMIHH, 14.1%; UIHH, 17.0%; and TOTHH, 13.7%. By the 4th iteration the correlations were all 1.00, and the mean absolute percent deviations from the 3rd iteration were all less than 0.30%. By the 8th iteration the mean absolute percent deviations from the 7th iteration were all less than 0.08%.

The fifth experiment represented a step back from the FSS intended for equation (10.20). In particular, the equation solved by this fifth FSS experiment was

$$N_t = f(E_t, L_{t-1}, N_{t-1}, c_t) \,. \tag{10.21}$$

The difference between the fourth and fifth experiments was that the fifth restarted each iteration of DRAM with the data on base-year land use and household location, whereas the fourth experiment used the output of each DRAM iteration as input to the next. Note that in both experiments there were no DRAM internal iterations, ('self-iterations'), because the original, or standard, solution method was used. In both experiments the MSA algorithm was used to try to achieve convergence. In principal, both experiments should not only have converged, but they should have converged to identical solutions. The fact that the fourth experiment did not produce a convergent solution suggests that the successive DRAM and NETWRK solutions were overresponding and thus, in effect, overshooting what might be called the 'envelope' of convergent solutions. It is the purpose of the MSA algorithm to cope with this problem. However, as discussed in chapter 7, the MSA algorithm does not give the best choice of step length or step direction, it just gives ones that are often able to provide a convergent solution path. As such, it is a very crude approximation to methods, such as Newton's, where appropriate derivatives would be calculated for each iteration in order to guide the search for an equilibrium solution. The general question of techniques for solving these model systems is an important topic for future research (Chapra and Canale, 1985; Reklaitis et al, 1983).

A sixth experiment was done, in a sense, to complete this set. In this configuration the MSA algorithm was used, DRAM self-iterations of the form 3-HH, 50-LU were done, and DRAM was restarted from the data on base-year land use and household location for each iteration. This was an

Equilibrium solutions to location models

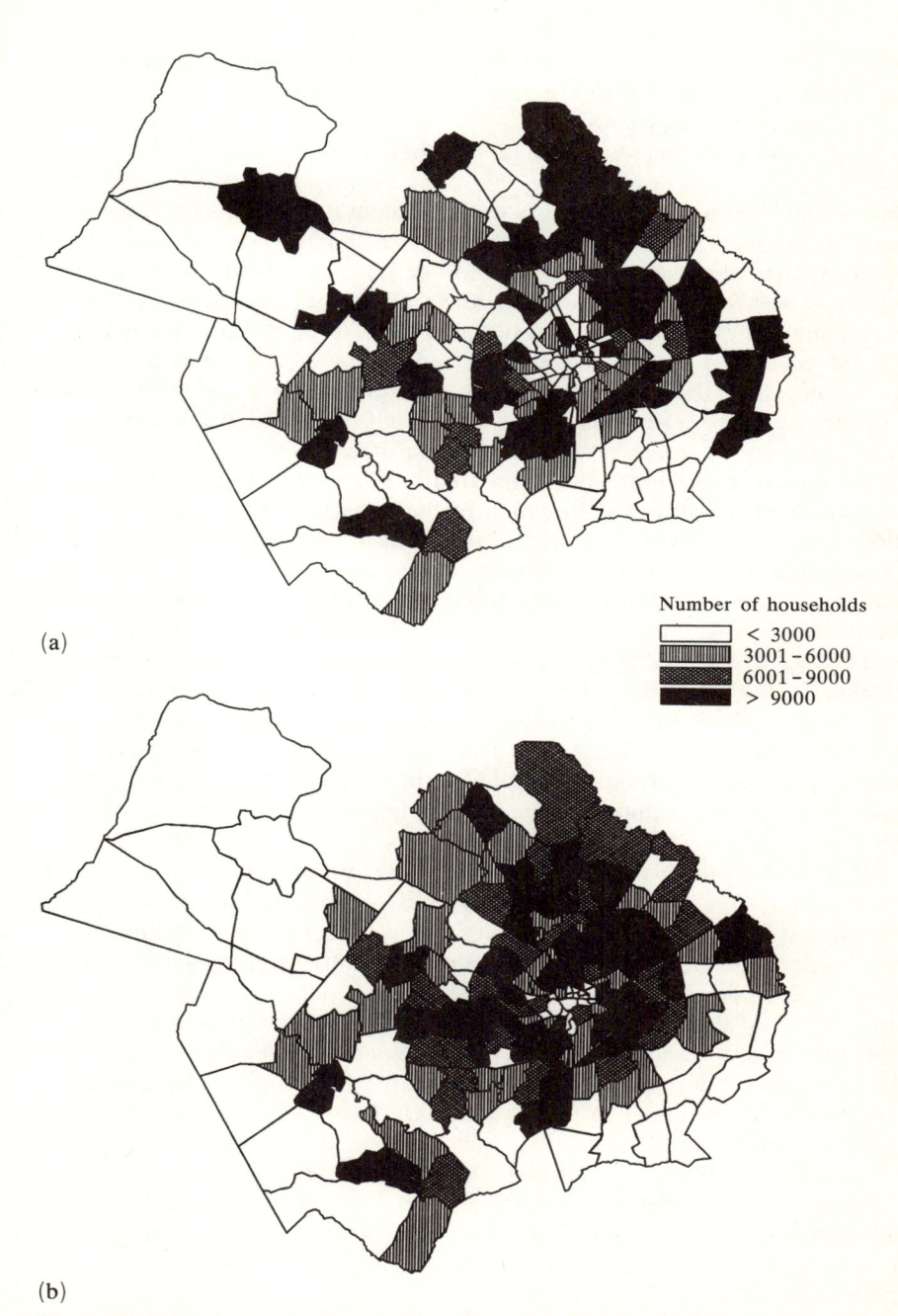

Figure 10.6. Washington, DC, region: results from the 8th iterations of (a) the fourth method and (b) the fifth method of full simultaneous solution, for 1985.

attempted solution of equation (10.20) again, as in the first experiment, with the addition of the MSA algorithm to aid convergence. The system did not converge.

If all these experiments are taken together, some fascinating possibilities are suggested. First, it is possible that these results are artifacts of the data set or that they result from errors in the computer program. The former is unlikely, as the 182-zone Washington data set is large enough to preclude the problems of discontinuities common to small numerical examples, and rather extensive checking and prior usage pretty much rules out the latter. What appears more likely is best described in a three-dimensional analogy, though the actual model system operates in many more than three dimensions (consider the numbers of locators, zones, and network paths). It appears, very generally speaking, that the self-iterations in DRAM which are necessary to solve simultaneously for household location and land consumption, move the equilibrium solution in a direction away from or, perhaps, orthogonal to the direction of the household-location and travel-time equilibrium. Thus, in a sense, one might think of the existence of one equilibrium solution between household location and congested (implicitly, by the household location) travel times (HLTT). This HLTT equilibrium can be calculated with the assistance of the MSA algorithm. The existence and stability of this equilibrium appear, from these numerical experiments, to depend on the use of fixed, base-year values of household location and land use for the calculation of the locational attractions [equation (10.6)]. Even the use of one step in the direction of the equilibrium between household location and land consumption (HLLC) appears to either eliminate the existence of the HLTT equilibrium or make it impossible to calculate with the methods used here.

It is too tempting to resist speculation on the possible correspondence of these results to real-world behavior. Clearly, there may be no correspondence at all. These results may be simply an aspect of the mathematics of the models. Yet, to a considerable degree, various forms of these models have been shown to describe human behavior at, let us say, the mesoscale. The notion of differential rates of adjustment to disequilibrium conditions provides a convenient framework in which to consider this matter. The HLLC equilibrium may perhaps be thought of in terms of phenomena of neighborhood character or neighborhood change. Of course, the term 'neighborhood' may not be strictly enough defined here, especially in view of the areal systems within which these models are customarily defined. These changes in neighborhood character take place, under the usual circumstances, rather slowly. Thus movement toward an HLLC equilibrium should, perhaps, be calculated much more slowly than movement towards an HLTT equilibrium.

Whereas a neighborhood life cycle can take twenty or thirty years, shifts in household location in response to changes in transport time (or cost) can occur much more rapidly, in a decade or less. Perhaps then, the HLTT

Equilibrium solutions to location models

equilibrium reflects these shifts in household location in response to changes in travel time. Such an equilibrium need not represent the same phenomenon as the change in neighborhood character implied by the HLLC equilibrium. This set of questions clearly implies the need for additional research, and will be discussed again later in the text (see chapter 11). Before continuing the general thread of this chapter, some rather interesting results from the sixth experiment mentioned above will be discussed.

10.6 Chaotic behavior of a deterministic system
10.6.1 DRAM and NETWRK: 1

Recall that the configuration of the sixth FSS numerical experiment involved the use of DRAM, including the self-iterative 3-HH, 50-LU solution procedure, and of NETWRK for the assignment of trips to the highway network using a stochastic multipath procedure. Note that the use of the term stochastic does not imply a random or probabilistic component of the system, rather it describes the use of a logit function to disperse trips both to the minimum path and to longer paths through the network between each $i-j$ zone pair. The two models are linked together and the MSA algorithm is used to attempt to help the system converge to an equilibrium solution. Finally, each iteration of DRAM begins with the same base-year distribution of land use and household locations and the same forecast-year distribution of employment at place of work.

An iteration of this system begins with DRAM being run by self-iteration to an HLLC equilibrium. The trip matrix which results from the DRAM solution is then assigned to the highway network. The flows on the network result in congestion of some of the links and consequently in a change in the time necessary to traverse them. The shortest paths through the congested network are then traced. The lengths of these paths, in travel time between each $i-j$ zone pair, become the elements of the travel-time matrix which is then used as an input to the next iteration of DRAM. In the 2nd iteration, DRAM is restarted with the forecast-year employment distribution (as in the first iteration) and with the base-year household location and land consumption. The only change in inputs from the 1st iteration to the 2nd iteration is the travel-time matrix. DRAM is again run to an HLLC equilibrium. The resulting trips are assigned to the network, and a new travel-time matrix is produced. The process is then repeated. As mentioned above, this procedure showed no signs of convergence after 8 iterations. The spatial patterns of total households for the 7th and 8th iterations of this experiment are shown in figure 10.7.

In and of themselves the spatial patterns in figure 10.7 do not seem particularly unusual. The only readily noticeable differences from the 7th to the 8th iteration occur in the central area. The spatial patterns for the central area only are shown in figure 10.8. Even there, the differences do not appear to be all that great, though there are some obvious changes of zones from '<3000' to '>9000'. However, zone 28 has 43459 households

274   Chapter 10

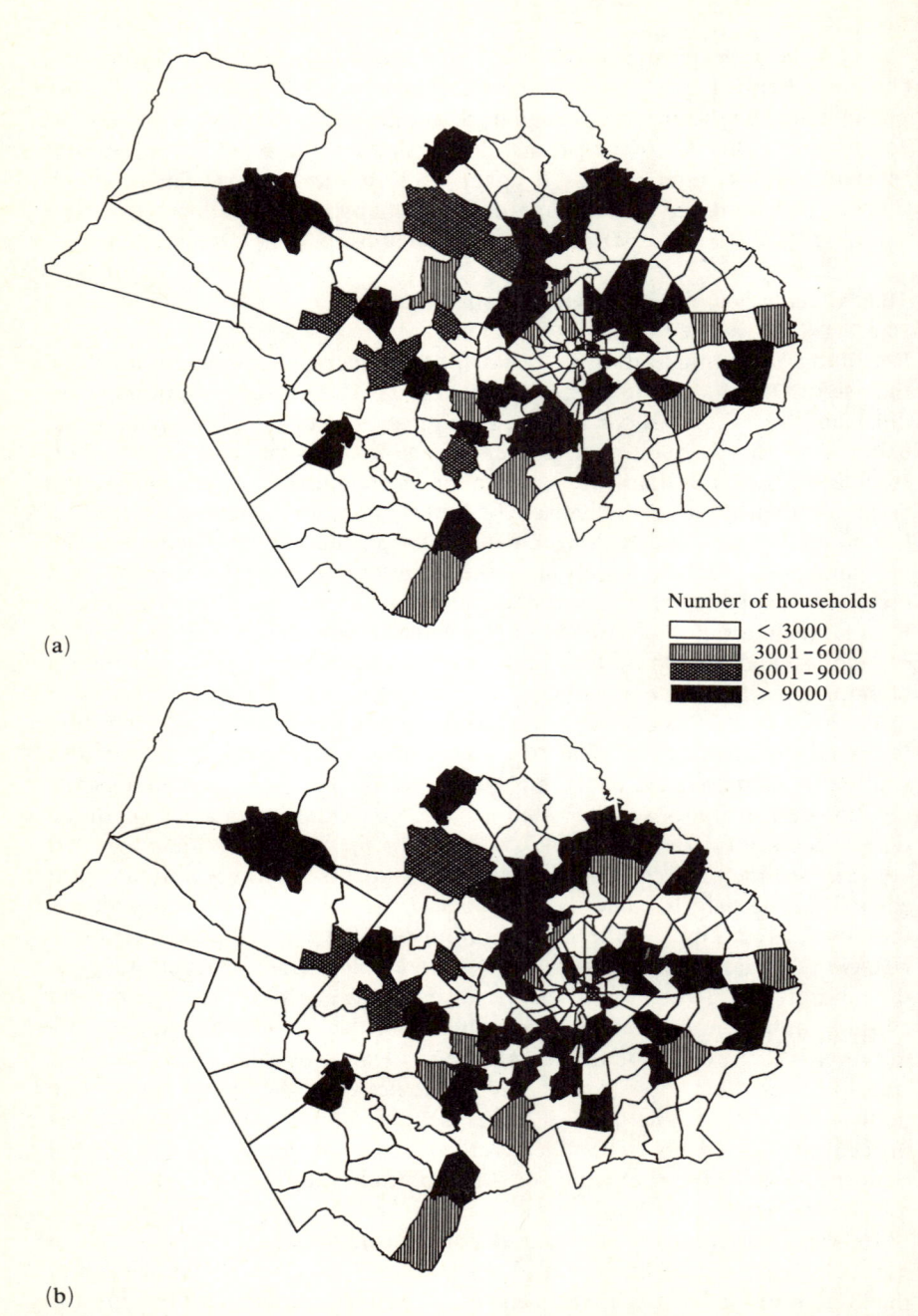

(a)

Number of households
☐ < 3000
▦ 3001–6000
▥ 6001–9000
■ > 9000

(b)

Figure 10.7. Washington, DC, region: results from (a) the 7th iteration and (b) the 8th iteration of the sixth method of full simultaneous solution, for 1985.

Equilibrium solutions to location models   275

in iteration 7, and 403 households in iteration 8. Zone 116 has 405 households in iteration 7 and 32 566 households in iteration 8. A few other zones are showing these enormous changes too, but these two will serve for the purposes of this discussion. First, this is not a situation of just two zones trading households. Not only are other zones involved, but the cycles do not correspond in that way. Table 10.1 shows the values of total households and of households in each income group for zones 28 and 116 for the base year and for 8 iterations of the DRAM-NETWRK system. It is apparent that the numbers of households in these two zones are not linked together. At some iterations both zones have only a few hundred households; at others both have tens of thousands of households; at yet other iterations one zone has few households when the other zone has many.

If the mechanics of these runs are recalled, it can be seen that the only difference in the DRAM inputs between iteration 7 and iteration 8 is the two different travel-time matrices. Both DRAM runs start with: (1) identical distributions of forecast employment (by employment type, by zone), (2) identical base-year distributions of households (by household type, by zone), and (3) identical base-year distributions of land use by type and zone. The mean zone-to-zone travel time for the matrix which is input to iteration 7 of DRAM is 51.93 minutes, and the corresponding value for iteration 8 is 51.99 minutes—not a very great difference. Comparison of the means of more than 33 000 travel-time values may obscure important individual differences, so a closer look was taken. As zones 28 and 116 were zones where large changes were seen, the 28th and 116th rows and columns of the

Table 10.1. Results (number of households by income group) for zones 28 and 116, obtained from 8 iterations of the sixth method of full simultaneous solution.

	Base year	Iteration							
		1	2	3	4	5	6	7	8
Zone 28									
LIHH	6 786	18		8	13 880	12 841	7	12 791	7
LMIHH	5 785	38	17	16	19 399	21 954	13	22 031	15
UMIHH	4 393	94	43	39	5 806	5 584	34	5 659	35
UIHH	3 695	1 100	436	384	2 978	2 956	340	2 978	346
TOTHH	20 659	1 250	505	446	42 063	43 335	394	43 459	403
Zone 116									
LIHH	2 845	7	9	7	8	9 373	7	8	8 886
LMIHH	3 567	14	19	15	16	20 583	15	16	18 098
UMIHH	3 252	34	52	38	42	3 744	37	41	3 803
UIHH	2 098	317	424	323	358	1 742	311	340	1 779
TOTHH	11 762	372	505	385	423	35 442	370	405	32 566

Note: LIHH, low-income households; LMIHH, lower-middle-income households; UMIHH, upper-middle-income households; UIHH, upper-income households; TOTHH, total number of households.

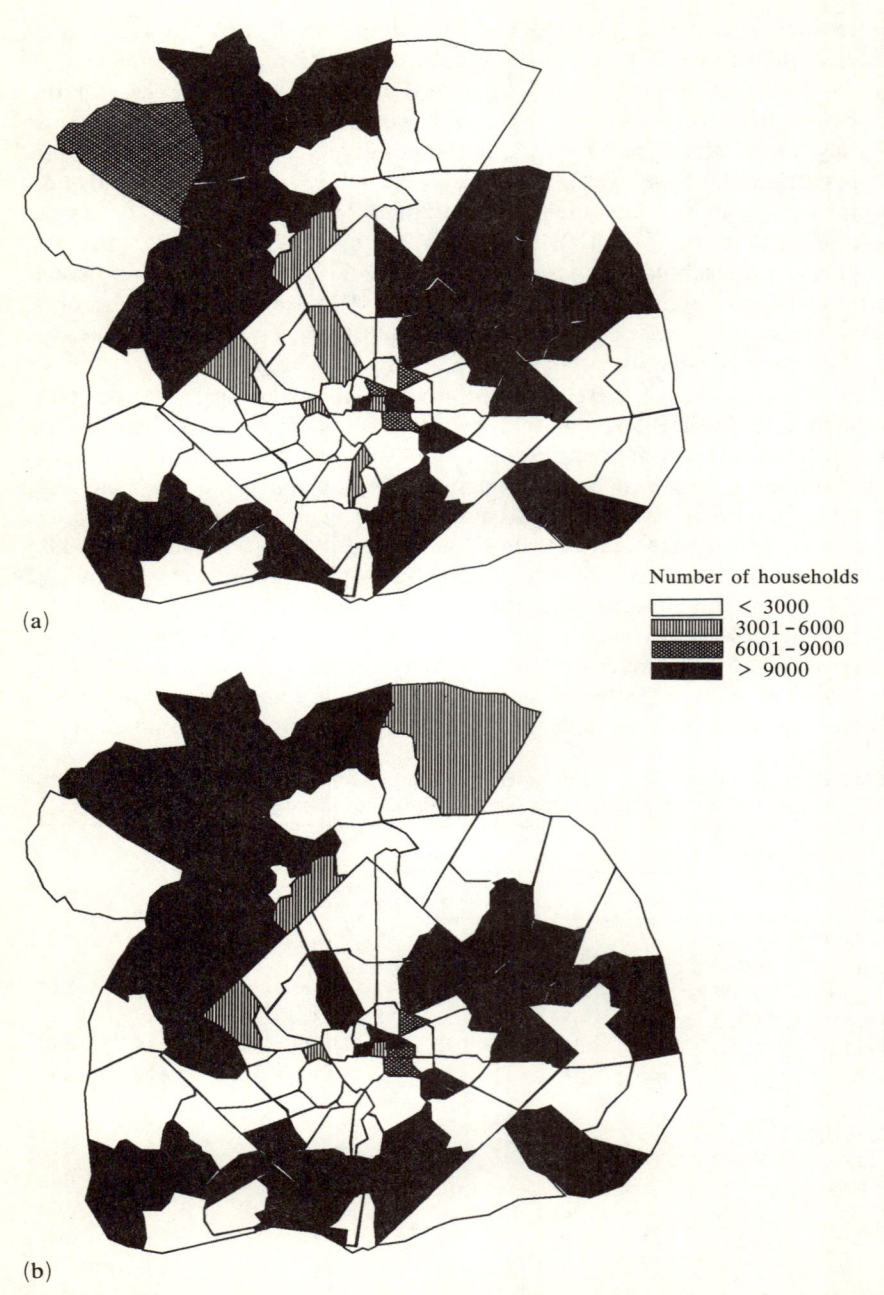

Figure 10.8. Washington, DC, central region: results from (a) the 7th iteration and (b) the 8th iteration of the sixth method of full simultaneous solution, for 1985.

travel-time matrix were examined. For the 28th column of the travel-time matrix the mean absolute difference between the iteration-7 and iteration-8 inputs was only 0.5%, and for the 28th row the corresponding value was only 0.3%. For the 116th column and row the values were 0.2% and 0.4% respectively. These means *did not* result from a few large deviations averaged over 182 observations. For the 28th row and column there were only 6 and 20 deviations, respectively, in excess of 1%. An examination of the diagonal elements, (28, 28) and (116, 116) showed that the differences between iterations were 0.0% in both cases.

Thus there was a situation where very modest differences in inputs resulted in substantial differences in outputs. This situation was *not* simply a matter of oversensitivity, rather it was an example of 'sensitive dependence on initial conditions', (that is, chaotic behavior in a deterministic system). The cause of this behavior, from a mechanical point of view, was the self-iterative solution of DRAM for the HLLC equilibrium. It has already been stated that the experiment did not converge overall. What is more intriguing are the correlations between the spatial patterns of total households from one iteration to the next. These are as follows:

Iterations	Correlation
1, 2	0.22
2, 3	0.77
3, 4	0.82
4, 5	0.84
5, 6	0.80
6, 7	0.78
7, 8	0.74

The correlations between alternate pairs of iterations are:

Iterations	Correlation
1, 3	0.39
2, 4	0.71
3, 5	0.77
4, 6	0.86
5, 7	0.90
6, 8	0.93

Thus it appears that although the process is clearly *not* converging to a stable solution, there may be a pair of alternate stable solutions between which it is switching. Perhaps iterations 1, 3, 5, 7 are tending towards one of these equilibria and iterations 2, 4, 6, and 8 are tending towards the other. To answer this question would require many more iterations of the model system and will have to await further research efforts.

A simple experiment was done to further understand what was being observed here. If as was implied above, the rather small iteration-to-iteration differences in travel times caused this chaotic behavior not in and of themselves, but via the agency of the 3-HH, 50-LU solution procedure, then those same travel-time matrices should result in very little difference in the output of a standard DRAM run. The travel-time matrices used as input to iterations 7 and 8 of the sixth experiment were used as input to two standard DRAM runs. The distributions of households which resulted were then compared. The simple correlations between the four household types and total households were all 1.00. The mean absolute percentage deviations were all less than 0.5%, with no single deviation being in excess of 1.9%. Thus it is not the sensitivity of the model in standard form which causes these curious results, it is the result of the repeated self-iterations of the 3-HH, 50-LU solution procedure.

10.6.2 DRAM and NETWRK: 2

Another possibility for this behavior, it might be argued, is that stochastic multipath assignment was used to assign the trips to the network. To examine this possibility, before a conclusion was drawn as to the overall meaning of the results, the experiment was rerun making use of user-equilibrium assignment. This time 12 iterations were run, by use of the same procedure as described above for the sixth experiment. Here too, the results were quite interesting, as the system started towards convergence and then moved away from it. The correlations for total households, for the 12 iterations were:

Iterations	Correlation	Iterations	Correlations
1, 2	0.20	7, 8	0.97
2, 3	0.67	8, 9	0.82
3, 4	0.80	9, 10	0.82
4, 5	0.84	10, 11	0.88
5, 6	0.75	11, 12	0.80
6, 7	0.82		

The alternate iteration correlations do not show the apparent movement toward different equilibria that they showed in the previous experiment. Here, each succeeding iteration of the system seems to yield a different spatial pattern of households. For all 12 iterations all inputs but the travel-time matrix are identical. Yet, as for the previous experiment (the FSS—sixth method) a number of zones seem to have two quite different solution regions. The zones take one or another of these solutions with no apparent regularity. In table 10.2 results are given for selected zones, for 12 iterations.

Two other points worth noting are related to the actual spatial patterns and the mean congested travel times. First, the spatial patterns produced in successive iterations, although different from one another, are not very different. Figure 10.9 shows the spatial patterns of total households for the 7th and 12th iterations. They are different, but there are more similarities than differences. Note, too, that these results are quite similar to the results from the sixth method shown on Figure 10.7(a). Thus although the choice of trip-assignment algorithm makes some difference, as it should, there is not a very great difference in the resulting spatial patterns. Second, it is worth noting that throughout the iterations the mean zone-to-zone travel time shows a very small but steady decline. The mean zone-to-zone travel times of the

Equilibrium solutions to location models 279

travel-time matrix inputs to iterations 7 to 12 of DRAM in this experiment were 52.91 minutes, 52.85 minutes, 52.84 minutes, 52.75 minutes, 52.69 minutes, and 52.65 minutes respectively. Again, these differences are so small as to be not worth mentioning were it not for the unusual equilibrium spatial patterns of households which result from the DRAM runs to which they are input.

An additional simple experiment was done to further confirm the uniqueness of this behavior. The travel-time matrices which had been used as input to the 7th and 8th iterations were averaged together, element by element. In a standard DRAM solution, the forecasts of household location which such a run would produce would, in general, be intermediate between the prior two DRAM results. For this additional HLLC run this was definitely not the case. Again, there was substantial switching of zones from very large to quite small numbers of households located therein. Thus even when the differences between the travel-time inputs were only half the difference between the original 7th and 8th iterations, the model still evidenced chaotic behavior when forced to an HLLC equilibrium.

The study of chaotic behavior of deterministic systems is a relatively recent development (Iooss et al, 1983), yet the frequency with which such behavior does appear, and the great number of research areas which are touched, has even led to a 'best-seller' on the topic (Gleick, 1987). The brief description given here of the apparent chaotic behavior suggests that (a) sometimes unexpected results (that is, nonconvergence) may be more than just an annoyance to be got around and, (b) there is much more work to be done here. A proper examination of these phenomena, now being planned, will require the use of a supercomputer to do the thousands of iterations necessary to produce a clear picture of what is really happening in this model system.

Table 10.2. Results (total households) for selected zones, obtained from 12 iterations of the seventh method of full simultaneous solution.

Zone	Base year	Iteration					
		1	2	3	4	5	6
9	9775	19042	339	287	285	296	290
28	20659	1250	600	472	45525	44950	416
45	8316	27227	4649	22428	21345	20692	21651
88	2231	20355	310	4147	3915	17166	3752
116	11762	372	589	382	389	410	389
		7	8	9	10	11	12
9	9775	24843	24709	283	24668	275	25467
28	20659	44266	43668	394	44100	41732	377
45	8316	6812	20622	21159	8140	21139	21182
88	2231	4015	17237	3810	4052	3923	3947
116	11762	394	393	368	391	391	367

280   Chapter 10

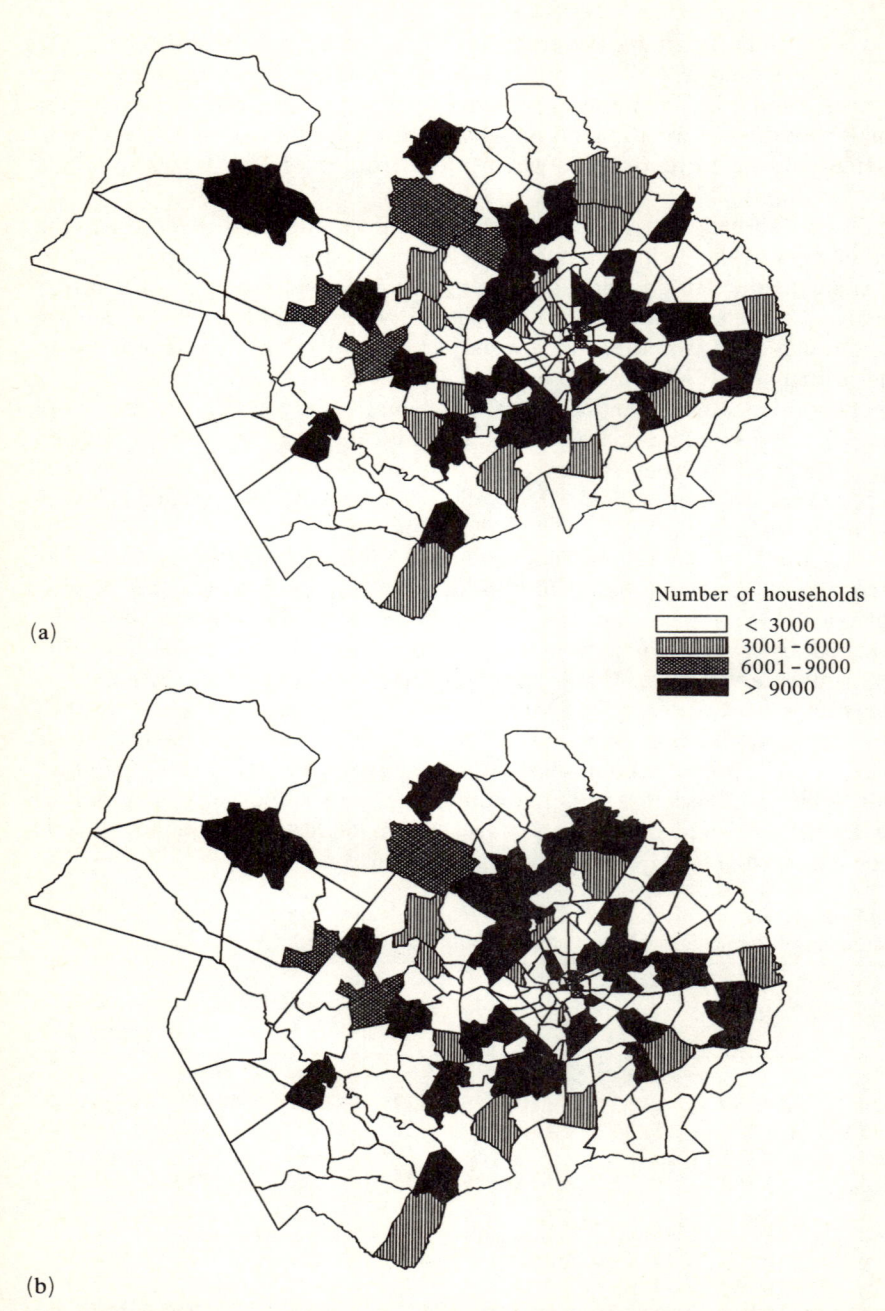

(a)

Number of households
- < 3000
- 3001–6000
- 6001–9000
- > 9000

(b)

Figure 10.9. Washington, DC, region: results from (a) the 7th iteration and (b) the 12th iteration of the seventh method of full simultaneous solution, for 1985.

10.7 Conclusion

Why, one might ask, are all these computer experiments necessary, when there are algebraic methods for analyzing systems of nonlinear equations that will yield answers to questions of solvability, uniqueness, and sensitivity to variables and parameters? Such analyses have been done for a restricted version of these models where most of the interactions examined in the experiments in this chapter were held constant (Miller, 1987). In the versions of the models described here such analyses might, at first glance, appear to be much more difficult, but still possible. There is, however, a factor which precludes such methods of analysis. This factor is the congested travel-time matrix. The congested travel times are the connection through which c_t is a function of N_t and E_t. These times are found by tracing paths through the network. The paths are over the network links, each of which may or may not be congested by the flows on it. Just as in the case of solving for user-equilibrium assignment, the critical matter is the numbers of paths. For the Washington network used in these experiments there are tens of millions of paths. A direct matrix–algebraic solution of a system where employment and household location are a function of travel time, and where travel time is a function of employment and household location via flows on links (paths), is therefore quite impossible. The numerical methods of the Frank–Wolfe and MSA algorithms for the trip-assignment problem and the Evans approach for combined location and assignment problems appear to be the only feasible methods for obtaining solutions to this class of problems at the present time. Thus the analyses in this chapter are only possible via numerical experiments.

Some discussion of the results of these experiments has already been given and need not be repeated here. In brief, there appear to be equilibrium solutions to portions of a linked household-location and trip-assignment model, but it does not appear possible to find an equilibrium solution to a fully simultaneous system. For fixed employment location and fixed travel time, it is possible to solve for an equilibrium solution to household location of several interacting household types and to land consumption. For fixed employment location, other household location, and land consumption it is possible to solve for an equilibrium solution to household location and travel time. It was not possible to solve for a simultaneous equilibrium solution to multiple household-type location, land consumption, and travel time. It is important to note that for the few combined location and assignment models that have been reported in the literature, the solution difficulties reported here would not have been encountered as virtually all such examples deal with only one category of locator. For the multiple-locator combined-model solution reported in chapter 9, the equilibrium between household location and land consumption was not attempted. What is new in this chapter is the attempt at HLLC equilibrium, and it is in the solution thereof that the surprises arise.

There was discussion in chapter 4, as part of the experiments with variations on the Herbert–Stevens formulation, of the role of the supply side in household location. In particular, the numerical experiments described there showed the importance of housing cost and/or housing stock estimates in preventing absurd solutions to mathematical programming formulations of the households-location problem. In particular, these supply-side variables had the effect of preventing, say, all the high-income households from locating in the single 'best' zone. If similar constraints were imposed on the solution of the HLLC equilibrium it might well be less likely that one would get the extreme solution fluctuations found in some of the numerical experiments in this chapter. This, too, is a matter to be examined in future research efforts, though, as mentioned in chapter 4, the estimation of housing price and/or housing stock is an extremely difficult undertaking.

The finding and identification of the potential for chaotic behavior in these models suggests a direction of research into some previously puzzling phenomena. In US cities there are regularly found instances of catastrophic change. Neighborhoods of a particular racial composition may experience slow change for years until, quite suddenly, there is an explosive shift in the predominant race or ethnic group. In other neighborhoods a long period of stagnation or slow decline may be followed by rapid 'gentrification' and thus a shift in the predominant household type, in terms of both income and household size. Though often examined, these phenomena have yet to be systematically understood. Clearly there is the potential for a great deal of additional work here.

Last, there is the more general issue of what these experiments suggest regarding location forecasts in general. It is obvious that the HLLC equilibrium solution is *not* a good forecast of household location. If anything, it may suggest the 'direction' in which locations would tend, all other things being equal. The HLLC equilibrium is *not* a market-clearing solution either, as there is no representation of the supply side included. Nor is the HLLC equilibrium a cost-minimizing or utility-maximizing solution. It represents a working out of the household-to-household preferences revealed in the coefficients obtained from model calibration, along with the preferences for various land-use compositions as revealed in the remaining coefficients shown in equation (10.6). It is clear that these preferences are never completely expressed in actual household location. There are mitigating factors such as house price and availability. There is also a certain locational inertia. Even in highly mobile societies only a small percentage of the households in an urban area change household location in any given year. This raises again the issue of rate of adjustment, that is, the length of time it is expected to take for a complete adjustment from any given disequilibrium pattern of household locations to the appropriate equilibrium.

Recall that, for all the solution methods discussed in this chapter, the first iteration of household-location calculations is identical to the standard nonequilibrium solution method. Thus each iteration may be seen as an additional step (or movement) towards the potential equilibrium solution. The HLLC equilibrium is the target, along one set of dimensions, towards which these iterations are moving. The HLTT equilibrium is the target along another set of dimensions. Progress towards these equilibria may be thought of as processes of adjustment in response to disequilibrium conditions. Each adjustment process may be proceeding at a different rate. It is precisely in this role that the equilibrium solution may be of use. The ability to solve for an equilibrium solution does not provide a forecast. It provides a target for use in models based on the concept of partial adjustment to disequilibrium conditions. The development and testing of such models will be the focus of the next chapter.

11
Preliminary development of dynamic spatial models

11.1 Introduction
In the early development of residence-location models it was common practice to gloss over the distinction between simulated periods of time, and the iterations which were computationally necessary to solve the models. The customary procedure for solving any of the various Lowry derivative models was that of successive substitution (Putman, 1979). Virtually all of these models located total households or total population (which may, in a subsequent calculation, have been disaggregated to population or household subcategories). If any measure of zonal attractiveness for residential locators was used, it was calculated from fixed, exogenous variables. Travel functions may have been nonlinear functions of, say, travel cost, but they too were fixed and exogenous. Although some of these models could have been solved by matrix algebra, most were not. The iterative solution methods used were usually convergent and gave what Lowry (1964) called an 'instant metropolis' as a solution. The temporal designation of such a solution, for example, 1980, 1985, and so on, was identified solely in terms of the temporal specification of the spatial distribution of basic employment and the regional population totals that were given as input.

The question of temporal specification, that is, how the output from a model is to be associated with a particular instant in historical time, is not yet resolved. Most authors do not address the issue. It is customary to assume that the regional control totals (for example, total population or employment by industry) which are provided as input to these models provide the temporal specification. Yet, there is no particularly good reason for this to be the case. Very few of these models have anything inherent in their structures which connects them to any particular real time. This is the case with DRAM if it is used independent of other models. EMPAL, however, uses the simplest method of connection to historical time, a lagged variable. This provides a degree of temporal specification to forecasts of the spatial distribution of employment produced by EMPAL which, in turn, provides (albeit to a lesser degree) a degree of temporal specification to forecasts of the spatial distribution of households produced by DRAM. Nonetheless, neither of these models individually or together has a formal specification of time incorporated in their equation structure. It is this deficiency which the work described in this chapter was intended to remedy.

In the second section of this chapter, the concept of system adjustment to disequilibrium conditions as a structure within which dynamic spatial models may be formulated is introduced. In the third and fourth sections of this chapter numerical experiments with the Archerville data are described and the behavior of alternate solution structures for achieving intramodel equilibria is explored. In the fifth section of the chapter these experiments are extended to the full-scale Houston and Washington data sets. In the last major section of the chapter, the sixth, some preliminary tests of the disequilibrium adjustment structure in which use of the Houston data set is made, are discussed.

11.2 Adjustment to disequilibrium as a modelling framework

In making a forecast of household location the DRAM model, as specified in equations (10.4)-(10.6), and discussed at length in chapter 10, in effect, creates an 'instant metropolis'. The employment and travel-time inputs are for time t, and the household and land-use inputs are lagged, for time $t-1$. Then, if it is assumed implicitly that locational behavior will be as it was in the year of the data used for the estimation of the parameters, the spatial pattern for households at time t can be calculated. The difficulty is that the model has no information for a zone with, say, 2000 households, as to whether that zone has recently declined from 3000 households, or increased from 1000 households. One way to provide that information would be to add a lagged variable for households to the structure of the model. This approach was taken in several recent planning-agency applications of EMPAL and DRAM (Lenk, 1988; Waddel, 1987; Watterson, 1986). This, in and of itself, still provides no information on the reasons for decline or growth in a particular zone. What is needed is not just lagged data but a structural mechanism to attempt to explain the positive or negative direction of zonal change.

There have been various past attempts to develop dynamic model structures for spatial interaction models. Indeed, one of the most widely used of early residence-location models, EMPIRIC, had a linear difference equation structure (Putman, 1979). This, however, was a strictly associative model, depending almost entirely on the output of multiple regression analysis to determine the model structure. In a much more interesting line of research, possible dynamic formulations of spatial interaction models have been investigated, including the work of Batty (1977), and the rather considerable work of Wilson (1985; 1987) and his associates (Beaumont et al, 1981; Clarke and Wilson, 1983). Another line of work in this area has been followed by Allen and Sanglier (1979) and Pumain et al (1984). Finally, another extensive line of work has been described by Bertuglia and his associates (Bertuglia and Leonardi, 1979; Bertuglia et al, 1985; Lombardo and Rabino, 1984). The differences between all these approaches are considerable, and many of the major issues remain to be resolved. There is, however, some work in the economics literature which offers a potentially useful organizing framework.

Consider first the following model of a market for some good (product) where there is a simultaneous determination of price and quantity (Bowden, 1978a):

$$D_t = X'_{1,t}\beta_1 + \alpha_1 P_t + u_{1,t}, \tag{11.1}$$

$$S_t = X'_{2,t}\beta_2 + \alpha_2 P_t + u_{2,t}, \tag{11.2}$$

and

$$Q_t = \min(D_t, S_t), \tag{11.3}$$
$$\Delta P_t = \lambda(D_t - S_t) + u_{3,t}, \tag{11.4}$$

where

D_t is the number of units of good demanded at time t,
S_t is the number of units of good supplied at time t,
$X_{1,t}$ is a vector of exogenous variables which determine demand,
$X'_{1,t}$ is the transpose of $X_{1,t}$ (similarly for $X'_{2,t}$),
$X_{2,t}$ is a vector of exogenous variables which determines supply,
Q_t is the quantity of the good actually transacted,
P_t is the price at which the transaction takes place,
$\alpha_1, \alpha_2, \beta_1, \beta_2, \lambda$ are parameters,
u_1, u_2, u_3 are disturbances terms with a mean of 0.

This formulation asserts that change in price is a function of the difference between supply of and demand for the good in question, plus or minus a random disturbance term (u_3) with a mean of zero. The parameter λ, determines the speed at which the system responds to differences in supply and demand, and adjusts to equilibrium. If $\lambda = 0$, then the system does not respond, or responds infinitely slowly, to imbalance between supply and demand. As $\lambda \to \infty$ the system response becomes infinitely fast, and imbalances between supply and demand cannot exist for finite time periods.

The process by which the quantities supplied and demanded adjust to achieve equilibrium in such models is called 'market clearing'. It is customary to consider that in an equilibrium model, market clearing takes place infinitely fast, as when in this model $\lambda \to \infty$. For any finite value of λ there will be a finite time required for this model to adjust to disequilibrium starting conditions, that is, for it to clear.

Consider next that for a particular set of exogenous variables $X_{1,t}$ and $X_{2,t}$ (note too, that individual vectors of exogenous variables might appear in both $X_{1,t}$ and $X_{2,t}$) there is a price P_t^* at which the market would clear in the absence of exogenous constraints. At this price the quantity supplied would equal the quantity demanded, thus one may write

$$D_t = X'_{1,t}\beta_1 + \alpha_1 P_t^* + u_{1,t}, \tag{11.5}$$
$$S_t = X'_{2,t}\beta_2 + \alpha_2 P_t^* + u_{2,t}, \tag{11.6}$$

and

$$X'_{1,t}\beta_1 + \alpha_1 P_t^* + u_{1,t} = X'_{2,t}\beta_2 + P_t^* + u_{2,t}, \tag{11.7}$$

and thus, for any period t,

$$P_t^* = \frac{1}{\alpha_1 - \alpha_2}(X'_{2,t}\beta_2 - X'_{1,t}\beta_1) + \frac{1}{\alpha_1 - \alpha_2}(u_{2,t} - u_{1,t}). \tag{11.8}$$

If the market were to always achieve a state of equilibrium, then P_t would equal P_t^* for all t. Equation (11.4) might be restated as

$$P_t = \mu P_{t-1} + (1-\mu) P_t^* + u_{3,t} , \qquad (11.9)$$

where

$$0 \leq \mu \leq 1 .$$

Here, the limiting case of instantaneous market equilibrium corresponds to $\mu = 0$. The case of no market response corresponds to $\mu = 1$. Thus the value of the parameter μ characterizes the responsiveness of the system to disequilibrium conditions. The effect of the disturbance term, u_3, in equation (11.9) is to say that even when both supply and demand adjust to the market-clearing price, P_t^*, the actual price level may differ from the market-clearing price by the random disturbance value. Note, too, that equation (11.9) can be rearranged to give

$$P_t = P_{t-1} + (1-\mu)(P_t^* - P_{t-1}) + u_{3,t} . \qquad (11.10)$$

This expresses present price as a function of prior price plus-or-minus an adjustment response to the difference between prior price and the present 'equilibrium' or potential market-clearing price. Note that for $\mu = 1$ there is no response, and present price equals prior price, and for $\mu = 0$ there is an instantaneous achievement of equilibrium, and present price equals the market-clearing price plus the disturbance term. For any value of μ between 0 and 1 this equation gives a partial adjustment to a moving equilibrium, which Bowden (1978a) calls a PAMEQ formulation. Note that a negative value of μ gives an overresponse to the disequilibrium condition, and a value of μ greater than 1, gives a perverse response, for example, it lowers the price when the market-clearing price exceeds the prior price.

It is possible to imagine other adjustment processes. Suppose that current price is a function of the difference between prior price and present market-clearing price and of the difference between prior and present market-clearing prices. This implies both an adjustment to disequilibrium conditions and an adjustment to the direction of change of the market-clearing price. This would give an equation of the form

$$P_t = P_{t-1} + (1-\mu)(P_t^* - P_{t-1}) + \mu(1-\gamma)(P_t^* - P_{t-1}^*) + v_t . \qquad (11.11)$$

Here v_t is the disturbance term, corresponding to $u_{3,t}$ in equation (11.10). The parameter γ determines the degree of responsiveness to the change in market-clearing price. If γ is restricted to being less than 1, then for the maximum value of γ the response to the change in market-clearing price vanishes. The effects of values of γ less than 0 could be perverse, depending upon the value of μ. Thus it is clear that the actions of μ and γ, taken together, serve to determine the responsiveness of the system to (1) disequilibrium conditions, and (2) the direction of change of the market-clearing price. Clearly a good deal more could be said here, but this suffices to introduce the general ideas.

Another economic model structure, which figures prominently in discussions of equilibrium adjustment processes, is the cobweb model (Van Doorn, 1975). Without describing the structure of such models in detail, the salient point for this discussion is that an important variable in the cobweb model is the 'expected market-clearing price'. Given the concept of such a variable, it becomes possible to discuss various model structures in terms of hypotheses about how such expectations develop and of what sorts of behavior they evoke. The simplest cobweb model structure has suppliers expecting that the market-clearing price will equal the actual price from the previous time period. Alternatively there is the structure called a 'cobweb model with learning' where the expected market-clearing price is a function of prior expectations, as well as of the observed error in prior expectations.

The equation structure of the basic model is

$$D_t = \alpha_0 - \alpha_1 P_t , \qquad (11.12)$$

$$S_t = \beta_0 - \beta_1 P_t^e , \qquad (11.13)$$

$$D_t = S_t , \qquad (11.14)$$

$$P_t^e = P_{t-1} , \qquad (11.15)$$

where

P_t^e is the price at which suppliers expect the market to clear.

Note that in this simplified model there are no exogenous variables such as those in equations (11.1) and (11.2), though this model could be extended to include them. Note, too, that here the quantity of good supplied depends solely on the expected price. In the cobweb model with learning, the expected price equation is extended to

$$P_t^e = P_{t-1}^e + \gamma(P_{t-1} - P^e_{t-1}) , \qquad (11.16)$$

with the condition that $0 < \gamma \leq 1$. There is an obvious similarity to the PAMEQ model structure of equation (11.10). In general, equations such as (11.16) are termed 'adaptive expectation models'. Such models, more generally, have expected or desired price as a function of a set of independent variables. Then the actual change in price is some fraction of the difference between current expected price and prior actual price, such as

$$P_t - P_{t-1} = \gamma(P_t^* - P_{t-1}) . \qquad (11.17)$$

There is a literature on these models and their estimation which can be traced via some of the sources cited here (Bowden, 1978a; Van Doorn, 1975), plus others (Fisher, 1983). There is also discussion of the similarities and differences between the adaptive expectations and partial adjustment models (Griliches, 1967), and of the parameter estimation problems associated with them (Bowden, 1978b; Maddala, 1983). For the purpose of this chapter these issues are not so important. It is important to note that the cobweb model of equations (11.12)-(11.15), or of equation (11.16) is an equilibrium model.

It is dynamic, because there are variables with different time subscripts included, but the assumption is that price adjusts, instantaneously, so that supply and demand will be equal. In the partial adjustment model of equations (11.1)-(11.4) supply usually does not equal demand. Actual price does not fully adjust to the market-clearing price. The assumption is that there are high costs to the producer of adjusting to equilibrium. The adjustment process may be considered to be one of balancing the cost of being out of equilibrium (for instance, unsold goods or lost sales) against the costs of adjusting to the equilibrium supply, demand, and price.

There are several major points to be drawn from this brief review. First, there is the notion of a potential market-clearing value of a variable (price, in these examples). Second, there is the idea of different response mechanisms (equation structures) which could determine the character and the extent of the response of the system to the difference between actual price and the hypothetical market-clearing price. Last, and in a sense, most important for the material to follow, there is the proposition that the response to disequilibrium will usually not be completed in each time interval. The system will be perpetually in a state of disequilibrium. Forecasting within this construct becomes a problem of (a) estimating the market-clearing (equilibrium) solution, and (b) estimating the extent to which the system moves from its prior state towards that equilibrium.

Clark (1983) cites several examples of how economists use disequilibrium model structures, including several models of housing disequilibrium. Perhaps of greater importance here is that the previously mentioned works (see the opening paragraphs of this chapter), by Wilson, and Bertuglia, and their associates, and summarized by Wilson (1987) may now be seen in a somewhat different perspective. Most of that work, usually done in terms of retail trade, follows from the following equation structure:

$$\frac{dW_j}{dt} = \varepsilon(D_j - k_j W_j) , \qquad (11.18)$$

where
W_j is the attractiveness (to retail trade, or sales) of zone j,
D_j is the revenue (sales) attracted to j,
k_j is the unit cost of operating a facility in j,
ε is a parameter.

The right-hand side of equation (11.18) is said to be a measure of imbalance between supply (of retail facilities) and demand (retail sales). In most of these analyses, W_j is taken to be the 'size' of retail facilities in j, and is usually measured in terms of retail employment. The parameter ε governs the rate of response (that is, the change in size of retail facility in j) to a supply–demand disequilibrium situation for j. A frequently used version of this equation adds a function of W_j as a multiplier of the right-hand side, to

yield
$$\frac{dW_j}{dt} = \varepsilon(D_j - k_j W_j) W_j \ . \tag{11.19}$$

The difference equation version of this is

$$W_{j,t} - W_{j,t-1} = \varepsilon(D_{j,t-1} - k_j W_{j,t-1}) W_{j,t-1} \ . \tag{11.20}$$

The demand for retail sales in j is usually estimated by a simple spatial interaction model of the form

$$D_j = \sum_i A_i e_i p_i W_j^\alpha \exp(-\beta c_{ij}) \ , \tag{11.21}$$

where the time subscripts $t-1$ are dropped for notational convenience, and where

e_i is the per capita retail sales demand,
p_i is the population of i,
c_{ij} is the cost of travel from i to j,
α, β are parameters,

and where

$$A_i = \left[\sum_k W_k^\alpha \exp(-\beta c_{ik}) \right]^{-1} \ .$$

Virtually all of the published research which followed, or was derived from, this approach then moves to a discussion of the equilibrium tendencies of this formulation, and to the trajectories which the components of the system will follow as they move toward the equilibrium solution.

It is clear that equation (11.20) may be written as

$$W_{j,t} = W_{j,t-1} + \varepsilon(D_{j,t-1} - k_j W_{j,t-1}) W_{j,t-1} \ . \tag{11.22}$$

Now it should be obvious that the system described by equation (11.22) does not have to reach an instantaneous equilibrium, and that the equation could represent an ongoing process of partial adjustment. As such, one has attractiveness of zone j at time t being a function of the prior attractiveness, at j at $t-1$, plus or minus an adjustment term. The adjustment term includes a measure of disequilibrium at time $t-1$, a response rate parameter ε, and a weighting which acts to amplify or attenuate the response.

Two major aspects of this equation structure seem questionable. First, attractiveness at time t must be related to revenues at time t. Attractiveness at time t clearly cannot be solely a function of attractiveness and revenues from the prior time period. Second, the action of the $W_{j,t-1}$ which multiplies the 'level' of disequilibrium is questionable. Should a zone with a large attractiveness value be more or less responsive to disequilibrium than a zone with a small attractiveness value? One might expect the larger centers to respond less rapidly to disequilibrium, and the smaller centers to respond more rapidly. These questions notwithstanding, some very interesting research has been done, cited above, into the paths these models take to reach equilibrium. The sensitivity to parameters, and their demonstration of catastrophe and bifurcation phenomena has been most informative, and it is consistent with the chaotic behavior observed in the experiments described above in chapter 10. Nonetheless, *an equilibrium solution is not likely to be a good forecast*. What is needed is a way to reformulate these experiments to be congruent with the PAMEQ paradigm to see whether better forecasts can be produced.

Last, note that Maddala (1983) refers to partial adjustment models with the same form as equation (11.10), but suggests that the adjustment parameter, μ, is related to two 'costs'. The first cost is that of being out of equilibrium with respect to some measure Y, and the second cost is that of making the adjustment to equilibrium. Maddala then assumes that both 'costs' have quadratic loss functions. In equation form, the cost of adjustment, C_A, is

$$C_A = \eta(Y_t - Y_{t-1})^2 . \tag{11.23}$$

The cost of being out of equilibrium, C_D, is

$$C_D = \phi(Y_t - Y_t^*)^2 , \tag{11.24}$$

then the overall loss function L, is,

$$L = \phi(Y_t - Y_t^*)^2 + \eta(Y_t - Y_{t-1})^2 . \tag{11.25}$$

For a given value of the observed variable, Y_{t-1}, in the prior time period, and of the equilibrium (or optimum) value of Y_t^* for the current time period, one may solve for the current value of the observed variable which minimizes loss. This is done by setting the derivative of L with respect to Y_t equal to zero, and solving for Y_t. Thus

$$\frac{dL}{dY_t} = 2\phi(Y_t - Y_t^*) + 2\eta(Y_t - Y_{t-1}) = 0 , \tag{11.26}$$

and, solving for Y_t, we obtain

$$Y_t = Y_{t-1} + \left[\frac{\phi}{\phi + \eta}\right](Y_t^* - Y_{t-1}) . \tag{11.27}$$

This is equation (11.10) again, though without the disturbance term $u_{3,t}$.

If one is to consider the use of such a framework for location models many questions must be answered. Which adjustment process is most suitable for which types of locator? How are hypotheses about adjustment processes to be tested, and how are their parameters to be estimated? In the next section of this chapter some very preliminary steps are taken, by way of a series of numerical experiments in which the Archerville data are used.

11.3 Some preliminary experiments: Archerville with fixed travel costs

Stated simply, the use of a disequilibrium adjustment process gives a solution for some variable at time t as a function of that variable at time $t-1$ and of the equilibrium (or optimum) value of that variable at time t (and sometimes of its equilibrium value at time $t-1$ as well). If the variable being analyzed is a single variable such as price, then, in effect, the time t value, again in the simplest case, may be seen to fall between the value at time $t-1$, and the equilibrium value at time t. If the process of calculating the equilibrium value is iterative, it may well be that the solution for the variable at time t will be given directly after some number of iterations of the equilibrium-solution technique. For a spatially disaggregated variable one will be dealing with vectors of variables at times t and $t-1$ and at equilibrium. Each activity type in each zone *may* be adjusting to disequilibrium at its own rate. The first set of experiments to be described was done for the purpose of observing the paths taken by the activities in various zones as they moved to the equilibrium solution.

In chapter 10 a series of numerical experiments with equilibrium solutions was described. These experiments, with use of the Washington data, were made with the intention of defining and analyzing different definitions of equilibrium. Here some of those same experiments will be repeated with the Archerville data, which will permit a more detailed examination of what is happening. In addition, several new experiments will be described.

A simple residential location model was described in chapter 2 and was tested with the Archerville data. The equation structure was

$$N_{i,h} = \sum_j E_{jT}[L_{iR}^{\gamma_h}(1+n_{i,1})^{\sigma_h}(1+n_{i,2})^{\rho_h}\exp(-\beta_h c_{ij})]$$
$$\left[\sum_k L_{kR}^{\gamma_h}(1+n_{k,1})^{\sigma_h}(1+n_{k,2})^{\rho_h}\exp(-\beta_h c_{kj})\right]^{-1} \quad (11.28)$$

where

$$n_{i,1} = \frac{N_{i,1}}{\sum_h N_{i,h}} = \frac{N_{i,1}}{N_{i,1}+N_{i,2}},$$

and where the variables are defined as

E_{jT} is the total employment, at place of work, in zone j,
c_{ij} is the travel cost from zone i to zone j,
$N_{i,h}$ is the number of type-h households, at place of residence, in zone i,
L_{iR} is the amount of residential land in zone i.

For this example, the Archerville data of chapter 2 will be used, and the assumed values of parameters will be

$\gamma_1 = 0.90$, $\gamma_2 = 1.50$,
$\sigma_1 = 2.00$, $\sigma_2 = 0.20$,
$\rho_1 = 0.90$, $\rho_2 = 3.00$,
$\beta_1 = 0.20$, $\beta_2 = 0.10$.

For the first equilibrium tests to be done here, it will be assumed that each locator knows the behavior of each other locator, and responds instantly, thus yielding a simultaneous solution of the type referred to as HH in chapter 10. The model is solved by successive substitution. The paths or

Preliminary development of dynamic spatial models 293

trajectories followed by each zone in moving from the base-year solution to the HH equilibrium solution are shown in figures 11.1(a) and 11.1(b) for low-income and high-income households, respectively. The trajectories begin at the base year. The first estimate of household location, at a value of 2 on the x axis, is the conventional, lagged-variable, nonsimultaneous solution. Each additional estimate corresponds to an iteration in the successive substitution process. Note that for both household types, in all zones, an equilibrium solution is reached after 5 or 6 HH iterations.

The process was then repeated for LU (land-use) iterations, where the residential land variable is updated from one iteration to the next. Figures 11.2(a) and 11.2(b) show the solution trajectories from this test. Note that the HH test trajectories differ from the LU test trajectories, and that the

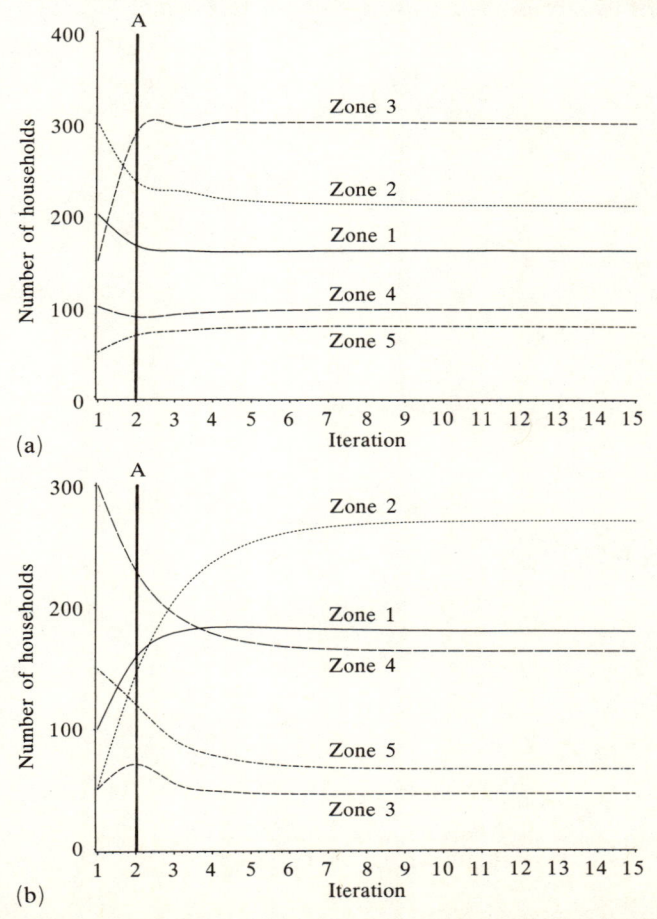

Figure 11.1. Trajectories of (a) low-income households and (b) high-income households for the HH run. Note: at A, the solutions are the conventional, lagged-variable, nonsimultaneous solutions.

equilibrium-solution values are quite different. This is, of course, what would be expected given the two different set of hypotheses represented. The LU tests represent the hypothesis that locating households know only the households distribution of the prior time period, but that they know instantly about changes in the land uses of each zone. In a very rough sense these land-use adjustments can be thought of as surrogates for the workings of the land market in each zone.

A third set of experiments was then run, where it was assumed that residential locators knew both the locational decisions of other locators as well as the changes in zonal land use. In effect, locators were provided with complete, instant information on the decisions of all other residential locators and their effects on land use. The numerical example of this model was solved, as for the prior two experiments, by an iterative technique.

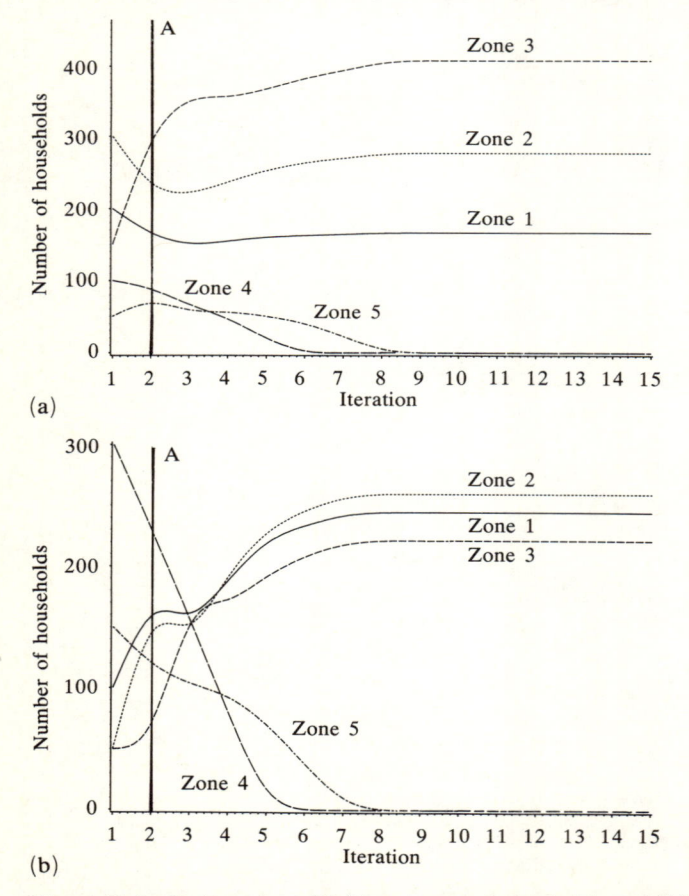

Figure 11.2. Trajectories of (a) low-income households and (b) high-income households for the LU run. Note: at A, the solutions are the conventional, lagged-variable, nonsimultaneous solutions.

Although it is not a difficult matter to join the two equilibrating processes, questions regarding the numbers and sequences of iterations are raised. It must be decided whether the HH or the LU iterations should be done first. Then, how many iterations of HH, say, should be done before the LU iterations are done? For example one might have 2 iterations of HH, (that is, calculate a first estimate of household location from the starting values of all variables) then update the households and calculate a second estimate of household location. This might be followed by 2 iterations of LU. The second estimate of households, along with the starting values of the other variables, is used to make a first estimate of, and thus update, the zonal land use. This first estimate of land use is used with the second estimate of households and the starting values of employment and travel cost, to produce a second estimate of zonal land use. Finally, if it is assumed that there are to be two 'grand' or overall iterations of the solution process, the second estimates of households and land use are used, together with the starting values of employment and travel cost, to repeat the entire process.

Numerous computer experiments were done, with different numbers of HH, LU, and grand iterations. All combinations produced equilibrium solutions. Different combinations produced different equilibrium solutions. This is not so surprising as it might sound. The rate of response of household locators to changes in the locations of other households is inversely proportional to the number of HH iterations. Similarly, the rate of response of household locators to changes in zonal land use is inversely proportional to the number of LU iterations. Suppose, for example, that 1-HH iteration is permitted, followed by 2-LU iterations. The locating households in this situation will have made less adjustment to the location of other households than to the changing land use in the zones. If, on the other hand, 20-HH iterations are permitted, followed by 2-LU iterations, then households will (probably) have fully adjusted to the location of other households. It is not proposed to use the numbers of iterations as any sort of measure of adjustment to equilibrium, but it should be clear that a change in the numbers of iterations has the effect of changing the degree of adjustment to disequilibrium of each household type. Although in each case they are somewhat different, the household trajectories for most of the combined HH/LU runs rather closely resembled those of the HH runs shown in figure 11.1.

At this point the question is whether there is a 'global' equilibrium for the combined HH/LU runs? The preliminary tests were worrisome. Different starting points gave similar but not identical solutions, even when the computer runs were done in double-precision arithmetic in order to eliminate any differences in solution due to rounding error. For a particular starting point, different numbers of HH and/or LU iterations gave somewhat different 'equilibrium' solutions. After some examination of the trajectories produced by these runs, it appeared that the differences in the implicit rates of adjustment produced by the differences in numbers of iterations (HH, LU, or 'grand') tended to force the system, in each solution test, to a local equilibrium. It seemed that perhaps a sequence of 'restarts' from subsequent local equilibria would eventually lead to a global equilibrium. This experiment was then run for several different configurations of HH and LU iterations with most gratifying results. A global equilibrium solution was found. Regardless of starting points or numbers of HH and LU iterations the identical, final, 'global' equilibrium was found. What was involved was to simply use the local equilibrium values of households, which were the result of a particular test run, as the input (starting) values for a subsequent run. Some combinations of HH and LU iterations took fewer reruns than did others. For example, from the same starting points a run with 1-HH and 1-LU iteration in 15 grand iterations (GI) required 7 full runs to reach the global equilibrium, whereas a 5-HH, 3-LU, 4-GI run required only three full runs to reach the same equilibrium values. Several combinations of iterations were tested, with the global equilibrium values used as starting values. In all such cases the iterations did not shift from the starting point, and the final solution was identical to the starting values. Thus not only could a global equilibrium be found, it appeared to be stable as well. The trajectories of the zonal household values for the three successive runs of the 5-HH, 3-LU, 4-GI run are shown in figures 11.3. Note that each run (and restart) consisted of 60 iterations ($5 \times 3 \times 4$), but in every case the model run converged at or before 30 iterations. Thus in figure 11.3 each successive run is shown for 30 iterations. For the three successive runs which it took to reach the global equilibrium, a total of 90 iterations is shown.

Before the next set of experiments are presented, one further point should be noted. The question was raised earlier in this discussion as to whether, in an iterative solution process, the results from some particular iteration would approximate the forecasts from the partial adjustment formulation. The answer to this clearly depends upon what degree of adjustment is expected for the given time period. To explore this notion, consider the difference between the starting values of households (by zone) and their equilibrium values to be full adjustment. Then consider the difference between the starting values and the 1st iteration values to be a partial adjustment. In a PAMEQ model structure the ratio of the partial adjustment to the full adjustment would be constant over all zones, though it might vary by type of locator. In the Archerville example this ratio varies from 0.6 to 4.0 for the LI households and from 0.3 to 0.9 for the HI households, along with one case where the signs of the full and partial adjustments were opposite. The use of 2nd-iteration or 3rd-iteration results for this comparison make very little difference in the results. Thus, from these preliminary experiments it can be seen that the results of any particular iteration in the equilibrium-solution procedure will not provide a useful estimate of a PAMEQ model structure solution. It should, however, be noted that the correlation of a 1st-iteration solution with the PAMEQ solution (regardless of the degree of adjustment, which in this case would be only a scalar transform) will be relatively high. This will be even more the case with urban data sets of many zones, where in either solution larger zones (in the base year) will be larger in the forecast. So too with smaller zones (in the base year) which will be smaller in the forecast. This may explain how traditional model forecasts can still be reasonably reliable even given the static, cross-sectional structures of the models.

In all the experiments described in this section the travel costs remained constant from the input, through the solution process, to equilibrium. In addition, the location of employment remained constant. Household location and residential land use were variable. The experiments showed that for fixed employment location, travel costs, model parameters, and regional totals, there was a unique equilibrium solution to household location and residential land use. In the next section of this chapter the experiments are enlarged to make the travel costs endogenous, and the equilibrium properties are reexamined.

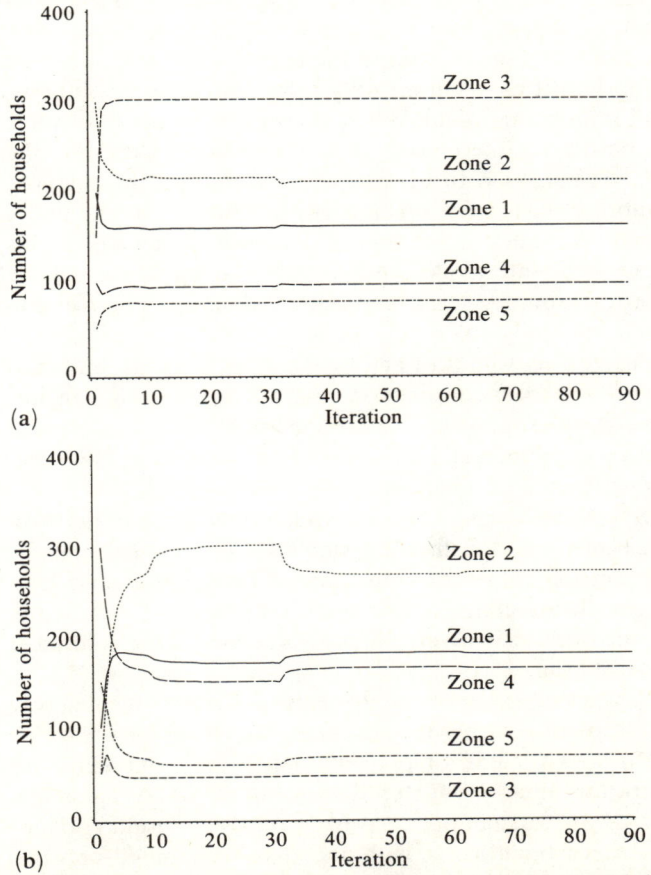

Figure 11.3. Trajectories of (a) low-income households and (b) high-income households for the 5-HH, 3-LU, and 4-GI run.

11.4 More preliminary experiments: Archerville with variable travel costs

The solution of location models with endogenously determined travel costs was discussed in chapters 8, 9, and 10. In chapter 8 experiments were done based on the Archerville data and a cubic link-cost–volume function, to solve for an intermodel equilibrium. The location models were not solved for intramodel equilibrium in the tests in chapter 8. In the experiments described in chapter 9, the full-scale Washington data set was used, along with the standard cost–volume function developed by the US Bureau of Public Roads. Again, intermodel equilibrium was achieved, without any attempt at intramodel equilibrium. In effect, the chapter 9 results demonstrated the effects of the shifting of travel costs from being exogenously determined to being endogenously determined. These tests explored the calculation of an equilibrium solution amongst the location and trip-assignment models whilst solving the location models for just one iterative step beyond their base-year inputs. In chapter 10, the question of intramodel equilibrium was taken up and alternative solution methods were explored, again making use of the Washington data. In the experiments in chapter 10 the intramodel equilibrium tests achieved only local equilibrium solutions, with no global equilibrium being found. When attempts were made to solve simultaneously for both the intermodel and intramodel equilibria, the system exhibited chaotic behavior. In the preceding section of this chapter, the intramodel equilibrium tests were repeated with a simpler model and a much smaller data set. A stable global equilibrium was achieved. In this section the combination of intermodel and intramodel equilibria will be examined, here too, with a simpler model and smaller data set. The solution process was as follows:

Step a With the same starting values of employment, households, and land use as were used for the Archerville experiments described earlier in this chapter, as well as the same exogenously determined travel costs, residential location was solved by a combined HH/LU run. At the same time, the work-trip distribution was calculated.

Step b The work trips were assigned to the Archerville highway network and the zone-to-zone travel costs for the congested network were calculated.

Step c The original starting values of employment, households, and land use were used, along with the congested-network travel costs from step b, residential location and work trips were then recalculated by a combined HH/LU residential model run.

Step d The work trips were assigned to the empty Archerville highway network to recalculate zone-to-zone travel costs for the (re)congested network.

Step e The congested-network link times from step b are combined with the congested-network link times from step d by using the MSA algorithm. (For the first combined-cost calculation, costs from step b are used. For subsequent iterations the combination is with the prior combined-network results.) Paths are then traced through the networks using the combined-congested link costs to produce zone-to-zone combined travel costs.

Preliminary development of dynamic spatial models 299

Step f As in step c, residential location and work trips are recalculated using the combined congested-network travel costs from step e.

Step g Check for an equilibrium solution by comparing the residential locations and zone-to-zone travel costs of successive iterations. If necessary, do another iteration beginning at step d.

For the Archerville data and the 5-HH, 3-LU, and 4-GI version of the residential model, and with the user-equilibrium assignment algorithm used for the loading of trips on the highway network, 6 iterations of the entire system were required to achieve equilibrium. This equilibrium solution was stable, in that restarting the model sequence either with the residential locations and trip matrices or with the zone-to-zone travel costs always gave the same solution. The solution also appeared to be independent of starting values of the model variables. Figure 11.4 shows the trajectories of the zonal households from their starting values through to system equilibrium.

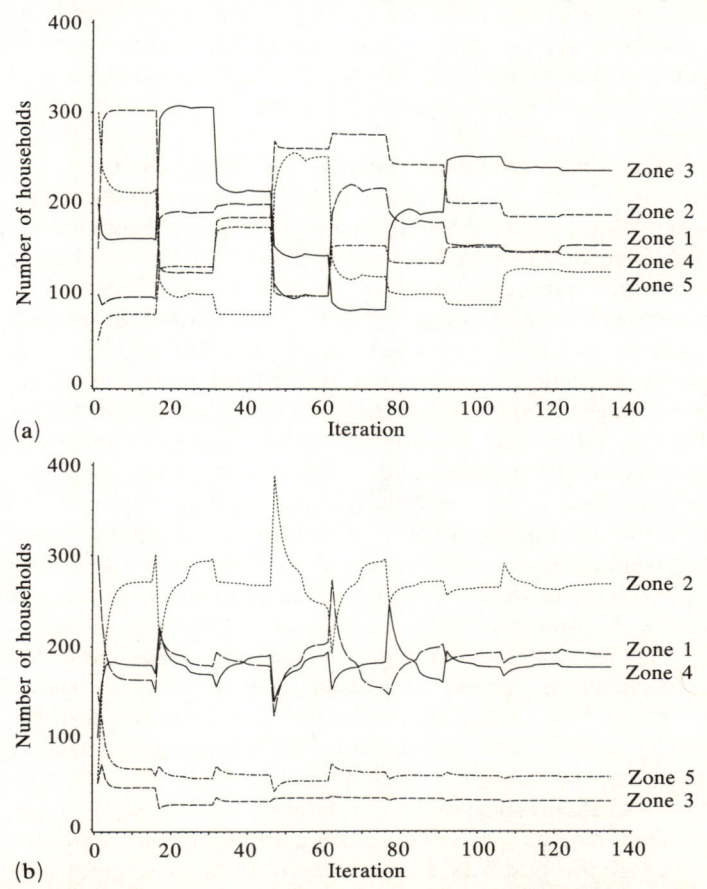

Figure 11.4. Trajectories for (a) low-income households and (b) high-income households for the combined LU/HH/transport run.

Note that in most cases, 15 iterations of the residential model were enough to achieve (or nearly so) equilibrium, thus only 15 iterations are shown within each of the six overall system iterations.

Now it should come as no surprise that the equilibrium solution for fixed transport costs is quite different from that for variable (endogenously determined) transport costs. The two solutions, at least for this example, are indeed quite different. Thus in considering the use of these equilibrium solutions as 'targets' in a PAMEQ formulation, it is clear that it will make a substantial difference as to which equilibrium is taken to be the target. A further concern is that the example given here converged to a unique, stable equilibrium with no complications, but the same solution procedure, albeit with a more complex model and a much larger data set, evidenced chaotic behavior during the experiments described in chapter 10. Now it is true that one may have completely stable simple forms of a system and then have chaotic behavior appear only as the system complexity is increased (Reiner et al, 1986). Nonetheless, it remains an open question as to what will happen when, in the next section of this chapter, the experiments again take up the full-complexity model and a large-scale data base.

11.5 Intramodel equilibrium experiments: Houston and Washington

Given the Archerville results described above, with evidence that a stable global equilibrium solution could be found for several model configurations, it seemed sensible to take a closer look at the results with the full-scale Houston and Washington data sets.

The first experiments were done with the 182-zone Washington data. A modified version of DRAM was used to solve for an HH equilibrium (as described in chapter 10). The model was run for 100 iterations. The zonal mean absolute percentage change was less than 0.1% for all household types when the run was terminated after the 100 iterations. Regionwide statistics can, however, be misleading. To understand better the behavior of the equation system as it moved towards equilibrium several graphic displays were prepared. In figure 11.5 the change is shown in the households of each zone from one iteration to the next for each household type. In all cases there is a good deal of zonal change in the first 30 to 50 iterations, with most zones having settled to a stable solution by the last iteration. However, there are still a few zones which are not completely stable, especially, for example, for low-income and lower-middle-income household types. Figure 11.6 shows the trajectories of household changes for zone 1, located in the center of the Washington region. It is quite clear that the households in this zone have not reached an equilibrium solution after 100 model iterations.

The experiment was then repeated for the Houston 199-zone data set. Again, the zonal mean absolute percentage change was well under 1% after 100 iterations. Although the model did not reach zero change, the regionwide percentage changes were small enough to suggest that the model probably had converged. The graphics, however, made it quite clear that convergence

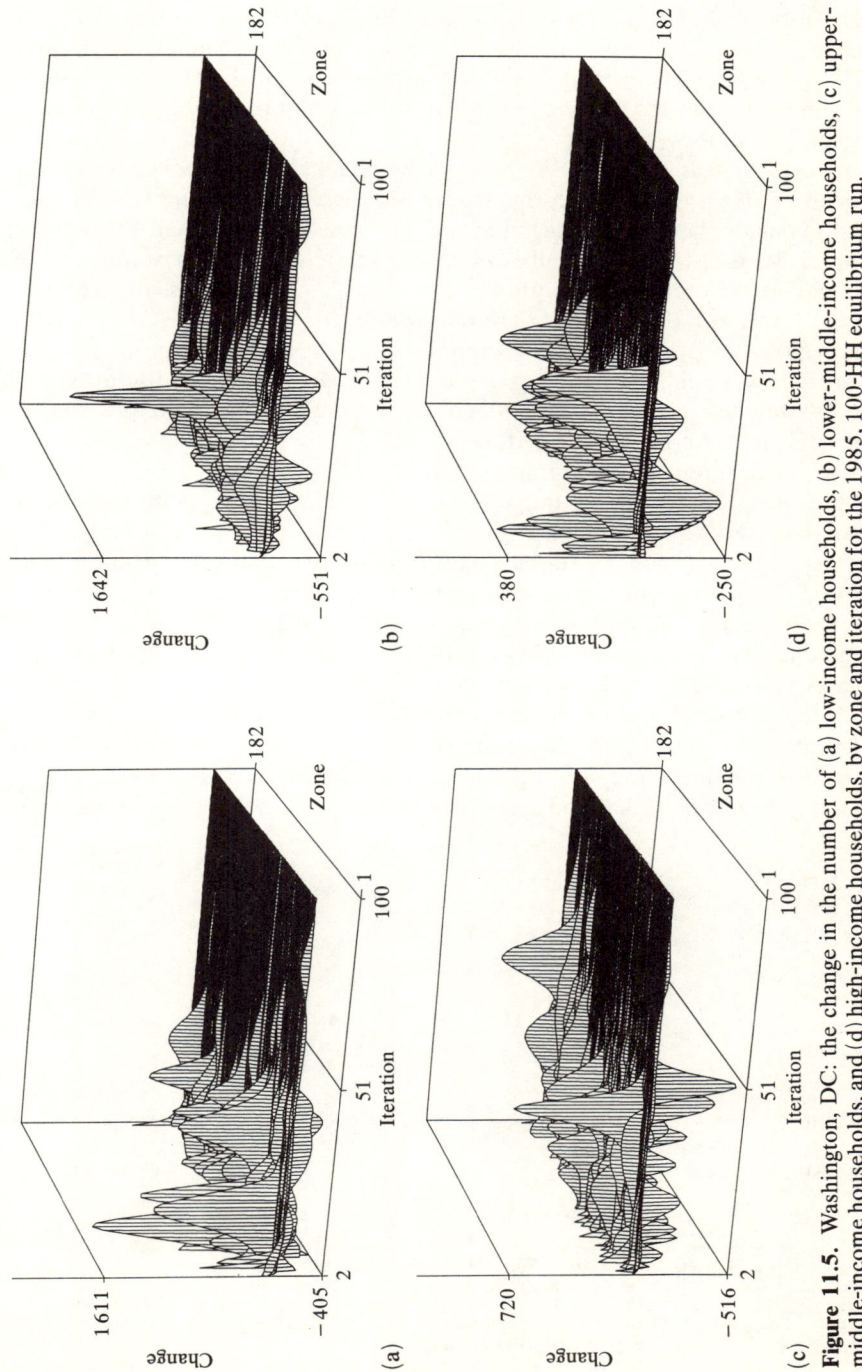

Preliminary development of dynamic spatial models

Figure 11.5. Washington, DC: the change in the number of (a) low-income households, (b) lower-middle-income households, (c) upper-middle-income households, and (d) high-income households, by zone and iteration for the 1985, 100-HH equilibrium run.

had not been achieved. A glance at figure 11.7 shows certain zones still not converging after 100 iterations. As an example the household change trajectories for zone 101, located in the western suburbs of the Houston region, are shown in Figure 11.8 (see over). An extra 50 iterations made no significant change in these results.

Attention was then shifted to the HH/LU simultaneous-solution procedure. Here, as described in chapter 10, a set of nested intramodel iterations is made, between households and land consumption. Remember from chapter 10 that it was difficult to achieve stable-equilibrium solutions for this model structure. The solution process in this set of tests involved ten outer iterations. Each outer iteration contained ten inner HH iterations followed by an LU update calculation. It is quite clear from figure 11.9 (see over) that while this procedure generally tends towards a stable solution for the Washington region, it is a much less stable process than the simple 100-HH iteration process. Again it is clear that certain household types in certain zones are not converging.

The Houston results for the 10-HH by 10-LU iteration solution showed similar patterns of apparent regional convergence. There, too, a graphic display of the solution iterations makes it quite clear that specific household types in specific zones are not converging. Figure 11.10 (see page 306) shows these results. Further examination of the unstable household types and zones suggests that not only are they not converging but that they may actually be evidencing chaotic behavior.

Even without repeating the experiments to include the travel costs in the solution process, as was done for the Washington experiments in chapter 10, it is clear that solving for intramodel equilibrium is a somewhat tricky process. It appears that the issue is not one of convergence in general.

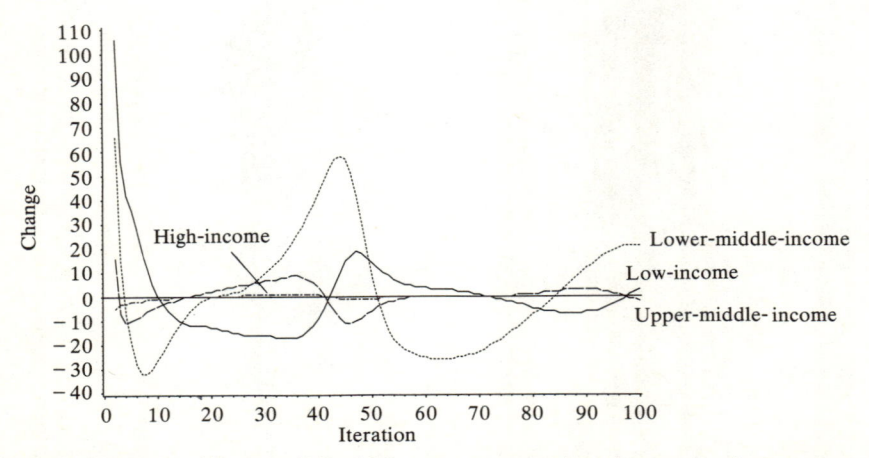

Figure 11.6. Washington, DC, 182-zone system: the change in the number of zone-1 households over 100 iterations, for 1985.

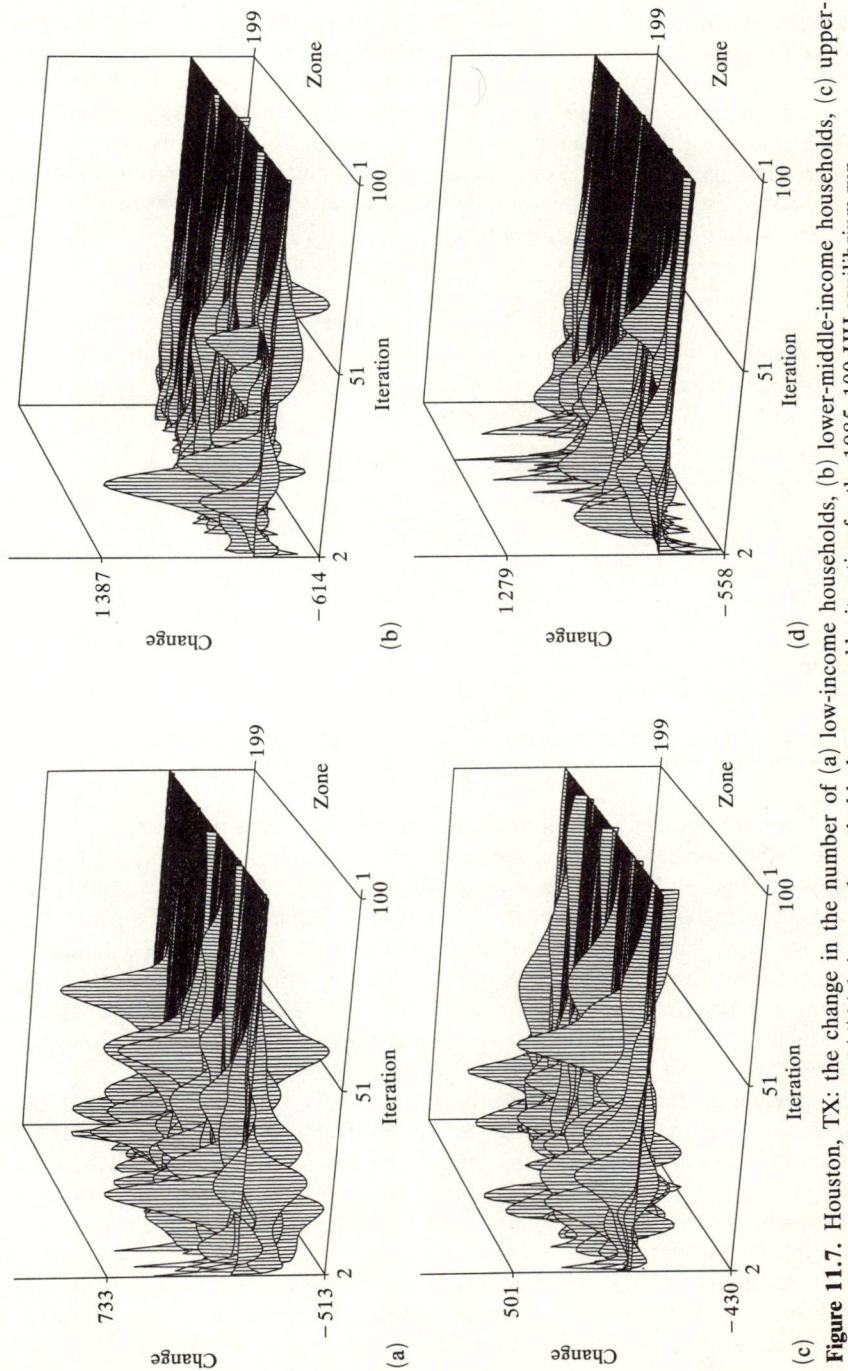

Preliminary development of dynamic spatial models

Figure 11.7. Houston, TX: the change in the number of (a) low-income households, (b) lower-middle-income households, (c) upper-middle-income households, and (d) high-income households, by zone and by iteration for the 1985, 100-HH equilibrium run.

In general the system is convergent. The problem is that specific locator types in specific zones, only a small portion of the total magnitude of the system (number of zones multiplied by number of locators), do not converge. For practical purposes this may not be an important matter. As a matter of research interest, the behaviors of the nonconvergent zones and locators are sure to see further investigation. More to the point of the original topic of this chapter, is the question of how these results relate to the use of the PAMEQ structure for spatial models.

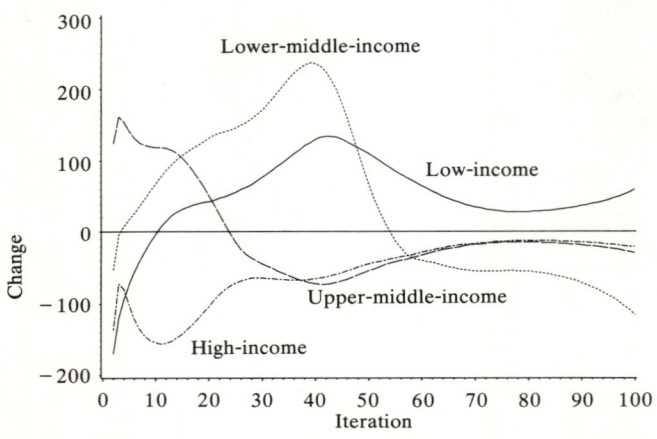

Figure 11.8. Houston, TX, 199-zone system: the change in the number of zone-101 households over 100 iterations, for 1985.

11.6 Very preliminary tests of an adjustment to equilibrium structure

As described earlier in this chapter, the PAMEQ structure requires both a base-year (prior) distribution of activities and an optimum (equilibrium) distribution for the current year. The results described above suggest that there are important issues to be resolved regarding not only the calculation of such equilibrium distributions but also the selection of the specific equilibrium structure from which the distribution is to be derived. The availability of data for the Houston region suggested the possibility of a preliminary investigation of these issues.

In particular there were data both for employment location and for household location for the 199 zones for 1975, 1980, and 1985. The 1975 and 1980 data had been used to estimate parameters for both EMPAL and DRAM. The parameter estimation process is described in Putman (1983a). The results were quite satisfactory in terms of goodness-of-fit, and of the signs and magnitudes of parameters.

Before the PAMEQ structure tests were begun, a simple test of the traditional forecasting method was made. With 1980 as the base year, EMPAL was used to forecast employment location in 1985. With 1980 as

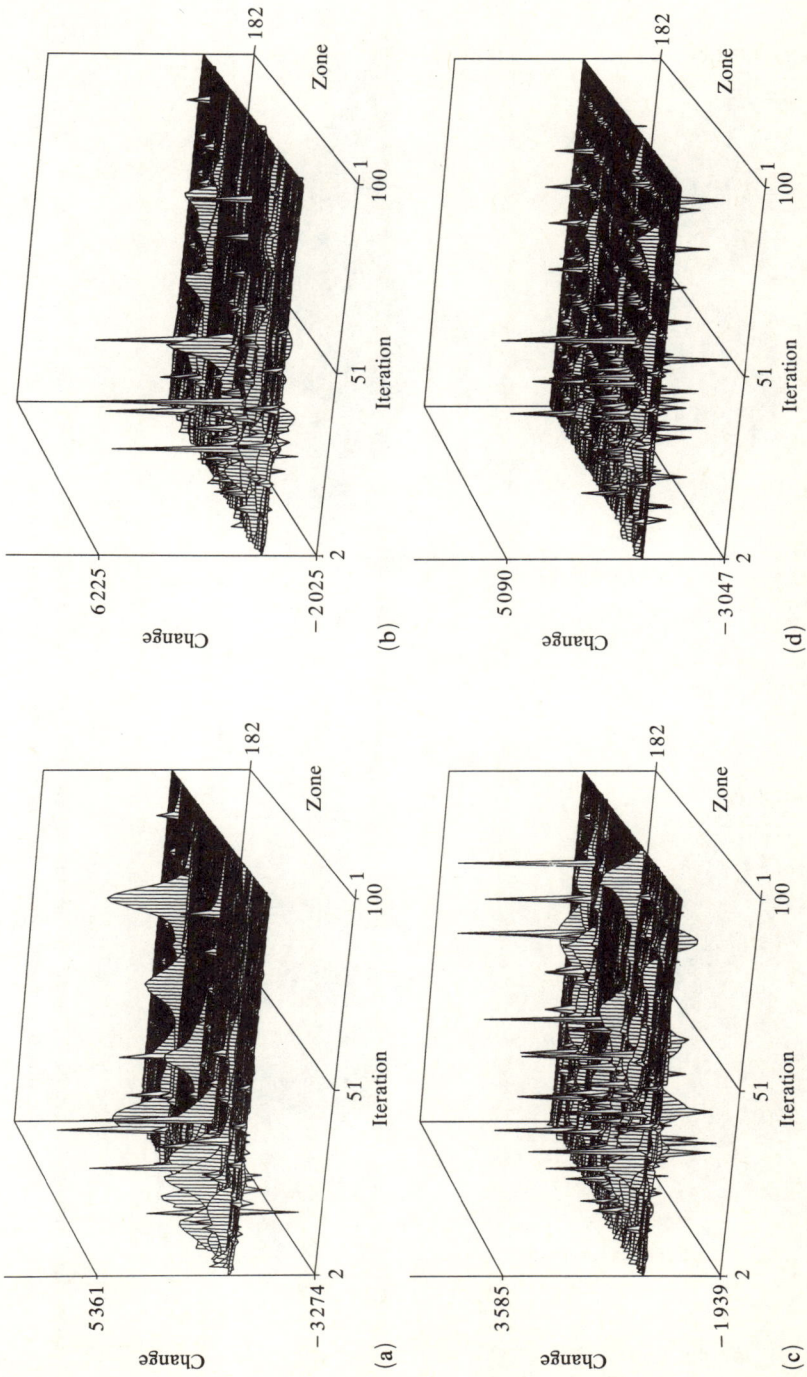

Figure 11.9. Washington, DC: the change in the number of (a) low-income households, (b) lower-middle-income households, (c) upper-middle-income households, and (d) high-income households, by zone and by iteration for the 1985, 10-HH, 10-LU equilibrium run.

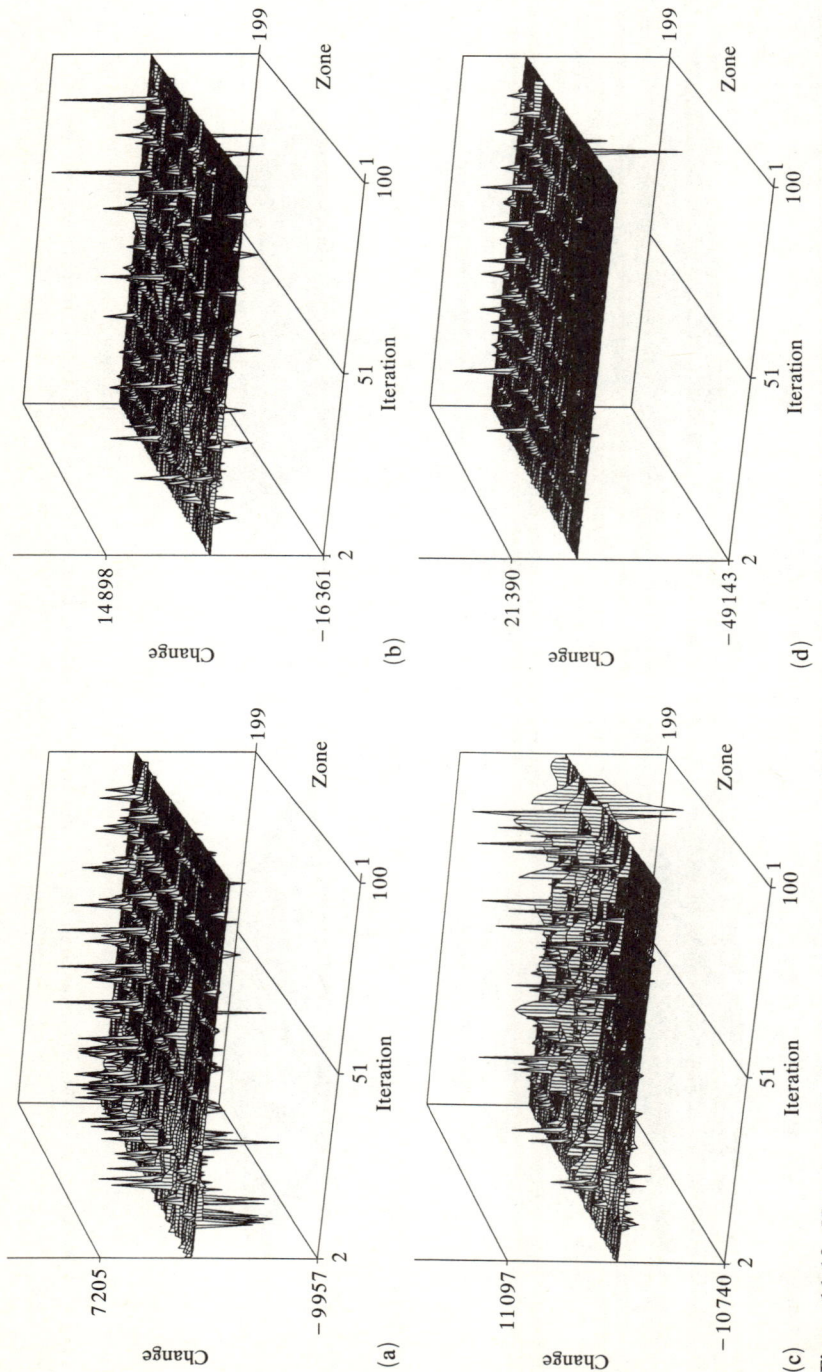

Figure 11.10. Houston, TX: the change in the number of (a) low-income households, (b) lower-middle-income households, (c) upper-middle-income households, and (d) high-income households, by zone and by iteration for the 1985, 10-HH by 10-LU equilibrium run.

the base year, and the 1985 employment forecasts from EMPAL as input, DRAM was used to forecast household location in 1985. These 1985 forecasts of employment and household location were then compared with the 1985 data. The correlations (Pearson's r) between the 1985 forecasts and data were as follows:

Employment type	r	Household type	r
Industrial	0.719	Low-income	0.954
Institutional	0.875	Lower-middle-income	0.965
Office	0.981	Upper-middle-income	0.919
Retail	0.928	High-income	0.944

From these results two points are clear. First, for activity totals the traditional method produces fairly good results. Second, by implication, only modest improvements are possible in correlations for most activity totals and it is therefore appropriate to examine correlations for changes in activity (that is, employment and households) by zone.

The 1980 and 1985 data were then used to calculate the 'observed' change in employment, by type and zone, and in households by type and zone. The 1980 data and 1985 forecasts were used to calculate 'estimated' changes in employment and households. It was to be expected that forecasts of changes in activity levels would be less accurate than forecasts of total activity levels, but the correlation results were even lower than what was expected. Thus began a careful investigation of the forecasts and the data.

Investigation of the forecasts began with a verification that the inputs, and so on, were correct. More complex issues arose with respect to the actual values used for some of the rather critical 'regional ratios'. This is a topic which is almost universally ignored in the modelling literature. Even so, when it comes to a comparison of model forecasts with observed data, this is a critical matter (Putman, 1981). The values of ratios such as employees-per-household could easily be obtained from the 1985 data. Other values, such as the conversion matrix of employee-type to household-type could not be derived from the 1985 data, as the data were not sufficiently detailed to permit the necessary cross-tabulations. Even so, no major problems could be found in the forecasting process. Thus it seemed necessary to conclude, albeit reluctantly, that the traditional forecasting procedure was quite incapable of reliably forecasting zonal changes in activity levels.

It was thought that the PAMEQ framework should provide the means of producing good forecasts of zonal changes, so two different 1985 equilibrium solutions were calculated. The first 1985 equilibrium was calculated with 100-HH iterations, and the second was a 10-HH by 10-LU equilibrium solution. Each of these was used, in turn, in the PAMEQ structure of equation (11.10). For each of these alternative equilibrium solutions a series of different estimates of zonal change in activity levels was made, with use of different values for the rate-of-adjustment parameter μ. None of these many different alternative estimates of zonal change showed any real improvement in correlation with 'actual' zonal changes.

At this point the only thing left to do was to reexamine the data. Almost immediately several problems became apparent. In the employment data there were two serious inconsistencies. First, at the most basic level there was a difference in the way in which certain types of employment were classified and reported in 1980, compared with the way in which it was done in 1985. Second, the way in which the detailed employment-type data were aggregated to the four major employment types used in EMPAL also differed between 1980 and 1985. There were two serious inconsistencies in the household data too. First, the 1980 household data were obtained from US Bureau of Census files, whereas the 1985 data were obtained from a variety of local (Houston area) sources, with particular reliance on public-utility connections. Second, the aggregation of the 1980 households to income groups was used to a considerable degree to determine (by proration) the distribution of the zonal households by income group in 1985.

Some attempts were made to correct for these inconsistencies. The zonal activity totals were afterwards shown by statistical analysis to have become reasonably consistent, with high correlations between the 1980 and 1985 data and between the forecasts and the 'observed' values for 1985. The problem remained, however, that the 'data' on zonal changes in activity levels were not sufficiently reliable to be used for calibration and/or evaluation of the PAMEQ formulation.

Even though a formal calibration was made impossible by the above-mentioned problems with the data, a number of rather interesting numerical experiments were done. Several different model constructs were used to calculate equilibrium solutions. Then these alternate equilibria were used with different response rates, (degrees of adjustment), to solve for the PAMEQ spatial patterns. Results from a few of those experiments are shown here to illustrate the behavior of the PAMEQ structure.

The results shown here were based on the 199-zone data for the Houston region. The 1980 observed data were used as the t-1 values. Several different 'optimal' solutions for time t were generated, two of which were used to produce the results shown here. The first optimal solution used here was the 100-HH solution. Figure 11.11(a) shows the PAMEQ solution for 0% adjustment. Thus the 1985 forecast is identical to the 1980 base-year distribution (multiplied by a scalar adjustment to match the changed regional household totals). In this case there is no spatial response to the disequilibrium represented by the difference between t-1 and t^*. Figure 11.11(b) also shows the forecast 1985 spatial pattern which results from a 30% adjustment to disequilibrium. Here it was assumed that the degree of adjustment to disequilibrium would be three tenths of the difference between the 1980 $(t-1)$ value and the 1985 (t^*) equilibrium forecast. The degree of adjustment was taken to be the same for all household types in all zones.

Figure 11.12(a) shows the 1985 forecast which results from assuming that the degree of adjustment is six tenths of the difference between the 1980

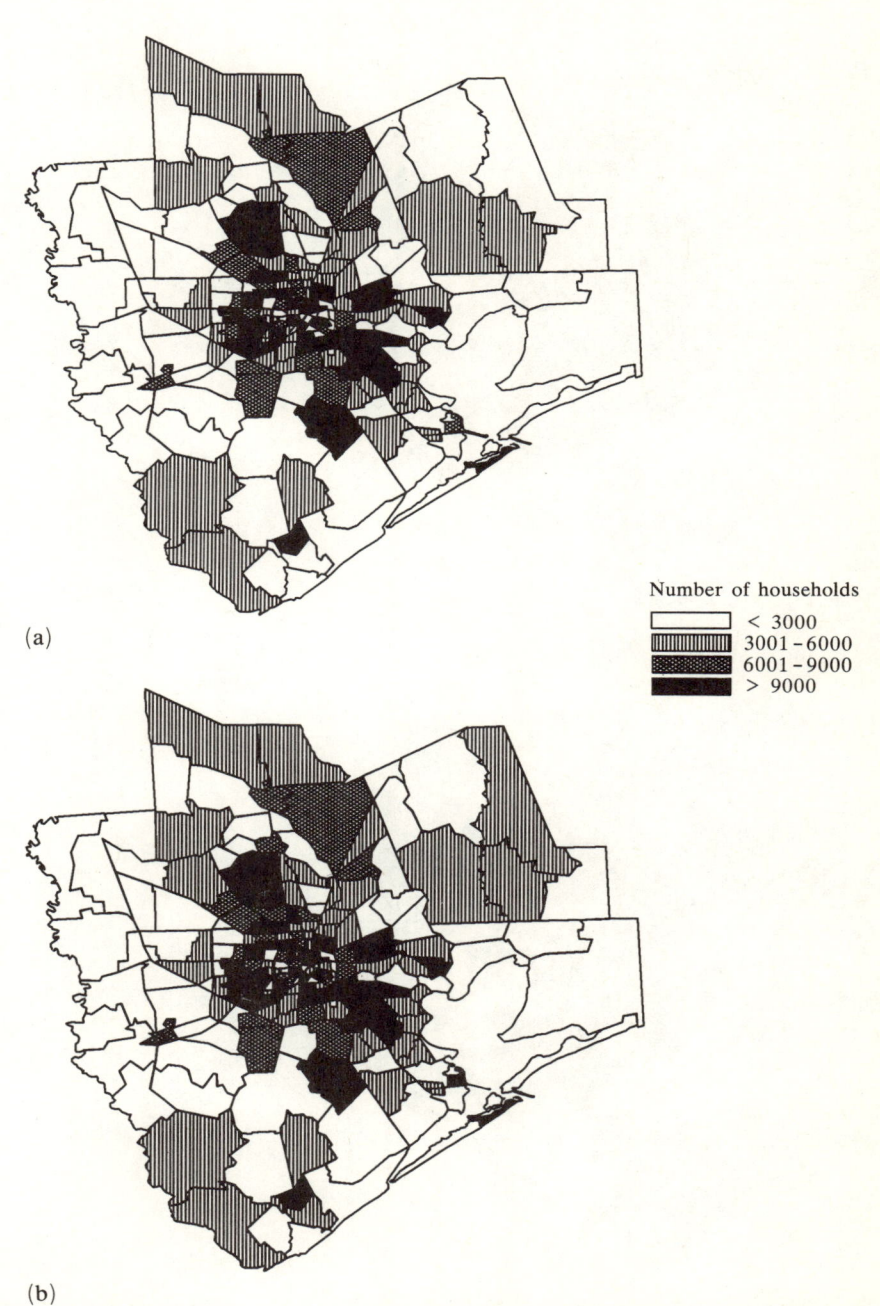

Preliminary development of dynamic spatial models

(a)

Number of households
< 3000
3001–6000
6001–9000
> 9000

(b)

Figure 11.11. Houston, TX: the forecast of total households in 1985 by the PAMEQ formulation test (100 HH) for (a) 0% adjustment and (b) 60% adjustment.

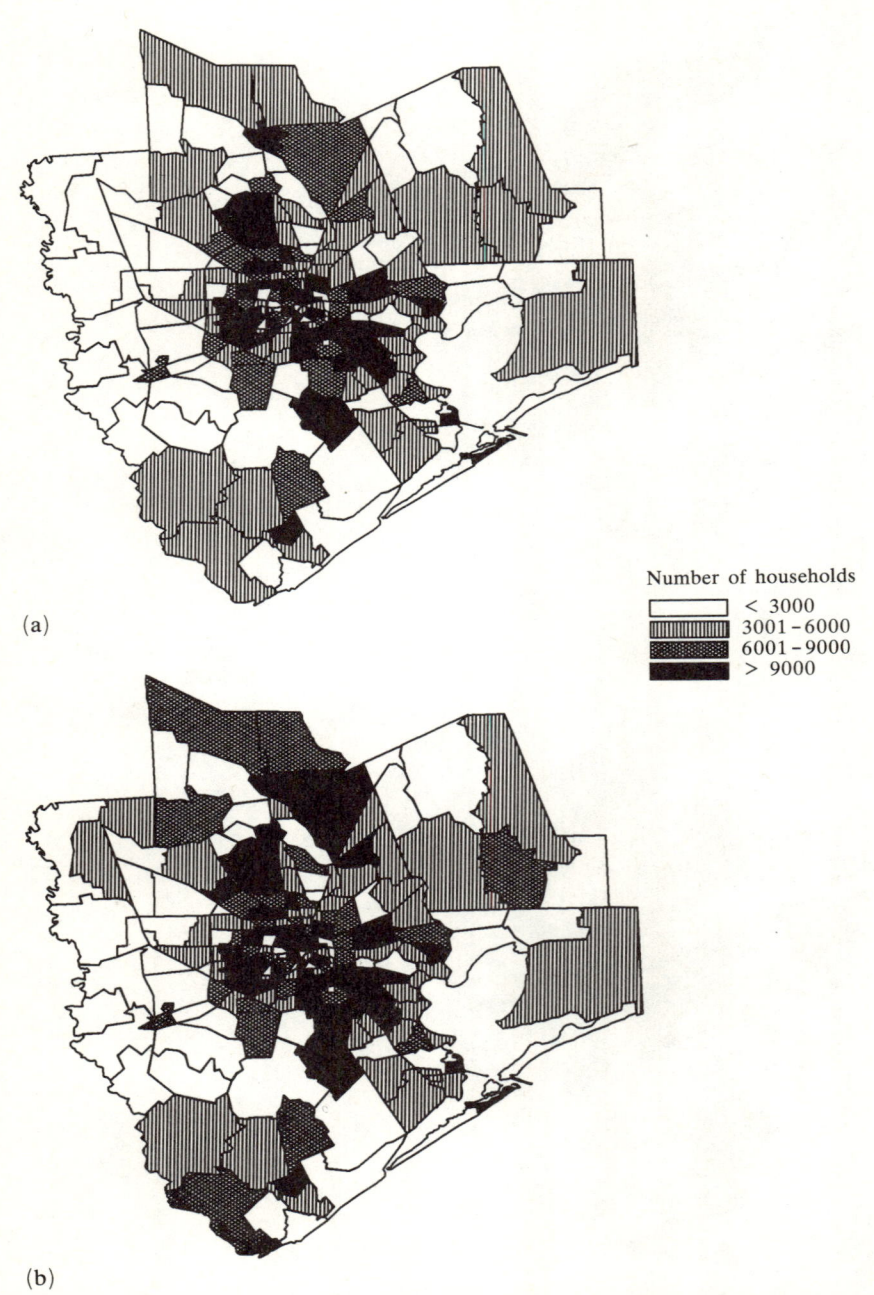

Figure 11.12. Houston, TX: the forecast of total households for 1985 by the PAMEQ formulation test (100 HH) for (a) 60% adjustment and (b) 100% adjustment.

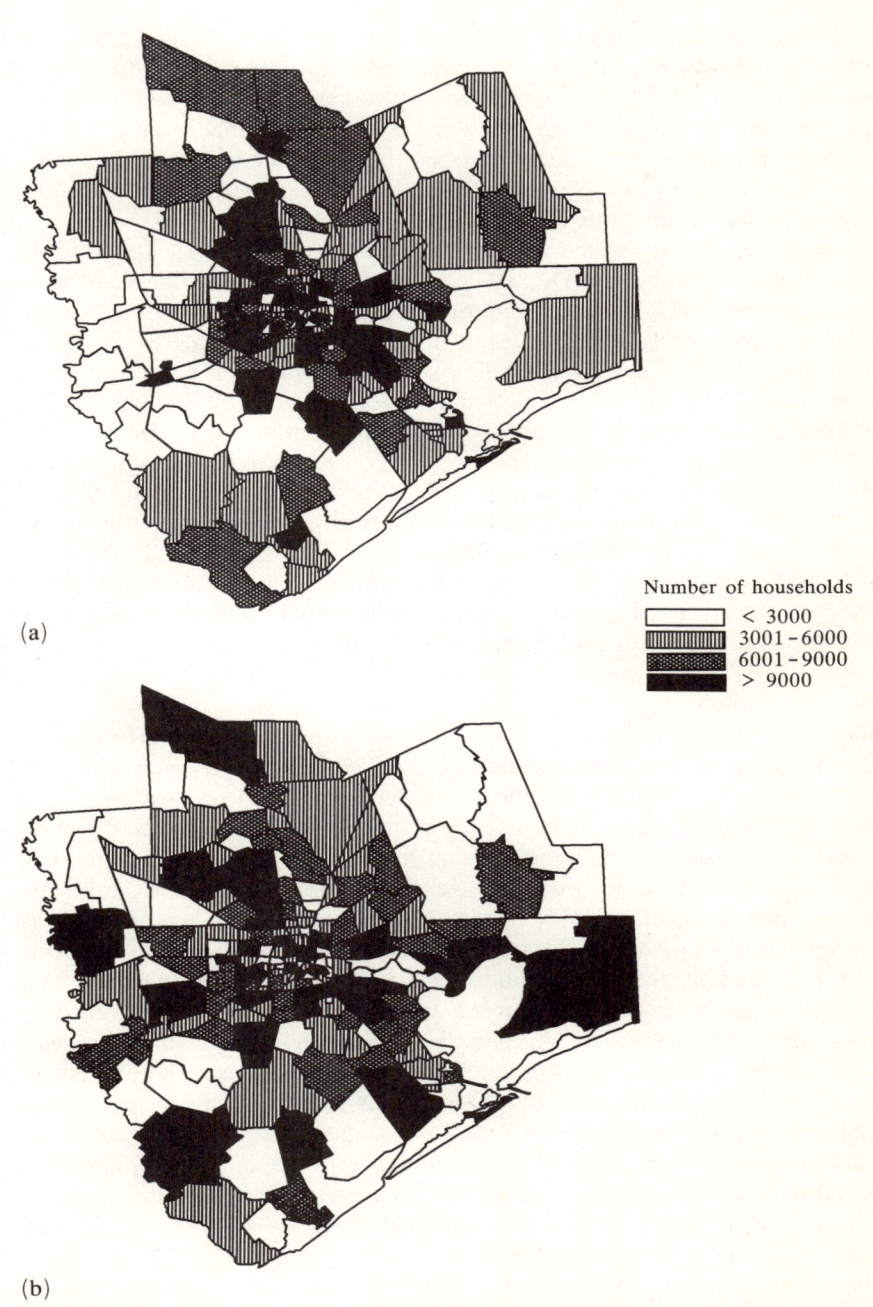

Preliminary development of dynamic spatial models 311

(a)

Number of households
- < 3000
- 3001–6000
- 6001–9000
- > 9000

(b)

Figure 11.13. Houston, TX: the forecast of total households for 1985 by the (a) conventional DRAM forecast and (b) the PAMEQ formulation test (HH/LU) with 30% adjustment.

$(t-1)$ value and the 1985 (t^*) equilibrium forecast. In figure 11.12(b) is the 1985 forecast which results from assuming full adjustment to disequilibrium. As can be seen, the 30% adjustment output looks rather like the 1980 base year (0% adjustment), with the difference that there is the beginning of some spatial dispersion. This tendency is, of course, continued in the 60% adjustment output and moves to a rather extensively dispersed pattern resulting from the full adjustment experiment (100% in a five-year period).

Figure 11.13(a) shows first the 1985 forecast spatial pattern which would result from a conventional DRAM forecast. It is worth recalling that this is also the pattern which results from one iteration of an equilibrium solution process. Figure 11.13(b) is the result of an assumed 30% degree of adjustment, but here to a different equilibrium solution from the one shown in figure 11.11. The 1985 (t^*) equilibrium forecast used here was the 10-HH by 10-LU solution described above. As can be seen here, the results of the HH/LU equilbrium give an unusually dispersed pattern of activities, even with only 30% adjustment. It is more than likely that future research will yield good reasons for preferring the HH equilibrium solution as an adjustment target, rather than the HH/LU equilibrium.

These experiments served to demonstrate the operationality of a PAMEQ structure for forecasts of changes in spatial patterns, while clearly leaving a good many unresolved issues for future investigation.

11.7 Conclusion

The original intent in this chapter was to demonstrate the potential usefulness of the concept of partial adjustment to equilibrium for dynamic models of spatial activity location. After the structure to be tested was set out it was clear that further exploration of equilibrium structures and solutions was necessary. To understand the process more thoroughly a series of numerical experiments were done with the Archerville data. These experiments showed that for each of three alternate sets of structural hypotheses it is possible to solve for a unique, stable, equilibrium solution. These solutions were independent of starting points. Each structural hypothesis produced a different, in some cases very different, equilibrium solution.

Despite the success of the Archerville equilibrium experiments, there was the evidence in chapter 10 that for full-scale data sets certain equilibrium structures might not be stable. As a consequence a new investigation was made of equilibrium structures for full-scale data sets. There it was found that according to regionwide measures of convergence, the HH and the HH/LU equilibrium structures did appear to converge after 50 to 100 iterations. Closer investigation showed, however, that there were individual locator types in individual zones which were not converging. This phenomenon, which appears to be chaotic behavior, seems to be the result of the large data sets and multiple, interacting, locator types. A good deal more investigation is necessary to achieve a complete understanding of the equilibrium behavior of these systems.

Finally, tests of the PAMEQ structure with real sets of data were not conclusive. The structure worked perfectly well, but there was no data with which the forecasts could be compared for purposes of evaluating their accuracy. An informal scrutiny of the results suggested that the HH-iteration structure provided a better target for the disequilibrium-adjustment process. The HH/LU structure seemed too unstable, and was shown to exhibit chaotic behavior. It may be that the HH procedure, in conjunction with variable travel cost, will yield an even better equilibrium target. There is a good deal of additional work necessary here, and empirical tests of this structure will have to await the availability of additional sets of data.

12

Conclusion: next research steps for dynamic spatial models

12.1 Introduction

It was clear at the beginning of this book, that many of the issues raised would not be resolved by the end of the book. That being given, perhaps one of the major accomplishments of the book is in laying the groundwork for the considerable additional work which will be required to develop a new generation of theoretically sound and operationally useful dynamic spatial models. In that vein, the purpose of this chapter is to identify and describe several of the major areas of concern.

Four general topics which require further research are discussed in this chapter. The first of these is the problem of the 'supply side' in location models, with particular emphasis on residential location and with the inclusion of supply–demand reconciliation as a part of the problem. The second topic is the problem of equilibrium structures and solution techniques. This includes both intermodel and intramodel equilibria. The third topic is the calibration problem. Here, too, there are questions of calibration both within and between models. The fourth topic is the broad, and crucial, one of the interplay between theoretical model building and the availability of data with which the models can be tested.

12.2 Supply–demand interaction in spatial models

The EMPAL and DRAM models, as well as the simple Archerville employment and residence-location models, which were used for numerical experiments in chapters 2, 8, 9, 10, and 11, are all 'demand'-driven models. Variables are included in the models which describe the relative attractiveness of each zone in the region being modelled. The quantity of locating activity allocated to each zone is directly proportional to the relative attractiveness of the zone. In the employment location calculations there are no supply-related variables at all. In the residence-location calculations, supply considerations enter via the land-use variables and their updating for each iteration or recursion.

DRAM was used for all of the intermodel equilibrium experiments described in chapters 8 and 9. DRAM was modified to enable the calculation of internal iterations to attempt to solve for intramodel equilibria in chapters 10 and 11. Both versions of DRAM include three land-related variables as a part of the zonal attractiveness calculation. These variables are: (1) residential land, (2) vacant developable land, and (3) percentage of developable land already developed. In each *recursion* of DRAM in the intermodel equilibrium experiments, or in each *iteration* of DRAM in the intramodel equilibrium experiments, the land-use variables are updated in response to the change in numbers of locators in each zone. The updating of the land-use variables is accomplished by a simple land-consumption model, LANCON, which was described in chapter 10. In LANCON, which for operational purposes, is embedded in DRAM, a decrease in the number of locators in a zone causes the release of land for other uses, whereas an increase in the number of locators results in the consumption of land at a rate determined by the equations of LANCON. There is no attempt made to represent competition between locators for the land in a particular zone. After the location and land-consumption calculations are completed, the sum of the land uses (of all types) is, if necessary, scaled back to the total developable land in the zone.

No locators are redistributed as a part of this process. Thus if the new locators in a zone, consuming land at the rates forecast by LANCON, 'use' more land than is available, the land they use is scaled down to what was available. The implicit assumption is that in circumstances where the demand for land exceeds the supply, the land market, unobserved in the model, adjusts so that activity densities increase. In subsequent iterations or recursions, the three revised land variables result in a change in the attractiveness of each zone to new locators. This, in turn, results in subsequent changes in activity location.

This, of course, is not the only way that land-use variables could be treated in a spatial interaction model. Many models simply have new land use computed in each recursion based on the prior average land use. They may or may not make use of the land variable in an attractiveness measure. In very few models has an explicit representation of rent or land price been attempted. One of the most interesting of such attempts was by Anas (1982). After extensive analysis, Anas concludes that for the marginal probability choice model (that is, choice of zone of residence) a rent or housing-price variable may not be essential. Of course, the experience from actual attempts to apply spatial interaction models is that it has usually been difficult or impossible to obtain reliable house-price data. Even more problematic is that in any recursive or iterative model structure it would also be necessary to forecast house price for future time periods. This task has proved to be virtually impossible.

The general problem of updating 'independent' variables is one which often causes difficulties for the development of models. It seems to be generally agreed in the field that for forecasts going beyond a five-year time horizon, a single-step procedure is inappropriate. Some model developers would perhaps say that even five years is too long a time period for a single-step forecast and have developed models with shorter temporal steps (Wegener, 1986). Although models operating on a one-year or two-year recursion period make for interesting speculation, the prospect of obtaining enough data to enable such models to be compared with reality is most unlikely. In many metropolitan areas it is not all that easy to obtain and/or develop multiple time point data sets with five-year intervals, with it often being necessary to estimate rather crudely some of the required variables.

In some cases, there is a need for supply-side variables to be produced as an output from whatever forecasting model(s) are to be used. Such estimates might be produced, after the fact, from the previously calculated forecasts of residence location. The question then becomes one of whether or not these supply-side forecasts of variables should become input to the residence-location forecasts. To answer this it must first be determined whether inclusion of such variables in the base-year calibrations of the model will improve significantly the ability of the model to forecast residential location. If not, then it may not be worth the effort of including the variables in the model. In this case, they would simply be estimated by, in effect, a postprocessing step, given as output, but not used as input to subsequent iterations or recursions. If the inclusion of such variables does result in an improvement in forecasting ability, then it will be necessary to determine how well these variables can themselves be forecasted for input to the subsequent recursions of the model. Unless these forecasts are reliable, it still may not be worth including the variables in the model.

As an alternative to using supply-side variables in the attractiveness term of a spatial allocation model, one might use them in some sort of explicit locational constraint. Traditionally, constraints have been used in spatial models to correct for obvious errors in the forecasts. An alternate use has been, in the form of such concepts as that of 'zonal holding capacities', to impose policy or the informed judgment of users on model outputs. In some cases such uses make good sense. Many spatial models are quite persistent in their attempts to locate residences in parkland or cemeteries. On the other hand, most use of supply variables, (for example, housing stocks) would not be for such clear-cut cases. In effect, the use of supply-side variables as constraints would still require the generation of forecasts of those variables for use as constraints.

For some authors the need to include supply variables, price, and an explicit representation of the land market seems almost an article of faith. For them, an economically sound model formulation 'must include a proper representation of the market'. Yet, there is very little empirical evidence to support the notion that such formulations, even where base-year data could be obtained for parameter estimation, would actually produce better forecasts than those which can be obtained from demand-oriented spatial models. Nonetheless, as discussed in chapters 4 and 5, there are some possible benefits to be had from studying mathematical programming formulations of location models, and the issue of the supply side is perhaps more important there, as will be discussed briefly in the next section of this chapter.

12.3 Supply–demand interaction in programming models

In chapters 4 and 5 there was an extended discussion of mathematical programming models for residence-location forecasting. A number of numerical experiments were also described. One clear conclusion from those numerical experiments was that for the optimizing or mathematical programming formulations, model solutions were both 'lumpy' and critically dependent on specific variables. For example, if houses in one zone cost

$200 000 and houses in another zone cost $199 999, then all other things being equal in the model, all the $199 999 houses would be occupied before a single $200 000 house would be occupied. This problem, it was shown, could be remedied by the addition of a dispersion term to the objective function of the model. By this means, in effect representing locators' imperfect information and differences of preferences, a distribution of location decisions around the 'optimum' was obtained. These solutions seemed likely to be more representative of actual behavior.

As was true for the spatial interaction models, supply-side variables can be used either as elements in the objective function or as constraints. In the first case they would, in their role in the formulation, be operating like a part of an attractiveness measure. Here, however, such variables could be more explicitly used. For example, if the objective function was one of maximizing consumer surplus, then house cost or price variables, site cost or price variables, or an apartment rent variable would be expected to show up in the objective function for residential locators with a negative coefficient. The same expectation would hold for a location-cost-minimizing formulation or a utility-maximizing formulation. Nonetheless, the inclusion of such variables in the objective function of one of these model structures still requires that the variables be forecast for successive time periods.

The use of, say, a stock variable such as number of housing units, or the like, as a constraint in a mathematical programming model poses similar problems. Further, the role of such supply variables as constraints is one of considerable importance in such models. A careful reading of the work with these models will often reveal the major role that such constraints play in developing the location forecasts obtained from the model (Prastacos, 1983). But, here too, these constraints will themselves have to be forecasted, just as in the early development work with the Herbert–Stevens model described in chapter 4, where before the model would produce useful forecasts the constraints had to be forecasted with a Lowry-type spatial interaction model. There was a circularity in this process which has yet to be resolved.

Regardless of whether the supply side of location is represented by actual *stock* variables such as housing units or structure types, or by *price* variables such as house price, land price, or rent, the first difficulty to be overcome is that of obtaining the data for the base year for model calibration. The second, and greater, difficulty is that of developing a method for updating those variables for successive forecast time periods. For stock variables this might involve modelling the processes of housing construction, conversion, and demolition. For price variables it would be necessary to model supply–demand balance, or a land market-clearing process of some sort. Efforts along these lines have, in virtually every case, met with enormous difficulty (deLeeuw and Struyk, 1975, Harsman, 1981, Ingram et al, 1972). If the goal is to forecast activity location it may make good practical sense to deal only with the supply side indirectly as is now done in EMPAL and DRAM, with perhaps some additional effort devoted to improving the land-consumption procedures. If, on the other hand, the goal is to produce supply-side variable forecasts of, say, housing stock or house price, it may be easier to produce them as a consequence of location-demand forecasts, than to try to model explicitly the operation of the land market. Even so, no one can deny the lure of attempting to model a process so central to human affairs. Only by scholars continuing attempts in that vein is there any hope that the process will be understood at a theoretical level and be proven empirically as well.

12.4 Location-transportation equilibrium: structures and solutions

Even in the earliest work on the integration of transportation and land-use models, one of the recurring issues was that of equilibrium solutions which, it was assumed, would represent some form of balance between the two very generally defined phenomena (Putman, 1973). Several attempts were made at that time to develop procedures which would produce a between models or, as it has been described in this book, an intermodel equilibrium solution. The high cost of computation in the early 1970s prohibited extensive numerical experiments, and even tests of what with hindsight turns out to have been the correct algorithmic approach could not be completed. It was not until the theoretical work of Evans (1976) that the general form of such approaches was clearly articulated. The practical focus of that approach, as discussed in detail in chapter 8, involves specific attention to link flows. In particular, if one is speaking of a model (or models) of activity location and trip distribution, combined with a model of trip assignment, the essential element in producing an equilibrium solution lies in developing a way of averaging or 'combining' link flows in successive iterations of the solution process.

In chapter 8, after exploring alternatives, it was shown that a simple procedure for combining link flows is the key to a rather robust intermodel equilibrium solution process. In chapter 9 it was shown that this approach works for a fairly complex model structure and for full-scale data sets. There are embedded issues of systemwide calibration, but generally speaking, solving for intermodel equilibrium is no longer an issue.

What is, however, very much an issue is the process of solving for intramodel equilibrium, and the extent to which intermodel equilibrium solutions and intramodel equilibrium solutions should be combined. The first part of this is the question of intramodel equilibrium. In the location models as well as in the assignment models there are still unresolved questions regarding equilibrium.

In chapters 4 and 5 the questions associated with network-assignment equilibrium were discussed. There it was shown that even though it is possible to give a rather clear definition of network equilibrium according to Wardrop's conditions. A major substantive issue is whether a better description of actual tripmaker behavior would be obtained by adding a dispersion term to the equilibrium defined by Wardrop's conditions. A great deal more

Conclusion

research is necessary on this matter. A major mechanical issue has to do with algorithms for solving equilibrium or stochastic equilibrium assignment. There are very serious questions about the ability of the algorithms now in popular use to achieve a correct equilibrium assignment solution. Even though equilibrium assignments may be preferable from a theoretical viewpoint, practical evidence of their superiority is scarce, and this may be because of inadequate solution procedures.

The situation with regard to equilibrium solutions for location models is much less clear. First of all, there is no clearly stated equivalent to Wardrop's conditions. In the absence of a price variable or some means of calculating location costs or benefits, no cost-based equilibrium condition can be stated. The closest one can come to such a statement for a demand-driven model, such as DRAM, is to say that an equilibrium solution has been reached when successive iterations of the solution procedure yield no changes in activity locations. This, however, is an algorithmic statement and provides no real theoretical insight into the process being modelled. In addition, there are many questions regarding the equilibrium solution process itself in such models. In chapters 10 and 11 experiments made with two such processes were examined.

The solution method called the HH process involved iterative recalculation of the numbers of each household type in a zone as a function of the number of households of each type in the zone as calculated in the previous iteration. In this process the land-use variables of each zone, as well as the matrix of zone-to-zone travel costs, remained constant at the values of the prior time period. The only 'reason' for change in household location from one iteration to the next was the degree of attraction, or repulsion, of each locating household type to each other household type. There was complete freedom of movement in this process, and, in effect, perfect information available to each household type about the movements of each other household type. Thus the equilibrium reached in such a model run would represent the preferred location of each household type with respect to the location of each other household type, while land-use and travel-cost variables were being held constant. The preferences were derived from a cross-sectional parameter estimation of base-year data. There is, of course, no way of knowing if these were the 'true' preferences, as some variables are clearly not included in the model structure. In any case, this solution process readily converges to a stopping point where no further shifting of household locations takes place.

The LU process is complementary to the HH process. Here the distribution of households from the prior time period, used in calculating zonal attractiveness, remains constant along with the zone-to-zone travel costs. The land-use variables change from one iteration to the next as the locations of households change in the current time period. Here, in effect, each locating household has perfect information about the land use in each zone and how

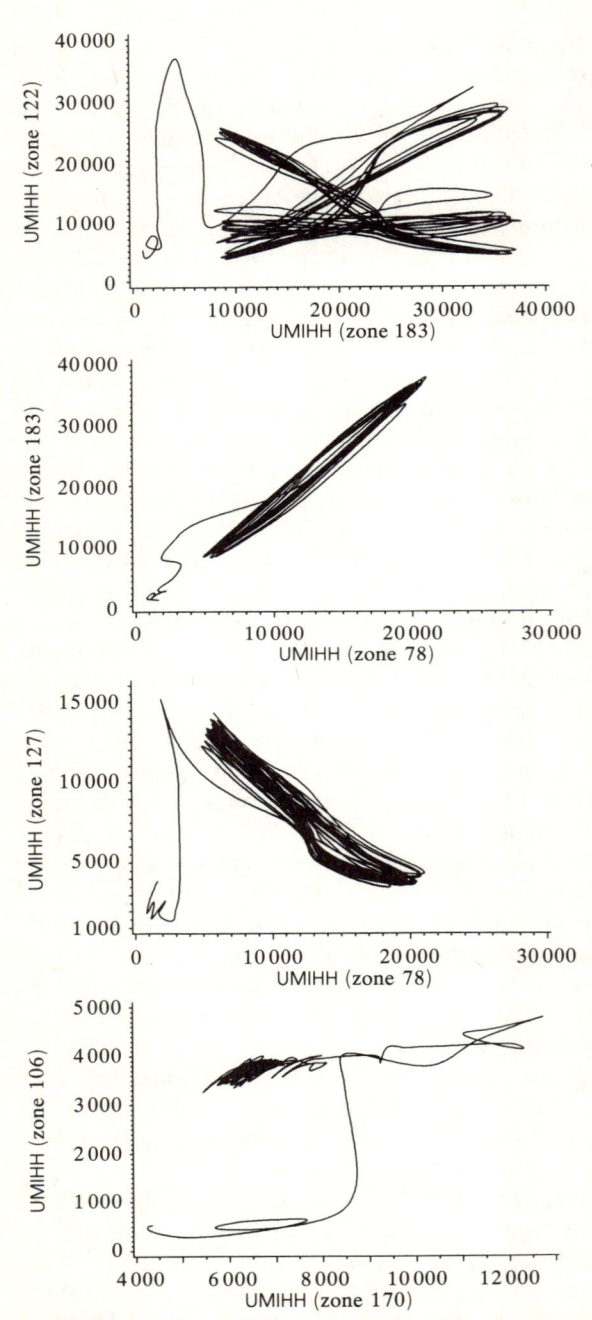

Figure 12.1. Houston, TX, 199-zone system: 1985 forecasts of the numbers of upper-middle-income households (UMIHH) in UMI zones, obtained from equilibrium experiments and evidencing chaotic behavior.

it changes with the location decisions of all households, but has only lagged information on the zonal distribution of household types and on travel costs. The solution of this structure was not so smooth as that of the HH structure, with convergence problems appearing as the system approached iteration-to-iteration changes of a few percent or less. It is not exactly clear why this should have been the case, but it appears to have resulted from the land-consumption calculations, and suggests the need for further work along that line.

Attention was finally turned to a combined HH/LU solution process. Here it is assumed implicitly that each household locator knows instantly and completely about the location decisions of each other locator and about the resulting changes in land use. Only the zone-to-zone travel costs remain constant at the levels of the prior time period. Here, after a number of iterations have moved towards convergence, the model system displays chaotic behavior. These chaotic patterns showed up when comparing the locations of specific household types in specific zones (Putman, 1990). In effect there is an unstable shifting of household composition of specific zones. There is a great deal of investigation to be done here. The chaotic behavior may be simply an artifact of the solution procedure. It is, at least in part, a consequence of a specific data set, as the phenomenon was observed with only one of the two rather different large data sets. The Houston equilibrium runs exhibited chaotic behavior, and the Washington runs did not. (Nor did it appear in tests of a very small sample data set.) This behavior may be indicative of real neighborhood effects in metropolitan areas. There is certainly evidence of what appear to be chaotic household location shifts in cities. The possibility that this particular behavior observed in the model solution represents a 'real-world' chaotic phenomenon offers a tempting research challenge. Figure 12.1 shows several phase diagrams of chaotic behavior in an HH/LU solution for Houston.

12.5 Dynamics of comprehensive integrated model structures

With the computation of intermodel equilibrium being reasonably well understood, but the computation of intramodel equilibrium still needing work, combining them into an overall model structure still presents solution-procedure questions. The PAMEQ structure described in chapter 11 assumes the possibility of solving for a comprehensive equilibrium, and involves not only the HH/LU structure, but also the inclusion of travel costs and employment location calculations as well. The numerical experiments of chapter 9 showed that this could be readily accomplished in the absence of the intramodel equilibrium calculations. The question remains, however, of what each of the several alternative equilibrium structures means, and which of them is the 'correct' one to use as the target equilibrium in the PAMEQ structure.

It is possible, for example, to start with the model structure given earlier, in figure 9.2. Before considering the question of intramodel equilibrium, it is necessary to consider the possibilities which swirl about the question of intermodel equilibrium. The chapter 9 numerical experiments with equilibrium involved progressing through the models, one after another. The equilibrium solution for the entire system of models was achieved by an MSA (mean successive averages) algorithm which combined successive estimates of network link flows before progressing to the next iteration through the entire model system. Mode split was not included in those experiments. Although it is unlikely that its inclusion would result in making the equilibrium procedure unsuccessful, the test remains to be done. Another test remaining to be done regards the connection of employment location (EMPAL) to residence location (DRAM). Perhaps it would be helpful to iterate between these models a time or two before passing on to mode split and trip assignment? This question leads into the one of intramodel equilibrium within an intermodel equilibrium structure. Should DRAM be solved for an HH equilibrium in each iteration of the intermodel equilibrium solution? If the DRAM intramodel equilibrium structure is expanded to include travel costs, then there is a direct overlap, rather than a simple nesting, of the intramodel and intermodel equilibrium structures and solution procedures.

All of this still leaves unanswered the question of what any one of these equilibrium structures might mean. Nor does it deal with whether or not omission of an explicit representation of the supply side and thus, of a market mechanism in the location models makes an important difference. With these questions unresolved, so must be the question regarding the proper target in the PAMEQ structure; so too the shape of the adjustment function and the response rates. Resolution of these questions will take further theoretical development and further numerical experiments. One key question along the path to further numerical experiments is that of model-system calibration, which will be discussed next.

12.6 Calibration of systems of models

The calibration, or parameter estimation, of models such as EMPAL and DRAM has received enough research attention as to be reasonably well understood (Putman, 1980; Putman and Ducca, 1978a; 1978b; Putman and Kim, 1984; Kim and Putman, 1984). The calibration of mode-split models has received considerable attention as well (Ben-Akiva and Lerman, 1985). The calibration of trip-assignment models requires considerably more work, as does that of the composite-cost calculation. This is not to mention the even more difficult calibration issues associated with response mechanisms and response rates in the PAMEQ structure. Some of this will be discussed below.

There is, as yet, no definitive answer as to which of the possible trip-assignment techniques discussed in chapters 5 and 6 is 'the best'. In the cases of both stochastic assignment and stochastic user-equilibrium assignment there are dispersion parameters to be estimated. Well-specified

Conclusion

methods of estimating these parameters have yet to be developed. The situation is made considerably more difficult by the many problems associated with the congestion, or volume-delay, functions which are used to calculate congested-link travel times or costs. The most commonly used of these functions is the one developed by the US Bureau of Public Roads [equation (6.9)], which has two empirically estimated parameters. Virtually all users of this function (who are virtually all users of traffic-assignment models) use, by default, parameters estimated decades ago. No adjustment is made for changes in facility or vehicle technology, nor for facility type (Fricker, 1989). Very little formal work has been done on the effect of the form and parameters of the volume-delay function on trip-assignment results. In addition, although not exactly a calibration problem per se, the entire matter is further complicated by the inherent difficulty of adequately describing abstract aggregated network links.

The level of spatial disggregation appropriate to the models described here is in the order of hundreds of zones per metropolitan region. For use at this level, road networks must be aggregated considerably from the 'on-the-ground' system. In the process of aggregating it becomes necessary to 'invent' specifications for the abstract aggregated links. This, in and of itself, is an almost intractable problem. No matter how well the aggregation is done it leads to a calibration problem with the model system. In particular, the total capacity of the abstract aggregated network will not match the total capacity of the 'on-the-ground' network. This results in a regionwide inconsistency between the capacity of the abstract network and the estimates of the model system of total number of vehicle trips. If the estimates of total trips in the system, even when known to be accurate, are assigned to the abstract network there is likely to be too much congestion (the abstract network invariably has less capacity than the tens of thousands of actual streets and highways). It then becomes necessary either to increase all the abstract link capacities or to scale down the estimates of the number of trips in the region.

From a practical perspective it is much easier to adjust the trips than it is to adjust the capacities of the thousands of abstract links. The question arises as to the extent to which the trips should be adjusted. No formal procedure exists for this problem. In some cases there is available an independently produced set of zone-to-zone travel times or costs. If this is the case then trip-assignment runs can be made with different scalings of the estimated number of vehicle trips. The resulting congested travel times can be compared with the independent estimates, to obtain a crude systemwide calibration. Clearly this is a calibration trouble spot. What is more, there are ramifications which trace back to the location model calibrations. If the zone-to-zone travel costs used in the location model calibrations were to be significantly different from the travel times or costs which resulted from trip assignments, then there certainly would be reason to consider whether the location model calibrations should be redone as well.

Further, there is a cluster of questions surrounding composite-cost calculations. First, the composite-cost function itself, [for example, equation (9.9)], contains one or more parameters. These parameters must be estimated, though, as mentioned in chapter 9, there are some questions regarding the structural form of the composite-cost function for use with location models having multivariate attractiveness functions, in addition to the question of how to estimate their parameters. Second, is, again, a question which goes back to the calibration of the location models. Should they be recalibrated, perhaps once again, by using composite costs in lieu of the highway costs which may have been used in a prior calibration? The need for a considerable amount of additional applied research here should by now be clear.

12.7 Additional calibration problems for PAMEQ structures

Some mention was made in chapter 11 of the potential problems of calibrating a PAMEQ structure, and references were given to the relevant literature. In using such a structure for integrated transportation and location models the biggest problems will not be with calibration of the PAMEQ structure, per se. The biggest problems will be those mentioned above, of determining what structure to use in solving for the target equilibrium and in estimating properly the parameters of that equilibrium structure.

Within the PAMEQ structure, however, a major issue is that of the degree of disaggregation of response mechanisms and response rates. The notion here is that different activity types are likely to have different rates of adjustment to disequilibrium. Further, within activity types there may be systematically different response rates for different zones or types of zones. There is some literature on parameter estimation for PAMEQ structures, but it does not begin to explore the complexities engendered by its use for location models, as is contemplated here (Bowden, 1978a). A particular difficulty here is the virtual certainty that response rates will be asymmetric with respect to the direction of adjustment. Thus, for example, the rate at which numbers of households increase in a particular zone when the imputed equilibrium value is higher than the actual value might be greater than the rate at which the numbers of households in that zone would decrease if the imputed equilibrium value were lower than the actual value. This implies that the data would have to be divided into, for each locator category, groups above and below equilibrium, and different response rates, and perhaps response functions, would have to be estimated for each group.

It should be readily apparent that there are substantial needs here for further research. Not only are these important questions of parameter estimation, but what are perhaps more fundamental questions of which parameters are actually to be estimated. In order for any of this work to pass beyond the purely speculative, a great deal of work with actual data will be necessary. Some highly data-dependent issues will be discussed next.

12.8 Theory development and data analysis

In much of what has been presented in this book, though there were many numerical experiments described, not too much discussion of data was given. Yet, no matter how clever the theory, no matter how sophisticated the equation structure and the computer program for its solution, they must all eventually face up to empirical validation. In the final analysis, it is only in terms of empirical validation that it is possible to judge the practical worth of any particular model structure versus that of any other. What is more, the process of attempting empirical validation often raises important theoretical issues itself.

One example comes from the calibration process. Consider the calibration of a location model such as DRAM. The procedure, statistically, is that of nonlinear curve fitting (see the example in chapter 3). If it is assumed that data are available and that a parameter-estimation computer program is available, then the best-fit parameters can be calculated. There will not be a perfect fit of model to data. The difference between the model and the data, the residuals, will be unexplained variation due to errors in model specification, errors in data, or random errors. The question is whether these residuals should be saved for further use in forecasting.

Use of the residuals to modify zonal attractiveness for a short-term forecast would almost certainly improve the reliability of the forecast (Putman, 1989). The difficult question is that of determining for how long a forecast period they will remain relevant. Were enough data to become available it would be worth experimenting with procedures for attenuating the influence of these residuals over, say, a fifteen-year forecast period. This could be done linearly with, for example, a 25% reduction (of the original value) in their magnitudes each five years in a recursive sequential forecast proceeding in five-year increments. If the base year were 1990, then the residuals would be used at 75% of their original values for the 1995 forecast, 50% of their original values for the 2000 forecast, 25% of their original values for the 2005 forecast, and not used at all for forecasting 2010 and beyond. Without additional data there is no way to evaluate this procedure.

Another question which comes from calibration, and which can only be answered with the availability of more data, concerns parameter variation over time. Implicit in most, if not all, spatial location model work to date is the assumption that the parameters are temporally invariant. This assumption is almost certainly false, though very little work has been done to attempt to prove or disprove it (Putman, 1977). Recalibration of models with updated data sets could shed some light on this question. It should be noted, nonetheless, that extremely good data consistency would be required from one time period to the next, lest the parameter differences be more reflective of different data sources and definitions than of differences in locational behavior. It would, however, be very nice to have a reliable updating procedure for parameters to use with models making long-range forecasts.

Another data-oriented area of concern is with various regional rates and ratios. In taking the employment, by zone and industry type, output from EMPAL and using it as input to DRAM, one must make a conversion. This conversion is done in DRAM, and amounts to converting from employees, at place of work, by zone and industry type, into heads of households, at place of work, by zone and household type. The data necessary to prepare what is in essence an employee-to-household conversion matrix are ordinarily available from census publications. The advantage of this 'mechanism' in the models is that the regional distribution of household types, usually by income, changes in direct response to changes in the distribution of regional employment. The disadvantage is the need to determine values for the conversion matrix for future forecast years.

Directly tied to the conversion matrix questions are questions regarding other regional ratios. Zonal population estimates depend upon conversion from households, by type, to population. The current models use regional values only for those conversions. There is the problem of updating the regional values for the forecast years. There is the question of disaggregating the rates to subregions or to zones.

Along with population per household is the question of employees per household, and of jobs per employee. The question of employees per household, in and of itself requiring consideration, also calls for an examination of the locational dynamics of multiworker households. *Explicit* in DRAM is the assumption that household location is jointly determined by place of work, travel cost, and residential zone attractiveness. *Implicit* is the assumption that after the number of employees at place of work is converted to the number of heads of households at place of work, a single place of work is associated with the residence-location decision of each household. The effect that changing patterns of employees per household will have on the location behavior of households has not been examined in this context.

Finally there is the panoply of data questions now, in effect, in the way of further work on PAMEQ formulations. What is needed is a complete set of employment, household, land-use, and travel-cost data for at least two time points (preferably five years apart). These data sets must be consistent with each other, but *not*, as is often the case, with one 'data set' actually being a set of estimates based on the other. Finding such a data set is proving to be a difficult task. In the absence of such a data set there is still much that can be done, but eventually it will have to be found. Only by demonstrating the performance of a model with real data, regardless of how sensible and/or sophisticated its formulation, can a definitive evaluation be made.

12.9 Conclusion

In a sentence, the conclusion is that there is no conclusion. A number of new approaches to integrated transportation, location, and land-use modelling have been described and tested. Some of them do indeed seem promising. Other approaches, perhaps equally promising, have been omitted. Notable

Conclusion

amongst these are microsimulation and the construction of location and transportation models as nested sets of choice models. Such nested choice models of location and transportation, in particular, should continue to be explored, as is being done by Anas (1982). Their omission here is a matter of judgment, preferences, and sheer lack of space in the book.

The purpose in the predecessor to this volume (Putman, 1983a) was to document the development of what was arguably the first operational integrated transportation and land-use model package. The purpose in this book is to set the stage for the next decade's research on the development of operational dynamic integrated transportation and land-use models. Only time will tell if that purpose was achieved.

Appendix

A1 Introduction

Owing to the nonlinear structure of the DRAM residential location model it is necessary to use a specialized parameter estimation procedure for its calibration (Putman, 1983a). The goal of a calibration effort is to develop estimates of the parameters of the equation(s) in a model which best fit(s) the general model structure to a specific data set. Most planners are familiar with this process in the context of multiple (linear) regression analysis. Calibration is analogous to regression analysis but different mathematics and a different computer program are used.

In multiple regression analysis, because of the linearity of the equations, it is possible to solve directly for the best-fit parameters, or regression coefficients. Although there are issues of statistical significance to be considered, only in the most peculiar of circumstances is there any concern about not being able to find the optimal, or best-fit, parameters in linear multiple regression. In nonlinear parameter estimation the situation is not quite so clear. First, for there to be a best-fit set of parameters there must be only one optimum on the mathematical surface defined by the equations and data set. Second, the computer program must be capable of locating that optimum.

In general terms this is an unconstrained nonlinear optimization problem. With regard to the presence of an optimum, experience with hundreds of parameter estimation efforts with data for more than forty different urban areas has never yielded a situation where an optimum could not be found. This is not to say that it is not sometimes difficult to locate the optimum. When there is high correlation amongst the independent variables of the model, the point giving the best fit can be a bit tricky to find. The computer program, called CALIB, which is used for this work, locates the optimum by a method called gradient search. In this method, the numerical values of the partial derivatives of the DRAM equations are used to guide a search procedure towards the optimum, or best-fit, parameters.

The measure used here for best-fit analysis is what is called the maximum likelihood criterion. In linear multiple regression analysis the best fit is measured by the R^2 criterion, with which most planners are familiar. The maximum likelihood criterion is more appropriate for the equation structure of the DRAM model. In addition to the best-fit criterion it is useful to know something about the statistical significance of the parameters obtained. In multiple regression analysis one is accustomed to seeing t-tests of the coefficients. The validity of these tests depends upon assumptions regarding the equation structure. These assumptions do not hold true for the nonlinear structure of DRAM, and thus the usual t-tests are not helpful. In lieu of these a set of asymptotic t-values is calculated. These can be interpreted in a manner similar to that which would be used for the usual t-tests, but their effectiveness cannot be guaranteed. In the negative direction they can show when a coefficient is not statistically significant. In a positive direction they can only show when a coefficient may be statistically significant, not whether it is for certain.

Appendix

A2 Calibration results
A2.1 DRAM, by household type

The data set for the Washington, DC, region was provided by the Metropolitan Washington Council of Governments. The data were for the year 1980 and were divided into 182 zones. The categories into which the locating group were placed were in terms of four household types, to approximate income quartiles. The total number of households in each income group in the entire region were, subject to round-off errors: low-income households (LIHH), 271 178; lower-middle-income households (LMIHH), 271 183; upper-middle income households (UMIHH), 271 174; upper-income households (UIHH), 271 188. The employment data for the area were disaggregated into four employment types (place of work) by zone. There were a total of 1 660 859 industry identified employees in the region, in the following groups: (1) industrial, 201 266; (2) government, 608 666; (3) wholesale and services, 548 462; (4) retail, 302 465.

The land-use data were provided in terms of residential and nonresidential acres, vacant acres—developable and not developable—and total acres, all by zone. The region includes a total of 1 367 942 acres, of which 325 281 are developed residential, 84 956 are developed basic, 34 951 are developed commercial, 616 402 are (unuseable) not developable, and 45 110 are in miscellaneous categories. The balance 261 242 acres, is vacant developable land. It should be noted that there is considerable discussion in the region, generally, as to whether agricultural land is to be considered developable. Several of the region's jurisdictions have enacted stringent land-use control laws, but it remains to be seen whether these will be adequate to resist the intense pressures to develop emerging in the region.

The equation structure for the current version of DRAM is given in equations (10.4)-(10.6). The calibration results for the Washington region are given in table A1. The first point to note is that the goodness of fit of the DRAM model to these data was quite good. The likelihood criterion can vary from zero to one. The lowest level is equivalent simply to taking the regional mean number of households (by type) per zone and using that as the estimated number of households (by type) for every zone. The highest level would only be achieved in the case of a perfect estimate of the actual numbers of households per zone. Thus the values of 0.74 and better achieved here are quite good. The values of R^2 were not used to guide the search for the best-fit parameters, but were calculated after the conclusion of the search in order to provide a basis for a comparison more readily understood by people unfamiliar with the likelihood criterion.

To turn next to the actual values of the parameters, the signs and magnitudes of the parameters are consistent with prior calibrations done for dozens of other regions. The negative values of the travel-time parameter are exactly as expected with the exception of the value for UIHH which most frequently will be lower (less negative) than the values for the other three income groups. This is because high-income households are usually able (and willing) to spend more (in time or cost) on travel than the lower-income household types. The value found here, a bit more negative than the value for the upper-middle-income households, implies that there are concentrations of high-income households in some of the more centralized population areas, whereas somewhat fewer of the upper-middle-income households live in such areas. Reference to the implicit average trip lengths (that is, the estimated length of work trips implied by comparing the spatial distributions of employment with the spatial distributions of households) shows that the upper-middle-income households are taking the longest trips, with slightly shorter trips being taken by the upper-income households. Both the high-income and upper-middle-income households are taking longer trips than the low-income and lower-middle-income households. Taken together with the travel-time parameter values, this confirms that a significant number of high-income households may be located in moderately 'dense' clusters of development, perhaps somewhat removed from the center of the region, but not so widely dispersed as are the upper-middle-income households.

Table A1. DRAM parameters for Washington (1980) region (182-zone system).

Variable	Low-income HH	Lower-middle-income HH	Upper-middle-income HH	High-income HH
Travel time	−1.23 (291.4)	−1.31 (339.1)	−1.31 (385.6)	−1.43 (470.5)
Vacant acres	−0.09 (124.6)	−0.07 (81.8)	−0.05 (58.0)	−0.05 (55.8)
Developed acres (%)	0.96 (26.9)	0.92 (27.7)	1.08 (36.1)	1.02 (33.4)
Residential acres (%)	0.75 (371.7)	0.75 (346.6)	0.73 (319.9)	0.71 (308.9)
Low-income households (%)	5.36 (261.5)	1.65 (72.9)	1.21 (54.5)	0.49 (21.6)
Lower-middle-income households (%)	3.80 (81.8)	7.26 (166.1)	2.48 (65.4)	1.78 (53.5)
Upper-middle-income households (%)	0.03 (0.6)*	1.73 (40.1)	6.62 (164.4)	3.80 (91.8)
High-income households (%)	−1.61 (71.4)	−0.96 (42.6)	−0.23 (11.6)	3.95 (226.6)
R^2	0.81	0.77	0.72	0.71
Likelihood criterion	0.86	0.81	0.74	0.74
Average trip length	19.4	19.2	21.5	20.9

* Not significant at 5% level.
Note: asymptotic t-tests in small type; HH, households.

As a matter of fact the only other slightly anomalous coefficient is that for high-income household attraction to intrazonal percentage of upper-middle-income households. Under ordinary circumstances it is expected that high-income household attraction to intrazonal percentage of high-income households will yield the highest parameter, and the next highest, but substantially lower, will be the high-income household attraction to upper-middle-income household concentrations, with a commensurately lower parameter value. Here, however, the results suggest that although the greatest absolute number of high-income households are found in zones where high-income households are in the majority, the second greatest attraction parameter, to concentrations of upper-middle-income households, is almost as great. The implication is that the upper-middle-income households are probably most spread out into the suburban areas, whereas significant numbers of high-income households remain more centralized. This may place them more closely to concentrations of lower-middle-income households than is the case in other urban areas. This implies, conversely, that those zones with the highest intrazonal percentage of high-income households will have rather smaller absolute numbers of high-income households, and equally great numbers of upper-middle income households.

The asymptotic t-test results are also quite satisfactory. Only one parameter of the thirty-two estimated is shown by a t-test to be not significant (which requires that the t-value be less than 2). As mentioned above, asymptotic t-tests cannot guarantee statistical significance of coefficients here, yet when the value of an asymptotic t obtained is ten times what is necessary for an ordinary t-test, it is probably safe to assume that the coefficient is statistically significant. Of the thirty-two parameters estimated only two had asymptotic t values less than 20. These parameters may not be statistically significant, but this is by a rather conservative standard, and there really is no way to tell for sure.

The data set for the Houston–Galveston area was provided by the Houston Galveston Area Council of Governments. The data were for the year 1980 and were divided into 199 zones. The locating group categories were in terms of four household types, approximating income quartiles. The total numbers of households, in each income group, in the entire region were (subject to round-off errors): LIHH, 228 058; UMIHH, 360 344; LMIHH, 291 741; UIHH, 222 458. The employment data for the area were disaggregated into four employment types, (place of work) by zone. There were a total of 1 518 438 industry identified employees in the region, in the following groups: (1) industrial, 504 620; (2) institutional, 140 020; (3) office, 548 887; (4) retail, 324 911.

The land-use data were provided in terms of residential and nonresidential acres, vacant acres—developable and not developable—and total acres, all by zone. The region includes a total of 4 891 544 acres, of which 279 847 are developed residential, 146 187 are developed nonresidential, 330 401 are (unuseable) not developable, and 194 711 are in miscellaneous categories (including streets and highways). The balance, 3 940 398 acres (about 80%) is vacant developable land.

Table A2 contains the results of the calibration results done for a 199-zone data set for the Houston–Galveston region. Virtually all of the comments made about the Washington results apply here as well. These results give a good sense of the parameter differences between the two regions as analyzed in terms of the DRAM equation structure.

Table A2. DRAM parameters for Houston–Galveston region (199-zone system).

Variable	Low-income HH	Lower-middle-income HH	Upper-middle-income HH	High-income HH
Travel time	−1.46 (297.4)	−1.44 (330.7)	−1.06 (224.6)	−1.12 (201.6)
Vacant acres	0.15 (110.3)	0.11 (87.5)	0.07 (74.3)	0.14 (111.9)
Developed acres (%)	0.71 (58.8)	0.63 (59.9)	0.77 (93.1)	0.99 (93.4)
Residential acres (%)	0.83 (267.2)	0.87 (299.6)	0.87 (340.7)	0.81 (253.5)
Low-income households (%)	4.22 (166.7)	−0.38 (15.7)	−1.34 (62.6)	−2.90 (98.5)
Lower-middle-income households (%)	3.07 (60.1)	5.84 (143.5)	1.63 (54.8)	2.23 (65.1)
Upper-middle-income households (%)	−0.48 (12.5)	0.60 (17.3)	3.94 (140.5)	1.06 (30.8)
High-income households (%)	−1.30 (45.3)	−0.94 (38.8)	−0.55 (29.0)	3.86 (195.4)
R^2	0.84	0.85	0.84	0.81
Likelihood criterion	0.91	0.90	0.87	0.89
Average trip length	23.5	24.2	33.4	33.2

Note: asymptotic t-tests in small type; HH, households.

A2.2 EMPAL, by employment type

Owing to the nonlinear structure of the EMPAL employment location model, as was also the case for the DRAM residential location model, it is necessary to use a specialized parameter estimation procedure for its calibration. The computer program, called CALIB, which was used for the DRAM calibrations, was also used for the calibration of EMPAL (Putman, 1983a).

EMPAL is a modified version of the standard singly-constrained spatial interaction model. There are three modifications: (1) a multivariate, multi-parametric attractiveness function is used, (2) a separate, weighted, lagged variable is included outside the spatial interaction formulation, (3) a constraint procedure is included in the model, allowing zone-specific and/or sector-specific constraints. The model is normally used for three to eight employment sectors whose parameters are individually estimated. The equation structure is given in equations (10.1) and (10.2).

Table A3. EMPAL parameters for Washington, DC, region (182-zone system).

Variable	Industrial	Governmental	Wholesale and services	Retail
Travel time (α)	1.45 (107.4)	3.08 (99.9)	2.07 (83.5)	1.99 (85.7)
Travel time (β)	−0.002 (6.8)	−0.39 (178.8)	−0.39 (120.7)	−0.80 (120.2)
Total employment	0.53 (195.9)	0.16 (112.7)	0.08 (32.8)	0.19 (21.4)
Total zonal acres	−0.25 (101.6)	−0.12 (35.2)	−0.13 (20.5)	−0.16 (15.0)
Lagged employment $(1-\lambda)$	0.22 (59.2)	0.06 (64.6)	0.03 (29.8)	0.05 (23.9)
R^2	0.84	0.96	0.95	0.90
Likelihood criterion	0.85	0.96	0.94	0.93

Note: see table A1.

Table A4. EMPAL parameters for Houston–Galveston region (199-zone system).

Variable	Industrial	Institutional	Office	Retail
Travel time (α)	1.36 (141.0)	0.97 (60.5)	1.53 (135.8)	1.35 (105.3)
Travel time (β)	−0.59 (329.9)	−0.73 (182.2)	−0.33 (248.3)	−0.41 (198.9)
Total employment	−0.53 (198.0)	−0.22 (48.6)	−0.02 (19.5)	−0.04 (21.9)
Total zonal acres	0.33 (45.8)	0.63 (56.0)	0.13 (36.4)	0.05 (8.6)
Lagged employment $(1-\lambda)$	0.18 (153.2)	0.56 (323.6)	0.14 (136.8)	0.17 (112.0)
R^2	0.77	0.81	0.95	0.76
Likelihood criterion	0.80	0.72	0.92	0.83

Note: see table A1.

The EMPAL calibration results for the Washington, DC, region are given in table A3, above. The goodness of fit of the EMPAL model to these data was quite good.

Turning next to the actual values of the parameters, we find that most of the signs and magnitudes of the parameters are consistent with prior calibrations done for dozens of other regions. The negative values of the travel-time parameter, β, are exactly as expected.

The asymptotic t-test results are quite satisfactory. No parameter of the twenty estimated is shown by a t-test to be not significant (which would require the t-value to be less than 2). As mentioned above, asymptotic t-tests cannot guarantee statistical significance of coefficients here, yet when the value of an asymptotic t is ten times what is necessary for an ordinary t-test it is probably safe to assume that the coefficient is statistically significant. Of the twenty parameters estimated only two had an asymptotic t-value of less than 20. The EMPAL calibration results for the Houston region are given in table A4.

A2.3 LANCON

The most recent versions of the EMPAL employment allocation model and the DRAM residential allocation model have a provision for the use of a new model to calculate land use. In earlier versions of EMPAL and DRAM, as is the case for virtually all existing land-use models, a simple extrapolation method is used for calculating land use. After new locations are calculated, a new value for land consumption is calculated in other models by simply applying the base-year average amount of land used per activity (for example, by households or employees). With the implementation of LANCON as an integral component of the EMPAL-DRAM model set, multiple regressions are used to estimate land consumption of several types.

For the Washington, DC, and Houston, TX, regions land-consumption equations were estimated for three classes of land use: residential, basic, and commercial.

The form of the regression equations used was developed under National Science Foundation (NSF) sponsorship as part of a study of "Intrametropolitan variation in rates of land consumption" (Putman, 1983b). The equations for estimating residential land consumption is given in equation (10.9).

Results of parameter estimation for Washington and Houston for this residential land-consumption equation are given in table A5.

In the same NSF sponsored study an equation was developed for basic land consumption as well. This was as given in equation (10.10). Results of parameter estimations for this equation for the Washington and Houston metropolitan areas are shown in table A6.

Last, the commercial land-consumption equation that was developed had the form given in equation (10.11). The results of the parameter estimation for commercial land consumption in Washington and Houston are given in table A7.

Appendix

Table A5. Residential land consumption in 1980—parameter estimates for equation (10.9).

	Value	t-statistic		Value	t-statistic
Washington, DC			Houston, TX		
k_0	−4.036	8.148	k_0	−1.912	4.040
k_1	−1.661	4.251	k_1	−0.127	3.079
k_2	−0.016	0.291*	k_2	−0.063	2.586
k_3	−0.138	2.675	k_3	−0.278	5.416
k_4	0.117	1.968	k_4	0.047	0.827*
k_5	0.628	10.208	k_5	0.108	1.497*
k_6	0.451	8.440	k_6	0.169	3.788
R^2	0.78		R^2	0.61	
F for whole regression	102.7		F for whole regression	50.5	

* Not significant at the 5% level.

Table A6. Basic land consumption in 1980—parameter estimates for equation (10.10).

	Value	t-statistic		Value	t-statistic
Washington, DC			Houston, TX		
g_0	−11.524	11.129	g_0	−6.800	5.536
g_1	−0.203	0.346*	g_1	−0.041	0.315*
g_2	1.227	15.536	g_2	0.548	3.780
g_3	−0.953	5.334	g_3	−0.620	5.234
g_4	0.575	6.398	g_4	0.910	11.053
g_5	0.151	1.763*	g_5	−0.022	0.152*
R^2	0.66		R^2	0.55	
F for whole regression	69.4		F for whole regression	46.8	

* Not significant at the 5% level.

Table A7. Commercial land consumption in 1980—parameter estimates for equation (10.11).

	Value	t-statistic		Value	t-statistic
Washington, DC			Houston, TX		
p_0	−10.914	15.451	p_0	−5.687	6.824
p_1	−0.590	1.217*	p_1	1.006	14.861
p_2	1.044	15.466	p_2	1.009	13.699
p_3	−0.340	2.513	p_3	−0.120	0.870*
p_4	0.358	5.455	p_4	1.078	14.085
p_5	0.007	0.090*	p_5	0.240	3.524
R^2	0.67		R^2	0.74	
F for whole regression	70.8		F for whole regression	111.5	

* Not significant at the 5% level.

All these parameter-estimation results compare favorably with those of the previous study. The signs and magnitudes of the regression coefficients were consistent with the previous results, as were the goodness-of-fit values and tests of statistical significance. These equations are incorporated in the current versions of EMPAL and DRAM and were used to compute land consumption in the model runs described in this book.

References

Akcelik R, 1978, "A new look at Davidson's travel time function" *Traffic Engineering and Control* **19** 459–463

Allen P, Sanglier M, 1979, "A dynamic model of growth in a central place system" *Geographical Analysis* **11** 256–272

Almaani M, 1988 *Network Trip Assignment and Aggregation Procedures in Urban Transportation Planning and Design* unpublished PhD dissertation, Urban Simulation Laboratory, Department of City and Regional Planning, University of Pennsylvania, Philadelphia, PA 19104, USA

Alonso W, 1960, "A theory of the urban land market" *Papers of the Regional Science Association* **6** 149–158

Alonso W, 1964 *Location and Land Use* (Harvard University Press, Cambridge, MA)

Anas A, 1973, "A dynamic disequilibrium model of residential location" *Environment and Planning A* **5** 633–647

Anas A, 1975, "The empirical calibration and testing of a simulation model of residential location" *Environment and Planning A* **7** 899–920

Anas A, 1982 *Residential Location Markets and Urban Transportation* (Academic Press, New York)

Arezki Y, Van Vliet D, 1985, "The use of quantal loading in equilibrium traffic assignment" *Transportation Research B* **19** 521–525

Avriel M, 1976 *Nonlinear Programming: Analysis and Methods* (Prentice-Hall, Englewood Cliffs, NJ)

Batty M, 1976 *Urban Modelling* (Cambridge University Press, Cambridge)

Batty M, 1977, "Operational urban models incorporating dynamics in a static framework", mimeograph, Department of Geography, University of Reading, PO Box 227, Reading RG6 2AB, England

Beaumont J R, Clarke M, Wilson A G, 1981, "The dynamics of urban spatial structure: some exploratory results using difference equations and bifurcation theory" *Environment and Planning A* **13** 1473–1483

Beckman M, McGuire C, Winston C, 1956 *Studies in the Economics of Transportation* (Yale University Press, New Haven, CT)

Ben-Akiva M, Lerman S, 1985 *Discrete Choice Analysis* (MIT Press, Cambridge, MA)

Benders J, 1962, "Partitioning procedures for solving mixed variables programming problems" *Numeriche Mathematik* **4** 238–252

Berechman J, Small K A, 1988, "Research policy and review 25. Modeling land use and transportation: an interpretive review for growth areas" *Environment and Planning A* **20** 1285–1309

Bertuglia C, Leonardi G, 1979, "Dynamic models for spatial interaction" *Sistemi Urbani* **1** 3–25

Bertuglia C, Leonardi G, Tadei R, 1985, "Dynamic analysis of transport–location interrelationships: theory and models", WP-27, Dipartemento di Scienze e Tecniche per i Processi di Insediamento, Facolta di Architettura del Politecnico di Torina, Viale Mattioli 39, 10125 Turin, Italy

Bertuglia C, Leonardi G, Occelli S, Robins G, Tadei R, Wilson A (Eds), 1987 *Urban Systems: Contemporary Approaches to Modelling* (Croom Helm, Andover, Hants)

Beveridge G, Schecter R, 1970 *Optimization: Theory and Practice* (McGraw-Hill, New York)

Bovy P, 1984, "Travel time errors in shortest route predictions and all-or-nothing assignments: theoretical analysis and simulation findings", memorandum number 28, Delft University of Technology, Department of Architecture, Institute for Town Planning Research, Julianalaan 134, PO Box 5, Delft, The Netherlands

Bovy P, 1985, "Error analysis of traffic assignment predictions", internal note RAV/1/85.1, Delft University of Technology, Department of Architecture, Institute for Town Planning Research, Julianalaan 134, PO Box 5, Delft, The Netherlands

Bovy P, Jansen G, 1983a, "An evaluation of traffic assignment predictions using route choice analysis and error decomposition", proceedings of the PTRC Summer Annual Meeting, Planning and Transport Research and Computation Co., Brighton, England, pages 101-113

Bovy P, Jansen G, 1983b, "Network aggregation effects on equilibrium assignment outcomes: an empirical investigation" *Transportation Science* **17** 240-262

Bovy P, Jansen G, 1983c, "Spatial aggregation effects in equilibrium and all-or-nothing assignments" *Transportation Research Record* **931** 98-106

Bowden R, 1978a, "Specification, estimation and inference for models of markets in disequilibrium" *International Economic Review* **19** 711-726

Bowden R, 1978b *Limited-dependent and Qualitative Variables in Econometrics* (Cambridge University Press, Cambridge)

Boyce D E, Chon K S, Lee Y J, Lin K T, LeBlanc L J, 1983, "Implementation and computational issues for combined models of location, destination, mode, and route choice" *Environment and Planning A* **15** 1219-1230

Boyce D E, Lundquist L, 1987, "Network equilibrium models of urban location and travel choices: alternative formulations for the Stockholm region" *Papers of the Regional Science Association* **61** 93-104

Boyce D, Janson B, Eash R, 1981, "The effect on equilibrium trip assignment of different link congestion functions" *Transportation Research A* **15** 223-232

Bradley S, Hax A, Magnanti T, 1977 *Applied Mathematical Programming* (Addison-Wesley, Reading, MA)

Branston D, 1976, "Link capacity functions: a review" *Transportation Research* **10** 223-236

Broad W, 1983, "Is science stymied by today's complexity?" *The New York Times* 28 June, page C1

Brotchie J, 1969, "A general planning model" *Management Science* **16** 265-266

Brotchie J, Sharpe R, 1974, "A general land use allocation model: application to Australian cities", in *Urban Development Models* proceedings of the third Land Use and Built Form Studies Conference (Construction Press, Harlow, Essex) pp 217-236

Brotchie J, Dickey J, Sharpe J, 1980 *Lecture Notes in Economics and Mathematical Systmes 180: TOPAZ* (Springer, Berlin)

Chapra S, Canale R, 1985 *Numerical Methods for Engineers* (McGraw-Hill, New York)

Clark W, 1983, "Structures for research on the dynamics of residential mobility", in *Evolving Geographical Structures* Eds D Griffith, A Lea (Martinus Nijhoff, Dordrecht) pp 372-397

Clarke M, Wilson A, 1983, "Dynamics of urban spatial structure: progress and problems" *Journal of Regional Science* **21** 1-18

Daganzo C, 1977a, "Some statistical problems in connection with traffic assignment" *Transportation Research* **11** 385-389

Daganzo C, 1977b, "On the traffic assignment problem with flow dependent costs—I" *Transportation Research* **11** 433-437

Daganzo C, 1977c, "On the traffic assignment problem with flow dependent costs—II" *Transportation Research* **11** 439-441

Daganzo C, Sheffi Y, 1977, "On stochastic models of traffic assignment" *Transportation Science* **11** 253-274

References

Dalton P, Harmelink M, 1974, "Multipath traffic assignment: development and tests", Systems Research and Development Branch, Research and Development Division, Ministry of Transportation and Communications, Downsview, Ontario, Canada

Davidon W, 1959, "Variable metric methods for minimization", report ANL-5990, Atomic Energy Commission, Argonne National Laboratory, 9700 South Cass Avenue, Argonne, IL 60439, USA

Davidson K, 1978, "The theoretical basis of a flow travel-time relationship for use in transportation planning" *Australian Road Research* **8** 32-35

de la Barra T, 1989 *Integrated Land Use and Transport Modelling* (Cambridge University Press, Cambridge)

de Leeuw F, Struyk F, 1975 *The Web of Urban Housing* (The Urban Institute Press, Washington, DC)

Dial R, 1971, "A probabilistic multipath traffic assignment algorithm which obviates path enumeration" *Transportation Research* **5**(2) 83-111

Dickey J, 1981, "Manual for FORTRAN (WATFIV) program for TOPAZ (technique for the optimum placement of activities in zones)", Division of Environmental and Urban Systems, Center for Public Administration and Policy, Blacksburg, VA 24061, USA

Eash R, Janson B, Boyce D, 1979, "Equilibrium trip assignment: advantages and implications for practice" *Transportation Research Record* **728** 1-8

Eash R, Chon K, Lee J, Boyce D, 1983, "Equilibrium traffic assignment on an aggregated highway network for sketch planning", Transportation Planning Group, Department of Civil Engineering, University of Illinois at Urbana-Champaign, Urbana, IL 61801, USA

Echenique M, 1985, "The use of integrated land use and transport models: the cases of Sao Paulo, Brazil and Bilbao, Spain", in *The Practice of Transportation Planning* Ed. M Florian (Elsevier, Amsterdam)

Ellerman D, Gibbons W, 1984, "Alternative volume delay and volume/capacity functions for the NETWORK model", CP-728 term paper, Urban Simulation Laboratory, Department of City and Regional Planning, University of Pennsylvania, Philadelphia, PA 19104, USA

Erlander S, 1977, "Accessibility, entropy, and the distribution and assignment of traffic" *Transportation Research* **11** 149-153

Evans S, 1973, "A relationship between the gravity model for trip distribution and the transportation model in linear programming" *Transportation Research* **7** 39-61

Evans S, 1976, "Derivation and analysis of some models for combining trip distribution and assignment" *Transportation Research* **10** 37-57

Ferland J, Florian M, Achim C, 1975, "On incremental methods for traffic assignment" *Transportation Research* **9** 237-239

Fiacco A, 1983 *Introduction to Sensitivity and Stability Analysis in Nonlinear Programming* (Academic Press, New York)

Fisher F, 1983 *Disequilibrium Foundations of Equilibrium Economics* (Cambridge University Press, Cambridge)

Fisk C, 1977, "Note on the maximum likelihood calibration on Dial's assignment method" *Transportation Research* **11** 67-68

Fisk C, 1980, "Some developments in equilibrium traffic assignment" *Transportation Research B* **14** 243-255

Floor H, de Jong T, 1981, "Testing a disaggregated residential location model with external zones in the Amersfoort region" *Environment and Planning A* **13** 1499-1514

Florian M, Nguyen S, 1974, "A new look at some old problems in transportation planning", paper presented at the Summer Annual Meeting of the Planning and Transport Research and Computation Co., University of Warwick, Coventry CV4 7AL, England

Florian M, Nguyen S, 1976, "An application and validation of equilibrium trip assignment methods" *Transportation Science* **10** 374–389

Florian M, Nguyen S, Ferland J, 1975, "On the combined distribution–assignment of traffic" *Transportation Science* **9** 43–53

Frank C, 1978 *A Study of Alternative Approaches to Combined Trip Distribution–Assignment Modelling* PhD dissertation, Department of Regional Science, University of Pennsylvania, Philadelphia, PA 19104, USA

Frank M, Wolfe P, 1956, "An algorithm for quadratic programming" *Naval Research Logistics Quarterly* **3** 95–110

Fricker J, 1989, "Two procedures to calibrate traffic assignment models", paper presented at the Second Conference on Application of Transportation Planning Methods, Orlando, FL; copy available from J Fricker, School of Civil Engineering, Purdue University, West Lafayette, IN 47907

Galster G, 1977, "A bid-rent analysis of housing market discrimination" *American Economic Review* **67**(2) 144–155

Geoffrion A, 1972, "Generalized Benders decomposition" *Journal of Optimization Theory and Applications* **10** 237–260

Gleick J, 1987 *CHAOS: Making a New Science* (Viking Press, New York)

Glover F, Karney D, Klingman D, 1974, "Implementation and computation comparisons of primal, dual, and primal–dual computer codes for minimum cost network flow problems" *Networks* **4** 191–212

Griliches Z, 1967, "Distributed lags: a survey" *Econometrica* **35** 16–49

Harker P, 1986, "The spatial price equilibrium problem with path variables" *Socio-economic Planning Sciences* **20** 299–310

Harris B, 1962, "Linear programming and the projection of land uses", Penn–Jersey paper number 20, Penn Jersey Transportation Study, Philadelphia, PA; copy available from B Harris, Department of City and Regional Planning, University of Pennsylvania, PA 19104

Harris B, 1966, "Basic assumptions for a simulation of the urban residential housing and land market", mimeograph, Institute for Environmental Studies, University of Pennsylvania, Philadelphia, PA 19104, USA

Harris B, 1972, "A model of household locational preferences", in *Karlsruhe Papers in Regional Science 1: Recent Developments in Regional Science* Ed. R Funck (Pion, London) pp 63–79

Harris B, Nathanson J, Rosenburg L, 1966, "Research on an equilibrium model of metropolitan housing and locational choice", interim report, Planning Sciences Group, Institute for Environmental Studies, Graduate School of Fine Arts, University of Pennsylvania, Philadelphia, PA 19104, USA

Harsman B, 1981 *Housing Demand Models and Housing Market Models for Regional and Local Planning* Swedish Council for Building Research, Stockholm, Sweden

Hearn D, 1984, "Practical and theoretical aspects of aggregation problems in transportation planning models", in *Transportation Planning Models* Ed. M Florian (Elsevier, Amsterdam) pp 257–287

Herbert J, Stevens B, 1960a, "A model for the distribution of residential activity in urban areas", Penn–Jersey paper number 2, Penn Jersey Transportation Study, Philadelphia, PA

Herbert J, Stevens B, 1960b, "A model for the distribution of residential activity in urban areas" *Journal of Regional Science* **2** 21–36

Himmelblau D, 1972 *Applied Nonlinear Programming* (McGraw-Hill, New York)

References

Horowitz J, 1984, "The stability of stochastic equilibrium in a two-link transportation network" *Transportation Research B* **18** 13–28

Houghton A G, 1971, "The simulation and evaluation of housing location" *Environment and Planning A* **3** 383–394

HRB, 1965, "Highway capacity manual", special report 87, Highway Research Board, Washington, DC 20418

Ingram G, Kain J, Ginn J, 1972 *The National Bureau of Economic Research Detroit Prototype Urban Simulation Model* (Columbia University Press, New York)

Iooss G, Helleman R, Stora R (Eds), 1983 *Chaotic Behavior of Deterministic Systems* (North-Holland, Amsterdam)

Jansen G, Bovy P, 1982, "The effect of zone size and network detail on all-or-nothing and equilibrium assignment outcomes" *Traffic Engineering and Control* **23** 311–317, 328

Janson B, Thint S, Hendrickson C, 1986, "Validation and use of equilibrium network assignment for urban highway reconstruction planning", paper presented at 65th Annual Meeting of the Transportation Research Board, Washington, DC; copy available from B Janson, Carnegie-Mellon University, Pittsburgh, PA 15213

Janson B, Zozaya-Gorostiza C, 1985, "The problem of cyclic flows in traffic assignment", Department of Civil Engineering, Carnegie-Mellon University, Pittsburgh, PA 15213

Karp R, 1982, "The computational complexity of network problems", in *The Mathematics of Networks* Ed. S Burr (American Mathematical Society, Providence, RI)

Kendrick D, Meeraus A, 1985 *GAMS, An Introduction* draft (The World Bank, Washington, DC)

Killen J, 1983 *Mathematical Programming Methods for Geographers and Planners* (Crown Helm, Andover, Hants)

Kim T, 1989 *Integrated urban Systems Modeling: Theory and Applications* (Kluwer, Dordrecht)

Kim Y-S, Putman S H, 1984, "Calibrating urban residential location models 5: a comparison of trip-end and trip-interchange calibration methods" *Environment and Planning A* **16** 1649–1664

Lapin L, 1981 *Quantitative Methods for Business Decisions* (Harcourt Brace Jovanovich, New York)

LeBlanc L, 1973 *Mathematical Programming Algorithms for Large Scale Network Equilibrium and Network Design Problems* PhD thesis, Department of Industrial Engineering and Management Sciences, Northwestern University, Evanston, IL 60201, USA

LeBlanc L, Morlok E, Pierskalla W, 1975, "An efficient approach to solving the road network equilibrium traffic assignment problem" *Transportation Research* **9** 309–318

Lenk F, 1988, "Small area forecasting in a slowly growing metropolitan area: the Kansas City experience, 1979–1988", paper presented at 1988 Conference of the Urban and Regional Information Systems Association, Los Angeles, CA; copy available from Mid-America Regional Council, 600 Broadway, Suite 300, Kansas City, MO 64104

Lewis H, Papadimitriou C, 1978, "The efficiency of algorithms" *Scientific American* **258** 96–109

Lombardo S, Rabino G, 1984, "Nonlinear dynamic models for spatial interaction: the results of some empirical experiments" *Papers of the Regional Science Association* **55** 83–101

Lowry I, 1964, "A model of metropolis", report RM 4125-RC, The Rand Corporation, Santa Monica, CA 90406, USA

Lundquist L, 1973, "Integrated location–transportation analysis: a decomposition approach" *Regional and Urban Economics* **3** 233–262

McBride J, 1985, "Testing NETWRK for bias due to zone selection sequence", CP-728 term paper, Urban Simulation Laboratory, Department of City and Regional Planning, University of Pennsylvania, Philadelphia, PA 19104, USA

McCormick G, 1983 *Nonlinear Programming* (John Wiley, New York)

McCullough D, 1977, "The American adventure of Louis Agassiz" *Audubon* **79** 3–17

Mackett R L, 1980, "The relationship between transport and the viability of central and inner areas" *Journal of Transport Economics and Policy* **14** 267–294

Maddala G, 1983 *Limited-dependent and Qualitative Variables in Econometrics* (Cambridge University Press, Cambridge)

Manheim M L, 1973, "Practical implications of some fundamental properties of travel demand models" *Highway Research Record* **422** 21–38

Miller H, 1987 *Urban Modelling Systems: Dynamic Properties and Equilibrium Tendencies* PhD dissertation, Department of City and Regional Planning, University of Pennsylvania, Philadelphia, PA 19104, USA

Murchland J, 1966, "Some remarks on the gravity model of trip distribution and an equivalent maximizing procedure", mimeograph LSE-TNT-38, London School of Economics, London WC2A 2AE, England

Murtagh B, Saunders M, 1983, "MINOS 5.0 user's guide", technical report SOL 83-20, Systems Optimization Laboratory, Department of Operations Research, Stanford University, Stanford, CA 94305, USA

Nakamura H, Hayashi Y, Mizamoto K, 1983, "A land-use/transport model in metropolitan areas" *Papers of the Regional Science Association* **51** 43–63

Pearson C, 1986 *Numerical Methods in Engineering and Science* (Van Nostrand Reinhold, New York)

Powell W, Sheffi Y, 1982, "The convergence of equilibrium algorithms with predetermined step sizes" *Transportation Science* **16** 45–55

Prastacos P, 1983, "The design and empirical estimation of a land use–transportation model for the San Francisco area", paper presented at the 30th North American Meetings of the Regional Science Association, Chicago, IL

PSCoG, 1986, "The DRAM85/EMPAL85 activity model system: development, structure, and application in the Puget Sound region", Puget Sound Council of Government, 216 First Avenue, South, Seattle, WA 98104, USA

Pumain D, Saint Julien T, Sanders L, 1984, "Dynamics of spatial structure in French urban agglomerations" *Papers of the Regional Science Association* **55** 71–82

Putman S H, 1973, "The interrelationships of transportation development and land development", report on contract DOT-FH-11-7843 to the Urban Planning Division, Federal Highway Administration, US Department of Transportation, Washington, DC 20590, USA

Putman S H, 1977, "Calibrating a residential location model for nineteenth-century Philadelphia" *Environment and Planning A* **9** 449–460

Putman S H, 1979 *Urban Residential Location Models* (Martinus Nijhoff, Boston, MA)

Putman S H, 1980, "Calibrating urban residential location models 3: empirical results for non-US cities" *Environment and Planning A* **12** 813–827

Putman S H, 1981, "Theory and practice in urban modelling: the art of application", paper presented at 13th Annual Conference of the British Section, Regional Science Association, van Mildert College, University of Durham, England; copy available from author

Putman S H, 1983a *Integrated Urban Models: Policy Analysis of Transportation and Land Use* (Pion, London)

Putman S H, 1983b, "Intrametropolitan variation in rates of land consumption: a preliminary analysis", final report on National Science Foundation Grant SES-8112792, Urban Simulation Laboratory, Department of City and Regional Planning, University of Pennsylvania, Philadelphia, PA 19104, USA

Putman S H, 1984, "Dynamic properties of static-recursive model systems of transportation and location" *Environment and Planning A* **16** 1503–1519

Putman S H, 1985, "Documentation for the CALIB program", Urban Simulation Laboratory, Department of City and Regional Planning, University of Pennsylvania, Philadelphia, PA 19104, USA

Putman S H, 1986, "Complexity in urban systems modelling: the effects of transit on urban form", in *London Papers in Regional Science 15: Integrated Analysis of Regional Systems* Eds P W J Batey, M Madden (Pion, London) pp 54–73

Putman S H, 1989, "A survey of recent applications of the EMPAL/DRAM/ITLUP model system: with comments on issues of model application", paper presented at the 2nd Conference on Application of Transportation Planning Methods, Orlando, FL; copy available from author

Putman S H, 1990, "Equilibrium solutions and dynamics of integrated urban models", in *New Directions in Regional Analysis* Eds L Anselin, M Madden (Belhaven Press, London) pp 48–65

Putman S H, Ducca F W, 1978a, "Calibrating urban residential models 1: procedures and strategies" *Environment and Planning A* **10** 633–650

Putman S H, Ducca F W, 1978b, "Calibrating urban residential models 2: empirical results" *Environment and Planning A* **10** 1001–1014

Putman S H, Kim Y-S, 1984, "Calibrating urban residential location models 4: effects of log-collinearity on model calibration and formulation" *Environment and Planning A* **16** 95–106

Putman S, 1990, "Equilibrium solutions and dynamics of integrated urban models", in *New Directions in Regional Analysis* Eds L Anselin, M Madden (Belhaven Press, London) pp 48–65

Reiner R, Munz M, Haag G, Weidlich W, 1986, "Chaotic evolution of migratory systems" *Sistemi Urbani* **2** 285–308

Reklaitis G, Ravindran A, Rogsdell K, 1983 *Engineering Optimization: Methods and Applications* (John Wiley, New York)

Robillard P, 1974, "Calibration of Dial's assignment method" *Transportation Science* **8** 117–125

Schlager K, 1965, "Land use plan design model" *Journal of the American Institute of Planners* **31** 103–111

Senior M L, 1974, "Housing market benefits: some reflections on Houghton's model and the Herbert–Stevens model and its variants" WP-56, Department of Geography, University of Leeds, Leeds LS2 9JT, England

Senior M L, Williams H C W L, 1977, "Model based transport policy assessment 1: the use of alternative forecasting models" *Traffic Engineering and Control* **18** 402–406

Senior M L, Wilson A G, 1974a, "Disaggregated residential location models: some tests and further theoretical developments" *London Papers in Regional Science 4: Space-Time Concepts in Urban and Regional Models* Ed. E L Cripps (Pion, London) pp 141-172

Senior M L, Wilson A G, 1974b, "Exploration and synthesis of linear programming and spatial interaction models of residential location" *Geographical Analysis* **7** 209-238

SEWRPC, 1966, "A mathematical approach to urban design: a progress report on a land use plan design model and a land use simulation model", Southeastern Wisconsin Regional Planning Commission, Waukesha, WI

Sharpe R, Brotchie J, 1972, "An urban systems study" *Royal Australian Planning Institute Journal* **10** 105-118

Sharpe R, Karlquist A, 1980, "Towards a unifying theory for modelling urban systems" *Regional Science and Urban Economics* **10** 241-257

Sharpe R, Brotchie J, Ahern P, Dickey J, 1974, "Evaluation of alternative growth patterns in urban systems" *Computers and Operations Research* **1** 345-362

Sharpe R, Wilson B, Pallot R, 1984, "Computer user manual for program TOPAZ82", Commonwealth Scientific and Industrial Research Organization, Division of Building Research, PO Box 56, Highett, Vic.3190, Australia

Sheffi Y, 1979, "A note on the turn and arrival likelihood algorithms of traffic assignments" *Transportation Research B* **13** 147-150

Sheffi Y, 1985 *Urban Transportation Networks* (Prentice-Hall, Englewood Cliffs, NJ)

Sheffi Y, Powell W, 1981a, "A comparison of stochastic and deterministic traffic assignment over congested networks" *Transportation Research B* **15** 53-64

Sheffi Y, Powell W, 1981b, "Equivalent minimization programs and solution algorithms for stochastic equilibrium transportation network problems", paper presented at Transportation Research Board, 60th Annual Meeting, Washington, DC; copy available from Y Sheffi, Department of Civil Engineering, Massachusetts Institute of Technology, Cambridge, MA 02139

Sheffi Y, Powell W, 1982, "An algorithm for the equilibrium assignment problem with random link times" *Networks* **12** 191-207

Taylor M, 1984, "A note on using Davidson's function in equilibrium assignment" *Transportation Research B* **18** 181-199

Tobin R, 1979, "Calculation of fuel consumption due to traffic congestion in a case-study metropolitan area" *Traffic Engineering and Control* **20** 590-592

Van Doorn J, 1975 *Disequilibrium Economics* (John Wiley, New York)

Van Vliet D, 1976, "Road assignment—III: comparative tests of stochastic methods" *Transportation Research* **10** 151-157

Van Vliet D, 1978, "Improved shortest path algorithms for transport networks" *Transportation Research* **12** 7-20

Van Vliet D, Dow D, 1979, "Capacity constrained road assignment" *Traffic Engineering and Control* **20** 296-305

Waddel P, 1987, "An overview of demographic forecasting in Dallas-Fort Worth", mimeograph, North Central Texas Council of Governments, PO Drawer C06, Arlington, TX 76005-5888

Watterson T, 1986, "The DRAM85/EMPAL85 activity model system: development, structure, and application in the Puget Sound region", mimeograph, Puget Sound Council of Governments, 216 First Avenue, South Seattle, WA 98104

Webster F V, Bly P, Paulley N (Eds), 1988 *Urban Land-use and Transport Interaction* (Avebury, Aldershot, Hants)

Wegener M, 1986, "Transport network equilibrium and regional deconcentration" *Environment and Planning A* **18** 437-456

Wheaton W, 1974, "Linear programming and locational equilibrium: the Herbert-Stevens model revisited" *Journal of Urban Economics* **1** 278-287

Wheaton W, 1977a, "A bid rent approach to housing demand" *Journal of Urban Economics* **4** 200-217

Wheaton W, 1977b, "Income and urban residence: an analysis of consumer demand for location" *American Economic Review* **67** 620-631

Wheaton W, Harris B, 1971, "Linear programming and residential location: the Herbert-Stevens model revisited", mimeograph, Department of City and Regional Planning, University of Pennsylvania, Philadelphia, PA 19104, USA

Wildermuth B, Delaney D, Thompson K, 1972, "Effect of zone size on traffic assignment and trip distribution" *Highway Research Record* **392** 58-75

Williams H C W L, 1977, "On the formation of travel demand models and economic evaluation measures of user benefit" *Environment and Planning A* **9** 285-344

Williams H C W L, Senior M L, 1977, "Model based transport policy assessment 2: removing fundamental inconsistencies from the models" *Traffic Engineering and Control* **18** 464-469

Wilson A, 1967, "A statistical theory of spatial distribution models" *Transportation Research* **1** 252-269

Wilson A, 1974 *Urban and Regional Models in Geography and Planning* (John Wiley, Chichester, Sussex)

Wilson A, 1985, "Spatial dynamics: classical problems, an integrated modelling approach and system performance" *Papers of the Regional Science Association* **58** 47-57

Wilson A, 1987, "Transport, location and spatial systems: planning with spatial interaction models", in *Urban Systems: Contemporary Approaches to Modelling* Eds C Bertuglia, G Leonardi, S Occelli, G Rabino, R Tadei, A G Wilson (Croom Helm, Andover, Hants) pp 337-426

Wilson A G, 1989, "The dynamics of urban economic development: an integrated model system", paper presented to the Congress of the European Regional Science Association, St Johns College, Cambridge; available from The University of Leeds, Department of Geography, Leeds LS2 9JT

Wilson A G, Coelho J D, Macgill S M, Williams H C W L, 1981 *Optimization in Location and Transport Analysis* (John Wiley, Chichester, Sussex)

Wolfe P, 1962, "The reduced-gradient method", unpublished manuscript, The Rand Corporation, Santa Monica, CA 90406, USA

Author index

Akcelik R 152, 153
Allen P 285
Almaani M 161
Alonso W 65, 72
Anas A 79, 80, 81, 315, 327
Arezki Y 158
Avriel M 48

Batty M 13, 285
Beaumont J R 285
Beckman M 121
Ben-Akiva M 258, 322
Benders J 174
Berechman J 4
Bertuglia C 5, 285, 289
Beveridge G 53
Bovy P 156, 157, 158
Bowden R 6, 285, 287, 288, 324
Boyce D B 6, 152, 153, 247
Bradley S 31, 112
Branston D 117, 151
Broad W 3
Brotchie J 95, 96, 98

Canale R 259, 270
Chapra S 259, 270
Clark W 289
Clarke M 285

Daganzo C 119, 126, 142, 145, 153, 155
Dalton P 126
Davidon W 56
Davidson K 152
de Jong T 4
de la Barra T 4
de Leeuw F 317
Dial R 126, 143, 146
Dickey J 96
Dow D 127
Ducca F W 322

Eash R 122, 156
Echenique M 4
Ellerman D 152
Erlander S 171
Evans S 107, 171, 177, 188, 259, 318

Ferland J 119, 125
Fiacco A 48, 112
Fisher F 6, 288
Fisk C 126, 141, 143, 144, 145, 146

Floor H 4
Florian M 119, 158, 171, 172, 173, 174, 176, 188
Frank C 174, 179, 186
Fricker J 323

Galster G 79, 80
Geoffrion A 174
Gibbons W 152
Gleick J 279
Glover F 175
Griliches Z 288

Harker P 179
Harmelink M 126
Harris B 70, 71, 72, 73, 79, 80, 82, 87
Harsman B 317
Hearn D 157
Herbert J 65, 66, 82
Highway Research Board (HRB) 152
Himmelblau D 53
Horowitz J 127
Houghton A G 83, 85

Ingram G 317
International Study Group on Land-Use/Transport Interaction (ISGLUTI) 4, 5, 6
Iooss G 279

Jansen G 156, 157, 158
Janson B 156, 158, 159

Karlquist A 4
Karp R 146
Kendrick D 56, 75
Killen J 31
Kim T 4, 6
Kim Y-S 234, 322

Lapin L 31
LeBlanc L 121, 126
Lenk F 285
Leonardi G 285
Lerman S 258, 322
Lewis H 146
Lombardo S 285
Lowry I 87, 284
Lundquist L 4, 247

McBride J 125
McCormick G 48
McCullough D i
Mackett R L 4
Maddala G 6, 288, 291
Manheim M L 236
Meeraus A 56, 75
Miller H 281
Murchland J 106, 143
Murtagh B 56, 173

Nakamura H 4
Nguyen S 119, 158

Papadimitriou C 146
Pearson C 260
Powell W 119, 122, 141, 145, 146, 149
Prastacos P 6, 317
Puget Sound Council of Governments (PSCoG) 13
Pumain D 285
Putman S H 1, 13, 14, 17, 52, 56, 119, 127, 132, 184, 185, 188, 190, 191, 202, 234, 238, 242, 256, 284, 285, 304, 307, 318, 321, 322, 325, 327, 328, 332, 334

Rabino G 285
Reiner R 300
Reklaitis G 48, 270
Robillard P 126

Sanglier M 285
Saunders M 56, 173
Schecter R 53
Schlager K 63, 64
Senior M L 70, 83, 85, 235
Sharpe R 4, 6, 95, 96, 98, 100
Sheffi Y 116, 117, 119, 122, 141, 142, 143, 144, 145, 146, 149, 185
Small K A 4
Southeastern Wisconsin Regional Planning Commission (SEWRPC) 63
Stevens B 65, 66, 82
Struyk F 317

Taylor M 152, 153, 154
Tobin R 158

Van Doorn J 288
Van Vliet D 116, 127, 142, 143, 145, 158

Waddel P 285
Watterson T 285
Webster F V 4, 190
Wegner M 4, 252, 315
Wheaton W 70, 79, 80
Wildermuth B 156, 158
Williams H C W L 235, 239
Wilson A G 5, 6, 13, 70, 83, 106, 235, 285, 289
Wolfe P 56, 174

Zozaya-Gorostiza C 159

Subject index

Accessibility measures 74
Adjustment to disequilibrium (see partial adjustment models)
Alternative model system configurations 190–191, 236
  effects of intrazonal travel costs 227–228
  effects of network detail 214–215
  effects of recursive solution methods 202
  effects of regional trip scaling 227, 228, 233
  equilibrium solutions 194–195, 242
  fixed activity location, variable transportation costs 192
  numerical experiments
    equilibrium solutions 242–251
    Houston, TX 203–208, 215–221, 227–232, 241
    sensitivity to intrazonal travel costs 227–230
    sensitivity to regional trip scaling 227, 228, 230–232
    variable activity location, fixed transportation costs 202–214
    variable activity location, variable transportation costs, single mode 214–226
    variable activity location, variable transportation costs, two modes 233, 241–242
    Washington, DC 208–209, 210–213, 221–226, 241, 243–247, 248–251
  variable activity location, fixed transportation costs 191–192
  variable activity location, variable transportation costs, single mode 192–193
  variable activity location, variable transportation costs, two modes 194
AON (all-or-nothing) trip assignment (see also trip assignment algorithms) 24–25, 116, 118–119
  compared with SM and UE trip assignment 27–28
  compared with SM trip assignment 26, 27
  deficiencies 25–26, 27
  incremental trip-assignment procedures 118–119
  tree-by-tree assignment 119, 127

AON (continued)
  iterative 'all-or-nothing' assignment 118
  numerical example 25–26
'Archerville'
  numerical examples 13–29, 42–44, 44–47, 53–55, 56–58, 73–78, 86–94, 100–111, 124–131, 166–167, 168, 169–171, 179–184, 292–300
  sample data set 8–13
Attractiveness measures, use in models 14, 15, 17, 18
  land use 14
  percentage of household types 17
  residential land 17
  total employment 14

Bid rent (see under Herbert–Stevens model)

CALIB (see under nonlinear programming: computer programs)
Calibration equations (see under DRAM)
Calibration of systems of models 322–324
  effects of network detail on 323
Chaotic behavior (see under equilibrium solutions to location models: numerical experiments)
Cobweb model (see under partial adjustment models)
Combined location and trip-assignment models (see also alternative model system configurations)
  Evans solution approach 178–179
    numerical example 179–184
    rate of convergence 181, 183–184
    sensitivity tests 181–184
  Florian solution approach 171–172, 174–176
    illustrative example 173–174
    numerical example 173–174, 176–177
  MSA, method of successive averages solution approach 185
    numerical examples 185–187
    rate of convergence 185, 187
  simple combined location and trip-assignment model 162–163

Combined location and trip-
  assignment models (continued)
  system optimal (SO) combined
    location and trip-assignment
    model 165
    numerical example 166–167,
      168
  user equilibrium (UE) combined
    location and trip-assignment
    model 169, 172
    numerical example 169–171
Complexity of urban models 3–4
Composite cost 233–241, 324
  sensitivity of 239, 240
Congestion
  effect on employment and residence
    location forecasts 28–29
  effects on zone-to-zone travel cost
    matrix 25–26
Congestion functions (see volume–
  capacity functions)
Conversion matrix, employee-to-
  household 9, 18, 203, 326
Convex combinations algorithms (see
  also Frank–Wolfe algorithm) 121,
  123, 146–149
Cyclic flows (see under Frank–Wolfe
  algorithm)

Discreteness of data 29
Dispersion term (see under TOPAZ)
Dissipation of unexplained variation
  246
Doubly constrained gravity model (see
  spatial interaction models: doubly
  constrained)
DRAM (disaggregated residential
  allocation model) 1, 191, 328
  as a spatial interaction model 17
  calibration equations 52–53
  formulation 17, 51–52, 237–238,
    254–255, 292
  model calibration examples
    329–332
  numerical example 18–19
  sensitivity tests 18–19
  temporal specification 17, 191–192,
    255, 259–260, 284, 285
Dynamic structures for spatial
  interaction models 285

EMPAL (employment allocation
  model) 1, 191, 332–333
  as a spatial interaction model 13
  formulation 13–14, 253
  model calibration examples
    332–334
  numerical example 13–16
  sensitivity tests 16
  temporal specification 14, 191,
    253–254, 284
EMPIRIC, residence-location model
  285
Empirical testing of urban models
  3, 325
Employment-location model (see
  EMPAL)
Employment-to-household matrix (see
  conversion matrix, employee-to-
  household)
Entropy measures 177–178
  Shannon's definition 178
Equilibrium solutions to location
  models 253, 258–259
  as forecasts of household location
    282
  as targets for partial adjustment
    models 291–292, 300, 321
  by solving a set of nonlinear
    simultaneous equations 259
  household behavior implicit in
    equilibrium solutions 264–265,
    294, 319, 321
  numerical experiments
    Archerville 292–300
    chaotic behavior 273–280, 282,
      300, 302, 320–321
    failure to converge 264, 265,
      267, 270, 272, 273, 277,
      302, 304
    full simultaneous solution (FSS)
      265–272, 298–300
    household location (HH)
      equilibrium 259–261, 262,
      292–293, 300–301, 302–304
    household location/land
      consumption (HHLC or
      HH/LU) equilibrium 261,
      263–264, 272–273,
      294–297, 302, 305–306
    household location/travel time
      (HHTT)equilibrium 270–273
    Houston, TX 300, 302–304, 306

Equilibrium solutions to location
 models (continued)
  numerical experiments
   land use (LU) equilibrium
    293-294
    uniqueness of equilibrium
     solutions 295-296
    Washington, DC 260-261,
     262-264, 265-272,
     273-280, 300-302, 305
   rates of adjustment to disequilibrium
    conditions 265, 272-273,
    282-283, 295
   successive substitution solution
    method 260
Evans solution approach (see under
 combined location and trip-
 assignment models)

Florian solution approach (see under
 combined location and trip-
 assignment models)
Flow-balance relationships 42-43, 44
Frank-Wolfe algorithm 58-59,
 123-124
 as a convex combinations algorithm
  124
 illustrative example 59-62
 numerical properties 139-140
 problem of cyclic flows 158-159
 rate of convergence 62, 123,
  158-159

GAMS (see under nonlinear
 programming: computer programs)
Generalized cost (see composite cost)
Gradient search technique 51, 52-53,
 328
 illustrative example 51-55
 likelihood function 52, 53
 rate of convergence 53
Gravity models (see spatial interaction
 models)

Herbert-Stevens model 65, 67-68
 application difficulties 70-73, 87
  estimating preference functions
   71-73, 79
  Hartford, CT 73
  New York, NY 80
 bid rent 66, 67, 69
 comparison of primal and dual
  formulations 69-70

Herbert-Stevens model (continued)
 computational difficulties 114
 consumer surplus 67
 deficiencies 70, 78-80, 94, 113
 dual problem 68-69
 illustrative example 66-67
 locational subsidies 68-69
 modified formulations
  Anas version 81-82
   deficiencies 82-83
  Houghton-Senior version 84-86
   dispersion parameter 86
   mean bid rent 85-86
  land version 83-84
  numerical examples 86-94
   sensitivity tests 91-93
  stock version 84
 numerical examples 73-79
  discontinuous distribution of
   households 75, 78-79
  residential budget 74
  sensitivity tests 77-78
 Pareto optimum 69
 preference functions 71-73, 74,
  75-76
 residential budget 66, 69
 utility-maximization behavior 68
Highway trip probabilities (see mode-
 split models: mode-choice
 probabilities)
Houston, TX
 data set 195-200, 220, 221,
  331-332
  deficiencies 308
 numerical experiments 138,
  159-160, 203-208, 215-221,
  227-232, 241, 300, 302-304,
  306, 320

Incremental trip-assignment
 procedures (see under AON)
Integrated transportation and land-use
 models 1, 4-6, 30 (see also
 ITLUP, LILT, MEPLAN)
 feedback mechanism in 30, 192
Interaction costs (see under TOPAZ)
ITLUP (Integrated Transportation and
 Land Use Package) 1, 2, 236
 'product-share' form 236-237

LANCON (land consumption model) 256, 315
   formulation 256–257
   model calibration examples 334–336
   temporal specification 256, 257
Likelihood function (see under gradient search technique)
LILT 4
Linear programming
   canonical form 37
      slack and surplus variables 37–38, 39, 40
      transforming a linear program into 37–38
   degenerate solutions 112
   dual problem 39, 41
      illustrative example 39–40
   dual variables 40–41
   feasible region 33
   graphical solution method 32–36
   primal problem 39, 41
   sensitivity analysis 111–112
   simplex algorithm 36–37, 38, 41, 43–44
      initial feasible solution 38
Link performance functions (see volume–capacity functions)
Location costs (see under TOPAZ)
Locational subsidies (see under Herbert–Stevens model)
Logit mode choice (see under mode-split models: logit formulation)
Logit trip assignment (see under SM)

MEPLAN 4
Minimum-cost flow problem (linear) 44, 45–46, 107, 163–165
   comparison with nonlinear minimum-cost flow problem 56–58
   numerical example 44–47
   relationship with spatial interaction models 107–108
   with maximum allowable flow on network links 47
MINOS-5.0 (see under nonlinear programming: computer programs)

Mode-split models (see also utility measures)
   logit formulation 20–22, 237, 238–239
      aggregate type 238
      disaggregate type 238–239
   mode-choice probabilities 20, 21–22
   MSPLIT computer program 194
   numerical example 20–24
   sensitivity tests 23–24
Model calibration (see also gradient search technique) 14, 328
   by solving the system of calibration equations 55
   Houston, TX 331–332, 333–336
   numerical examples 329–336
   Washington, DC 329–331, 333–336
Moore algorithm (see under shortest-path problem)
MSA (method of successive averages) (see under combined location and trip assignment models and under SUE)
MSPLIT (see under mode-split models)

NETWRK (see under trip-assignment algorithms)
Network nodes
   load nodes 164
   nonload (transshipment) nodes 164
Node-balance relationships (see flow-balance relationships)
Nonlinear minimum-cost flow problem 55–56
   comparison wtih linear minimum-cost flow problem 56–58
   numerical example 56–58
Nonlinear programming 48–62
   computer programs 56
      CALIB 56, 328
      GAMS 56, 75, 114
      MINOS-5.0 56, 173
   constrained 55
      illustrative example 55–58
   unconstrained 48
      illustrative examples 49–50, 50–51
      inflection point 48
      necessary conditions for a local optimum 48
      stationary point 48
      sufficient conditions for a local optimum 48–49

Subject Index

Optimization problems 31
  classification by constraints 32, 55
  classification by objective function 31-32, 55

Parameter variation over time 325
Partial adjustment models (see also equilibrium solutions to location models)
  adaptive expectations models 288
  cobweb model 288-289
  PAMEQ (partial adjustment to moving equilibrium 285-287, 288-289, 321-322
    calibration problems 324
    numerical experiments 304, 307-312
    relationship with iterative solution methods 296-297
  quadratic loss functions 291
  response rate parameter 286, 287, 289, 290
  retail trade model 289-291
Preference functions (see under Herbert-Stevens model)
Probit trip assignment (see under SM)

Regional control totals 15-16, 203
Residence-location models (see DRAM, EMPIRIC, Herbert-Stevens model)
Residential budget (see under Herbert-Stevens model)
Retail trade model (see under partial adjustment models)

Scaling factor 14-15
Schlager model (land-use design-plan model) 63-64
  application difficulties 64
  deficiencies 65
Shortest-path problem 41-42
  Moore algorithm 116
  numerical example 42-44
Simplex algorithm (see under linear programming)
Simultaneous model solution (see equilibrium solutions to location models)

SM (stochastic multipath) trip assignment (see also trip assignment algorithms) 26, 116-117
  compared with AON trip assignment 26, 27
  compared with AON and UE trip assignment 27-28
  deficiencies 27
  dispersion parameter 26-27
  logit formulation 26, 117, 142
    compared with probit, UE, and SUE trip assignment 143-145
    independence-of-irrelevant-alternatives 142-143
  numerical example 26-27
  probit formulation 117
    compared with logit, UE, and SUE trip assignment 143-145
Spatial interaction models
  doubly constrained 98, 106, 178
  equivalent optimization problem 106-107
  relationship with minimum-cost flow problem 107-108
  singly constrained 13, 233-234
Systems of urban models (see alternative model system configurations)
Successive substitution solution method (see under equilibrium solutions to location models)
SUE (stochastic user equilibrium) trip assignment (see also trip assignment algorithms) 119-120, 122, 140
  compared with logit, probit, and UE trip assignment 143-145
  compared with UE trip assignment 145, 154
  dispersion parameter 142
  relationship with SM trip assignment 120
  relationship with UE trip assignment 141-142
  solution procedures 146-151
    MSA (method of successive averages) 149, 150-151
    with logit path choice 141-142
      deficiencies 146
    with probit path choice 141

Supply-demand interaction
  in programming models 316-318
  in spatial interaction models
    314-316
System optimization (see under trip-assignment algorithms)

Temporal specification (see also DRAM: temporal specification and also EMPAL: temporal specification) 284
TOPAZ (technique for the optimal placement of activity in zones) 95-96
  absolute activity formulation 98
  computational difficulties 114
  deficiencies 113
  Dickey version 96-98
  dispersion (entropy) term 106-108, 111, 112
  interaction (transport) costs (or term) 95, 96, 97
    calculated from a doubly constrained gravity model 98, 99, 106-108
  location costs (or term) 97, 101
  numerical examples
    comparison of location cost and transport cost minimization 104-105
    incorporating dispersion (entropy) term 106-111
      sensitivity tests 110-111
    location and transport cost minimization 104-106
      sensitivity tests 105-106
    location cost minimization 102-103
    transport cost minimization ('transportation' problem) 103-104
  TOPAZ82 formulation 100
  sensitivity analysis 110-111, 112
Transportation-land-use interrelationships 5, 29
Transportation problem (see minimum-cost flow problem)
Travel cost functions (see volume-capacity functions)

Trip-assignment algorithms (see also AON, SM, SUE, UE)
  capacity-constrained 117-119
    trip reassignment 118-119
  comparison of trip assignment algorithms 161-162
  effects of network congestion on 153-155
  effects of network detail on 155-157
  effects of starting points on 158
  empirical verification of trip-assignment algorithms 157-158
  NETWRK computer program 124, 193, 258
  numerical experiments with AON, SM, UE, and SUE algorithms 124-138, 149-151, 159-161
    Houston-Galveston highway network 138, 159-160
    Minneapolis-St Paul highway network 135-137
    San Francisco highway network 132-135, 149-151
    sensitivity tests 126, 128, 129, 132-134
    Washington, DC, highway network 160
  problem of determining link capacity 157
  SO (system optimization) criterion 122
    relationship with UE trip assignment 167-169
  unconstrained trip assignment 24, 116-117
Trip distribution models 19-20, 233, 236-237
Trip-end constraints 19, 103
Trip matrices 19-20
Trip reassignment (see under trip-assignment algorithms: capacity-constrained)

UE (user equilibrium) trip assignment (see also trip assignment algorithms) 27, 28, 119-122, 147-149
  compared with AON and SM trip assignment 27-28
  compared with SUE trip assignment 145, 154

Subject Index

UE (continued)
    deficiencies 141, 160–161
    relationship with AON trip assignment 120
    relationship with SO criterion 167–169
    UE criterion 120, 122
    uniqueness of UE solution 126–127
Utility measures
    use in mode-split models 20–22
    use in SM (stochastic multipath) trip assignment 26

Volume–capacity functions 24, 55–56, 117, 151–153
    comparison of alternative volume–capacity functions 152–153
    effects on trip-assignment algorithms 152–153

Volume–capacity functions (continued)
    Davidson function 152–153
    link-flow constraints 153
    US Bureau of Public Roads (BPR) volume–capacity function 121–122, 151–152, 153, 323

Washington, DC
    data set 195–197, 199, 201, 220, 221, 329
    numerical experiments 160, 208–209, 210–213, 221–226, 241, 243–247, 248–251, 260–261, 262–264, 265–272, 273–280, 300–302, 305

Zone-to-zone travel cost matrix
    asymmetry of 26
    calculating 15
    calculating with congestion 24